OPIUM

OPIUM
Reality's Dark Dream

THOMAS DORMANDY

YALE UNIVERSITY PRESS
NEW HAVEN AND LONDON

For information about this and other Yale University Press publications, please contact:
U.S. Office: sales.press@yale.edu www.yalebooks.com
Europe Office: sales @yaleup.co.uk www.yalebooks.co.uk

Set in Minion Pro by IDSUK (DataConnection) Ltd
Printed in Great Britain by TJ International, Padstow, Cornwall

Library of Congress Cataloging-in-Publication Data

Dormandy, Thomas.
Opium: reality's dark dream/Thomas Dormandy.
p. cm.
ISBN 978-0-300-17532-5 (cl : alk. paper)
1. Opium abuse—History. 2. Opium trade—History. 3. Opium—History. I. Title.
HV5816.D67 2012
362.29'309—dc23

2011033287

A catalogue record for this book is available from the British Library.

10 9 8 7 6 5 4 3 2 1

FOR JOHN AND IN PRECIOUS MEMORY OF DAISY

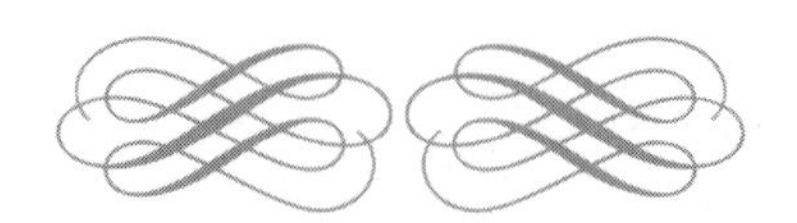

Contents

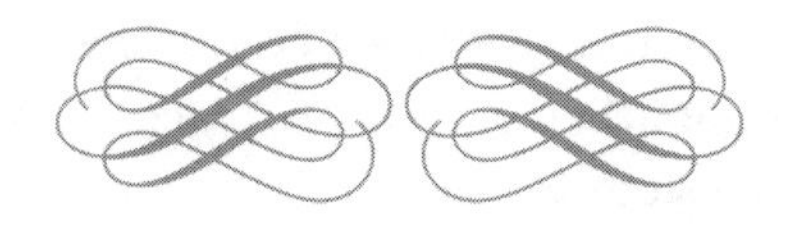

Illustrations

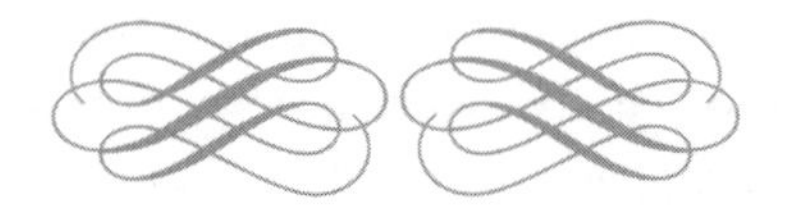

Introduction

Sir William Osler called it 'God's own medicine'. His friend, the eminent surgeon and addict William Stewart Halsted, declared it to be the vilest curse on society. Conflicting opinions about a drug are common; but such extremes could refer only to one.

Age and survival alone make opium unique. Fermented grape juice may be older; but it is the emergence of opium which is the more remarkable. A potpourri of rotting fruit and a few hours of sunshine are all that are necessary to start alcoholic fermentation. Prehistoric man or woman may have recoiled from their first sip but the after-effects must have drawn them back for another – and then perhaps another. The harvesting and processing of poppy juice could never have been so accidental. Cultivation was back-breaking, as it still is. It required considerable expertise, as it still does. Processing was lengthy and complicated. It could not have evolved without painstaking experimentation. Fermented grape juice could be made to taste divine. Natural poppy juice always tasted foul. And yet, not only was poppy juice harvested and processed by Stone Age man but, once discovered, it never disappeared. Even during the Dark Ages when cultivation ceased in Europe, it flourished in the Islamic world.

* * *

But age, fame and survival are not alone in making opium unlike any other drug. Even before the discovery of endorphins, the body's own morphine-like products, many regarded it in some mysterious way as 'natural'. Though it could gravely distort behaviour, it still seemed part of normal brain function. Or if, like Descartes, one believed in that elusive appendage, the human soul, nestling in the pineal gland, it worked through it . . .

A hundred years before Descartes the sixteenth-century doctor – or charlatan – Paracelsus was often obscure but he had moments of lucidity, even of illumination. In one of those he compared the action of his laudanum to the rekindling of the candles of faith and hope in the soul which the ill winds of doubt and despair had extinguished. And in 1750 George Young, author of the first modern monograph on the drug, attributed its properties not, 'as some of my simple-minded colleagues believe, to external fumes, vapours, auras and effluvia which attack the brain . . . Its wondrous effects cannot be wondered at since its Active Principle . . . is naturally carried within our own tissues . . . being exalted by Nature to stimulate us when necessary.' His insight was not explored for three hundred years; but that opium was in some way uniquely native to man was already widely believed.

Indeed, to millions of users opium would never be a drug. Baudelaire compared it to 'a dear and old woman friend . . . full of caresses and alas, full of deception.' Coleridge chafed at his enslavement but could not function without it. To his rival, Thomas de Quincey, it was 'my one and only true companion.' Both Coleridge and De Quincey lived into their mid-seventies, a good age in those days. Some have speculated – as some still do – that the 'naturalness' of the substance could explain perplexing individual variations. Natural hormones like insulin or thyroxine too can save the lives of those deficient in them while in similar doses kill those who are not.

* * *

No major development in the history of opium was without wider repercussions. In Western Europe the advent of affordable laudanum ushered in the Romantic Age. About the same time the Industrial Revolution was partly responsible for the emergence of tuberculosis in its modern form. A unique relationship between drug and disease developed. Though often painless at first, in its later stages the illness could be excruciating. Only one medicine relieved the suffering. Tuberculosis and opium together shaped the art of Keats, Shelley, Novalis, Thoreau, Weber, Chopin, Chekhov and the lives of countless others less famous but no less important to themselves.

In 1805 the isolation of the active principle, morphine, made accurate dosage possible. A few decades later the introduction of the subcutaneous syringe for administering the extract marked the beginning of modern therapeutics. The development of the more powerful synthetic morphine derivative, heroin, transformed drugs into big business. Big business means big crime (among other things) and crime does not come any bigger than war. The opium wars between China and Britain (with contributions by other Western powers) still tend to be glossed over in school books; but they have helped to shape the world of today.

Big crime also calls for big crime prevention, or at least attempts at it. Opium and its products have spawned the most expensive international anti-crime organisations in history. Is it also the most wasteful, misguided and corrupt? Opinions differ. The past never provides clear answers; but pointers surface uninvited.

* * *

Sir William Osler in his praise and William Stewart Halsted in his condemnation were both right. So were other experts who have passionately contradicted each other. Opium and its derivatives have destroyed, degraded, corrupted and killed individuals, families, communities and even whole nations. Yet many doctors before Osler have said that without the blessed effects of opium on their patients they could not have faced their profession. In Sir Joseph Lister's wards at King's College Hospital, London, the most advanced surgical unit in the world in its day, every patient was on doses of morphine which would elicit gasps today. The motives were mixed. Without the doping no dignified ward round would have been possible and Sir Joseph's celebrated puns would have been wasted in the uproar. But the practice also eased the suffering of the inmates as nothing else would. That too mattered. Contrasts are everywhere. If the diabolical Dr Fu Manchu was a characteristic figure of the opium saga, that good and godly woman, Dame Cicely Saunders, was another.

* * *

Historical treatises dealing with the drug and its derivatives, morphine and heroin, have tended to focus on their depredations. But histories should try to chronicle the lights as well as the shades. With opium this has never been easy. Jean Cocteau was an addict and an enthusiast. To express an opinion about the drug, he wrote, is like trying to express an opinion about human nature – not human nature at its most placid and rational but human nature at its most grotesque, tragic, creative and holy. Like other devotees, he never quite succeeded himself.

No history of opium can be perfectly objective, and historians should perhaps declare their personal luggage. The present writer's happy childhood was marred by a series of operations – or would have been if in those far-off days kindly anaesthetists had not administered an injection called 'premedication'. One of the objects of the injection, administered exactly two hours before surgery, was to calm patients. It transported the present writer into a land of indescribable bliss, never experienced before or since, remembered and cherished between operations long after memories of pain and discomfort had faded. Fortunately he did not realise until many years later that the injection contained a hefty dose of a morphine derivative. Had he known, he would today be an addict or, infinitely more likely, the memory of an addict sadly passed away in his prime. Otherwise his experience has been as a doctor; and he is not animated by any reforming zeal. Except perhaps one.

Over the past century professional historians have woven cultural, economic, social, industrial, anthropological and a score of other specialised strands into the fabric of their specialty – but not health and disease. Nothing less than the Black Death which exterminated a third of Europe's population seems to merit their attention. Wide-ranging books about the Victorians may dwell lovingly on abstruse theological disputations or the art of Marie Lloyd but leave unmentioned general anaesthesia, antisepsis, immunisation, the upsurge of syphilis or indeed any medical topic. In different ages tuberculosis, smallpox, scarlet fever, bladder stones, the complications of childbirth and even the common cold exercised more minds and changed more lives than any minor war or international treaty, but even in multi-volume general histories they rarely rate more than a paragraph. The reverse is also true. Specialist medical historians sometimes describe advances and reverses in health in splendid detail but as if they had happened on the moon.

The history of opium cannot be recounted in this way. The use and abuse of the drug critically influenced non-medical events, and non-medical events critically shaped the use and abuse of the drug. Even today the poppy is a major player on the international stage, sustaining or destroying governments, triggering wars. Seemingly unrelated events in the past like the end of the religious conflicts in Europe or the building of the Transcontinental Railways in the United States are part of its history.

* * *

That history naturally divides into two parts. By several names the juice of the capsule of the white poppy has been 'opium' for thousands of years. Laudanum has been one of its *noms de guerre*, a formative influence on European literature and the arts. There have been others. That is the theme of Part I of the present book. But since the mid-nineteenth century 'opium' has often meant not the juice but its active principle, the alkaloid morphine. A hundred years later morphine was refined into a more 'heroic' derivative, heroin. These changes affected the nature of addiction, transformed the public perception of it and created the most powerful criminal organisations the world has ever seen. They are the subject of Part II.

PART I

THE JUICE

CHAPTER I

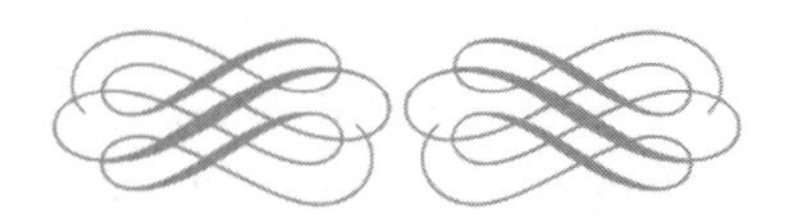

Petrified buns

THE ALPINE WINTER of 1854 caused much hardship. Children lost their ears and noses to frostbite and the frail and elderly died. But the cold also provided the strong and industrious with exceptional opportunities. Little steamers had started to criss-cross Lake Zurich during the previous summer; and the low water level now created the right conditions for building landing platforms. Such a facility would do wonders for the tourist trade of the small lakeside community of Meilen. But almost at once the diggers ran into difficulties. Jutting from the bottom of the lake were a dozen almost immovable wooden poles. They had obviously been implanted there by humans; but what kind of humans and when and for what purpose? The answers were soon revealed. The mud around the poles yielded a profusion of stone and wooden artefacts as well as fossilised objects, including petrified apples, raisins, hazelnuts and what looked like buns about three inches across. Until then all prehistoric finds in Europe had been accessories of death – graves, weapons, necessities of the afterlife. The Meilen excavators had uncovered the stuff of fairy tales, the remains of a Stone Age lake village where people actually lived.[1]

The discovery made headlines. Soon an expert team from the University of Zurich under the leadership of Professor Ferdinand Keller had difficulty in keeping thieving archaeologists at bay. But nothing could stop the flurry of sensational newspaper reports, inevitably followed by prehistoric romances (a now happily defunct literary genre), calendars, children's toys, dioramas and expensive forgeries. One of the sensations, as revealed by Professor Keller, was that the petrified buns contained recognisable fossilised poppy seeds. The professor even identified the seeds as belonging to the cultivated white rather than to the wild red variety. When fresh and processed, he suggested, they would have had a mild opium-like effect.

His interpretation was disputed by jealous colleagues. Professor Maurice Rochat of Nancy argued that even unfossilised the seeds of different kinds of poppy, none of them bigger than a pinhead, were impossible to distinguish. But Professor Keller, a patriotic Swiss, had already started to venerate his lake-dwelling ancestors and stuck to his guns. His forebears were obviously a prudent lot – why else would they build their dwellings on stilts in lakes and swamps – as well as being honest and hard-working. They deserved the occasional break. No breathtaking works of art emerged as would a few decades later in Lascaux in France and Altamira in Spain; but that in no way excluded artistic achievements. Lake water is not a hospitable milieu for paintings and the treasures could have perished.[2] Alcoholic fermentation may also have been discovered; but the people of Lake Zurich were no drunks, certainly not. The building of houses on stilts must have called for planning and ingenuity. The cultivation of the white poppy also required a high degree of expertise. It gradually transpired that the villages flourished early in the sixth millennium BC; that is, the late Stone Age. The lake dwellers may therefore have been the first humans to taste opium.

* * *

The people of the ancient civilisations of the Near East probably came next. In a Sumerian ideogram of about 4500 BC – roughly the age of Abraham – the poppy is called *Hul Gil* or the 'plant of joy', the first known such expression of high regard. Besides praising it, the tablet briefly instructs how the juice of the pod was to be collected at dawn and how it should be converted into a drink. The Assyrians who followed the Sumerians seem to have had special iron scoops to score the pods and let the sap ooze out. They called the juice *aratpa-pal*, the possible root of the otherwise obscure *papaver*.[3] The Persians, history's first empire-builders, recorded using poppy extracts for medicinal purposes. But it was the Valley of the Nile which the classical world – and historians ever since – have identified as the cradle of opium.

Though sizable on the map, the area which was regularly flooded by the great river and could therefore sustain life was smaller than present-day Belgium. Not much land to waste; but for thousands of years sustaining life included the cultivation of the poppy. The fields around Thebes (today's Luxor) on the Upper Nile in particular acquired a reputation for the richness of their growth as early as the third millennium BC; and the 'divinely elevating effect' of Theban opium was still the toast of Britain's Prince Regent in early nineteenth-century Brighton.

Two remarkable papyri are especially useful for understanding Egypt's fascination with the poppy. The discovery of both is linked to the name of Edwin Smith, generally described as an American farmer and excavator resident in Egypt in the 1860s. In fact Smith was probably more a maker than a digger-up of antique remains; but he was no ignoramus. He at once recognised the value of two scrolls offered to him by more proactive grave-robbers and paid what he later described as a fair price for them. The first, known today as the Edwin Smith papyrus, is preserved in New York.[4] The second and even more valuable was named after Georg Moritz Ebers, a Berlin Egyptologist who never went near Luxor but bought the treasure from Smith and translated the hieroglyphics into German.[5]

In these ancient texts, about three thousand years old, every medical and surgical ailment was covered. The longest sections dealt with the heart, regarded as the site of confluence of all body fluids – blood, tears, urine and sperm – and where opium was believed to act. (The brain, by contrast, was considered a squashy filling, unnecessary in the next world and therefore discarded before mummification.) Separate chapters covered pregnancy, gynaecological complaints, intestinal diseases, parasites, eye and skin problems, snake bites, decaying teeth, burns, fractures, deformities and methods of contraception. Everywhere the emphasis was on practical management and no sharp dividing line was drawn between afflictions of the body and turbulences of the soul. It was this unitary approach which justified the use of poppy juice in a surprising range and variety of ailments.

It also makes some passages familiar to ears attuned to the modish term 'psychosomatic'. It is only the term which is relatively new. In Egypt as in most other ancient civilisations healing depended on establishing spiritual contact with a god as much as on any medical ministrations to the body. When lucky and deserving, the god would take the complaint on himself and, if need be, consult more specialised deities. In one passage of the Ebers papyrus the victim of a severe headache identifies with the god Horus as an intermediary and it is Horus who exclaims: 'My head, my head, oh my head is bursting'. 'Which part of your head?' asks a sympathetic fellow god, Thoth. 'Upper part of my forehead and my right temple,' replies Horus. 'This is grave. I shall have to solicit the advice of Ra,' says Thoth. Ra, an all-wise power usually identified with the sun, is sympathetic and threatens the demons who have been such a plague with terrible punishment. 'I shall cut off thine trunks, you naughty spirits!' But that, it seems, is not enough: the demons continue to afflict Horus (and the patient). Ra recommends a well-tried poultice made of the skulls of catfish. Horus applies the poultice; but, though it eases the pain, the relief is temporary. Then Ra concludes that only a special extract made from poppies will relieve the suffering and orders his acolytes to prepare a concoction. Horus swallows it and lo! his headache disappears. So does the patient's.

Poppy juice sweetened with honey was also a remedy for the looseness of bowels, a common complaint in Egypt as it still is. In depression it lightened the spirits. It eased the suffering of patients afflicted with the Bilharzia parasite and bladder stone, both still prevalent. In the painful dressing of wounds and the setting of fractures it assuaged the agony. It must have been a godsend in the drilling of jawbones for dental abscesses, an operation which, judging by skeletal remains, seems to have been performed not infrequently.[6] Because even mixed with honey the draughts often tasted bitter and smelt nauseating, administration by suppositories and enemas was often preferred. The analysis of dreams as a diagnostic tool was highly developed by the priests of Karnak and poppy juice was sometimes administered to induce stirrings of the not-yet-so-named subconscious. The potions also helped in most forms of dying. Nothing suggests that ancient Egyptians exalted the stiff upper lip; but, like most people of the Fertile Crescent, they valued dignity and a brave and noble countenance. Nor was the afterlife forgotten. No pharaoh would be buried without a few dried poppies and implements of poppy harvesting to serve him on the journey and in future abodes. At the other extreme of life, a special preparation mixed with milk was recommended to quieten infants with 'colic'.[7] With so many uses, the plant was held in affection: it was incised on funerary monuments and painted on the walls of temples.

Poppy juice, like other remedies, was rarely if ever prescribed on its own. Medical treatment on the banks of the Nile always included prayers, incantations, healing amulets, purifying rites and changes – temporary or permanent – in lifestyle. Remedial and sacred baths would be performed before embarking on complex treatments; and parts of the body would be shaved and tablets inscribed with messages to the gods would be applied. But such communications too would often be soaked in poppy juice; some of the divinities addressed clearly appreciated a well-prepared concoction. Sexual practices were scrutinised and regulated, circumlocutions in some periods being as inventive as in Victorian Britain. The fact in particular that opium could counteract carnal urges did not escape the notice of the pharaonic healers. Incest was a grave sin in the Old and Middle Kingdoms (though not among royalty and under the later dynasties) and, when feared, poppy juice was an effective prophylactic. Unlike most modern textbooks, the papyri recognised that some ailments were incurable and should therefore be treated with prayers rather than with medicines; but even in such cases poppy juice was prescribed to ease the passing.

Inevitably over a period of thousands of years medical practices both in Egypt and in Mesopotamia changed. A high degree of specialisation developed in the Nile Valley under the XVIII dynasty in the fifteenth and fourteenth centuries BC, a time of intense artistic and spiritual ferment.[8] Written note taking of individual patients became formalised. Medical clichés would blossom even in hieroglyphics: indeed pictograms were well suited to convey platitudes. Many case histories would mention consultations with specialist dentists, bone surgeons, gynaecologists, gastroenterologists, oculists and experts on drugs and potions. The last were among the most highly esteemed. They advised on the processing of the poppy, an expertise requiring long training as well as inborn aptitude.

* * *

The blessings of Egyptian and Mesopotamian civilisation were spread by the Phoenicians and other seafaring people. Their traders sold special knives for incising the capsule of the poppy; and, sailing along the Mediterranean and the Atlantic coast, found profitable markets. In the 1930s digging up Minoan remains on the island of Crete archaeologists unearthed a charming female figurine with a come-hither smile (usually described as archaic) and crowned with a tiara of poppy capsules. The lady, initially labelled the Goddess of Healing, seems to have adorned a windowless opium den which, buried under volcanic ash, survived unchanged for three thousand years. Even more surprising was the collection of bronze needles found on mainland Greece in the 1970s and dating from before 1500 BC. Mystifying at first, they were probably prehistoric *yen-hoks*, the pins used by Chinese mandarins centuries later to pierce wads of opium over charcoal grills and let the wisps of smoke billow to their nostrils. Some of the heads of the prehistoric needles are hollow and fashioned into the shape of poppy pods.

Phoenicians also carried and sold copies of medical texts enshrining the discoveries of Egyptian and Mesopotamian doctors. Hippocrates revered the physicians of Thebes and Karnak; and even Galen, notoriously sparing of praise other than of himself, would write with approval of the cures performed by Amenhotep, son of Hapu. The Greeks too began to ponder what the secret of the poppy might be. And how did one particular species of the plant – and no other – acquire its magical properties?

CHAPTER 2

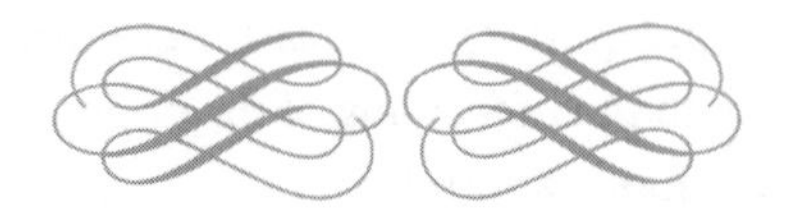

The magical seepage

TO THE SECOND QUESTION there is still no certain answer. Some botanists believe that the opium-yielding white poppy evolved naturally, the result of mutations in response to quirks of climate and geography. But others have suggested that *Papaver somniferum* was the result of deliberate selection by generations of prehistoric cultivators. That may sound far-fetched but would not be unique. Though spread over millennia, the cumulative ingenuity of these pre-human plantsmen equalled anything their modern descendants have achieved.

Poppies are a bounteous tribe: no less than 28 genera and over 280 species flourish in the temperate and subtropical zones of the northern hemisphere. Many have been – and still are – cultivated for their beauty. The Welsh poppy, the blue or Syrian 'tulip' poppy, the charmingly reticent off-white Alpine poppy, the subarctic Iceland poppy, the lush purple California poppy are wonderfully cheering and ornamental. In the wild the flower is a single bloom; but double blooms and varieties with serrated and fringed petals have been bred. Among the most exquisite though difficult to grow are the Pink Chiffon and the Oriental Paeony, both natives of Asia Minor and for centuries the adornment of Byzantine empresses.

In all this rich multiformity it is only the *P. somniferum* and the *P. bacteatum* which produce opium in significant amounts, the former being the easier to cultivate. It is an annual with a growth cycle of 120 days and requires a moderately rich, well-cultivated soil. In the wild it is most likely to flourish in earth that has been recently dug, ploughed or, most effectively, torn apart by shrapnel. Blood is a good fertiliser. The best climate is warm to temperate with not too much rainfall during the early stages of the development cycle. Sandy loam is ideal since it retains moisture and nutrients and is not too hard to penetrate for the delicate first roots. The opium poppy is a 'long-day' photosensitive plant: that is, it needs to grow through a period of long days and short nights, preferably with direct sunlight for at least twelve hours a day.

All this may sound demanding; but in most parts of the world the plant is easy to grow. It does not require regular irrigation or expensive chemicals; it has few pests and therefore does not depend on insecticides or fungicides. This is what makes it in poor countries an irreplaceable cash crop. The seeds are naturally sown by the pod blowing in the wind and shaking its contents like a pepper pot; but when sown deliberately they are dropped in rows of shallow holes made by a stick called the dibber. About 5,600 grammes are usually sown per acre. The timing depends on weather conditions and forecasting and requires intuition and experience; but other cash crops like peas, beans or tobacco can be planted alongside the poppy. All highly satisfactory for the poor farmer and his numerous brood.

The seeds germinate quickly and within six weeks the plant is established. It first looks like an ugly young cabbage, consisting of a bare stem or peduncle and secondary branches called tillers. Both stem and tillers are covered with velvety hairs. Eventually the plant can reach a height of between 90 and 150 centimetres, the tooth-edged leaves appearing on alternate sides. As the single buds at the ends of stem and tillers develop, the stem and tillers bend under their weight; but as the buds mature the stems too strengthen and begin to point upwards. About ninety days after germination the flowers start to bloom.

The event is magical though the poppy does not display its full glory at once. For about thirty-six hours it looks crumpled like a butterfly emerging from its chrysalis. But then, suddenly, the four petals separate and expand, each slightly overlapping one neighbour. Their delicacy and freshness have enchanted artists since time immemorial. They are the first flowers children paint. The traditional opium poppy is usually described as white but is more ivory-coloured; and there are rarer pink, crimson, faintly purple and multi-hued varieties. The colour fades delicately from the base to the edge of the petals. Even in countries where the plant is a weed, the moment of flowering is relished, at least by visiting townies. Inside the flower a ring of spikes, so-called anthers, marks the site of the future pod. Swarms of insects descend and get on with their job.

Except for poppies deliberately cultivated to last, the flowers bloom only for four or five days. The petals then drop and expose the pod, the size of a large pea. Within a week or two it grows to the size of a hen's egg and acquires a bluish-green and slightly waxy appearance. From the top sprouts a crown of excrescences, so-called stigmas. The roots of the fallen petals are marked by scars near the bottom. The outer skin protects the ovary which has a three-layered wall and seed-producing membranes. A single pod may produce more than a thousand seeds. When mature the seeds detach themselves and may remain loose in the pod for a day or two. Then they disperse through small holes that open up below the crown. As generators of an 'opium-like' effect poppy seeds have long been regarded by doctors and chemists, but not by the uninformed public, as inactive. In such matters the uninformed public is usually right. Blameless athletes have found to their consternation that a mansize poppy-seed bagel can give a positive urine test for opium; and folk wisdom has always attributed a sedative effect to seed infusions.

But pharmacologically the seeds are a by-product. The fame of the opium poppy rests on its sap. The name opium is Greek, from οπιον, meaning the fresh juice of the poppy. The Greeks also used the term *meconium*, from μηκώνιον, the poppy. Hippocrates also referred by this name to the watery bowel discharge of the newborn: he apparently thought that the two looked similar. Meconium in the latter sense is still used in neonatal medicine. By whatever name, the raw opium contains sugars, proteins, water, ammonia, plant wax, gums, fats, sulphuric and lactic acid and a range of alkaloids. It is the last which matter.

Alkaloids are a busy family of naturally occurring nitrogen-containing bases. They are produced by bacteria, fungi, plants and animals as well as by humans and include essential hormones (like adrenaline), important neurotransmitters (like serotonin), poisons (like strychnine), stimulants (like nicotine, caffeine and cocaine), natural 'drugs' (like quinine) as well as the opium alkaloids. The last alone number more than fifty compounds, divisible for practical purposes into two groups. Those centred on the phenanthrene nucleus include morphine, codeine and thebaine. Those centred on other chemical structures include noscarpine and papaverine and have minimal pharmacological action. All seem to be synthesised in the so-called 'lactifers', the cells which produce the sap, over a period of about a week while the pod is ripening. The starting material is probably quite ordinary, perhaps an albumin-like protein, but the detailed steps of the synthesis are still unknown. Once the pod is ripe the process stops and the substances are quickly – that is within a few days – broken down into inactive fragments. Their natural role, purpose or any advantage they might confer or might have conferred at some point in evolution remain a mystery. The poppy or its sap is not, so far as is known, consumed by any species other than man.

* * *

The two great blessed or cursed stupefiers of mankind, opium and alcohol, may have been discovered about the same time but in alcoholic fermentation a single observation, the discovery that fresh saliva could initiate fermentation even in the cold and dark, was needed. The processing of poppy juice leading to opium could not have been the result of one such happy discovery: timing was all. Harvesting has always begun, as it still does, about two weeks after the petals have dropped. The farmer would then examine the pods and the crowns. By now the pods have darkened from bluish-green to a dull brown. When the points of the crown are standing straight out or pointing upwards, the pods are ready. Since not all pods mature at the same time the field has to be inspected more than once every day for several weeks.

A specialised implement with three or four parallel sharp blades is used for the tapping. The manoeuvre is sometimes known as 'scoring' or 'lancing' and it is where most expertise is required. If the cuts are too deep, the sap will ooze out too quickly and dribble uselessly to the ground. An even deeper incision may lead to the pod weeping internally, causing it to shrivel and die. If, on the other hand, the cuts are too shallow, the flow will be too slow and the sap will harden prematurely, sealing the cut like a scab. The ideal depth is usually about 1–1.5 millimetres. The modern knife can be set for a certain depth but even a modern knife cannot decide which depth is the right one.

Timing is another decision that needs judgement. The sun encourages the milky sap to trickle out; but if the tapping is done too early, when the sun is still high, the heat dries the first drops and seals the wound. If the tapping is done too late, there may be insufficient time for all the sap to ooze to the surface. The aim is to let the drops emerge slowly overnight and coagulate on the surface. Different fields require different timetables.

When opium – that is the sap of the pod – first appears, it is a cloudy white and fairly mobile liquid. It is also biologically inactive. Only on contact with air for a few hours do the critical precursor alkaloids oxidise and acquire pharmacological potency. The liquid also becomes dark brown, viscous and sticky; and it develops a distinctive and delicate perfume, not unlike that of freshly mown hay. It is only then that the expert harvester scrapes the secretion off the pod with a short-handled blunt knife. To prevent the blade from getting covered in gum he licks it between plants. This is the basis of the harvester's reputed addiction to his crop.[1] It is a myth. Properly treated the pod will continue to secrete for some days and can be scraped five or six times. An experienced farmer works his way backwards across the field, tapping lower pods before taller ones to prevent the sap from spilling. The scrapings are collected into a conical container which hangs around his waist. Prudent cultivators mark the larger and more generously secreting pods with a coloured yarn. This will direct them back to the plants which will be gathered whole and sun-dried. It is their seeds which will be used for next season's planting.

A generous pod in the fields around Thebes in the time of the pharaohs might have produced up to 70–80 milligrammes of sap, an acre of poppy yielding approximately 16 kilogrammes of raw opium. Not all poppy fields were as fertile nor all farmers as skilful as those of ancient Egypt; but the basic pattern would be recognised by growers today anywhere in the world. The raw sap would still need time-consuming processing: it would indeed be the after-treatment as much as the native produce which would determine quality and price. But that is for another chapter.

CHAPTER 3

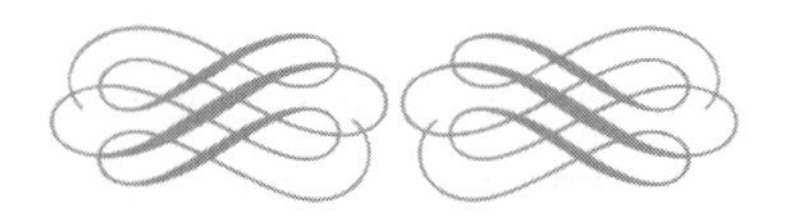

A gift of the gods

THROUGHOUT THE CLASSICAL WORLD – roughly twelve hundred years from Homer in about 800 BC to the fall of the Western Roman Empire in AD 456 – the poppy and its juice were cherished as gifts of the gods. Homer relates the visit of Telemachus, son of Ulysses, to King Menelaus of Sparta. It is not a happy occasion. Remembrance of the sons, brothers, fathers and friends the assembled company had lost under the walls of Troy makes them dejected. Many still suffer from the after-effects of the privations they had endured themselves. Helen, beautiful daughter of Zeus, takes pity on them. Many years ago she had been given a drug by Polydamna, mistress of the god Thoth of Egypt. Homer refers to it as nepenthe (νεπένθε) or 'no sorrow'.[1] Helen pours it into the wine and the effect is miraculous. Pains ease. Grief subsides. Sad memories remain but no longer hurt. At least they do not hurt as much as before. No more tears are shed even by those who had witnessed many deaths.

What was the drug which banished sorrow, calmed tempers and lulled those who drank it into slightly inane but happy conviviality? Some scholars have argued in favour of cannabis or even mandrake. But Jean Cocteau, an expert both on opium and on the classics, had no doubts. Only poppy juice would have had the effect described by the poet. The passage also suggests that in Homer's day the drug was still a somewhat exotic luxury. A thousand years later Pliny the Elder speculated that it was Helen's physical charms rather than her potion which achieved the transformation; but Romans of Pliny's class were incorrigible cynics.[2]

Similar uncertainty surrounds many other potions celebrated by Greek and Latin poets. What was the drug which Chiron the Centaur bequeathed to his adopted son Aesculapius, 'gentle artificer of drugs which dispel sorrows'?

Whatever it was, its effect sent the poet Pindar into iambic raptures.[3] What ointment did Patrocles administer to his friend Eurypylos 'which slayeth the pangs of despair and stops bleeding'?[4] What was the brew which Aphrodite gave to her son Aeneas to lighten his grief? To what 'well-tried herb' did Sophocles refer when he mentioned a concoction which would mollify sorrow? And what was the 'honeyed juice' consumed by the Lotus-eaters and offered to Ulysses' crew? The white lotus does contain pharmacologically active alkaloids – at least one with a Viagra-like action – but the clinical description of the effect of the Lotus-eaters' brew, 'to browse on the lotus and to happily forget all thoughts of return', is pure opium.[5] But the 'honeyed' adjective does not fit a usually unpleasant-tasting liquid. Or tastes may have changed. Or, as so often when trying to interpret the science of classical poets, there is a limit to how far speculation can go. They wrote poetry, not manuals of home doctoring.

Yet opium or 'poppy juice' was undoubtedly widely used in Athens in the fifth century BC, the city's golden age, and for a variety of purposes. Henbane, the poison of choice for state executions, was probably laced with opium. The equanimity with which Socrates continues to chat after downing the lethal drink and as his limbs become numb may testify not only to his composure but also to the Greek executioner's art. His slightly puzzling last injunction to Crito, to sacrifice a cock to Aesculapius, may refer to his admiration for the wonders of medicine even in the chilling context of a judicial murder. Opium was also used for suicides. Hannibal of Carthage came near to bringing republican Rome to its knees; but eventually his luck ran out and he was defeated at Zama (in today's Tunisia) in 202 BC. Banished by his ungrateful countrymen, he spent years trying to build coalitions against Rome among the remnants of Alexander's empire in Asia Minor. He more than once proved his skill as a commander; but in the end Rome prevailed and demanded his extradition. In Libyssa in what is today Turkey and was then part of the small kingdom of Bithynia he unscrewed the handle of his dagger and drank the contents of the phial of Theban poppy juice given to him by his mother. Or so the legend goes: it is doubtful if even the best Theban opium would remain active for thirty years. But he died instantly and the place became a shrine.[6]

And of course opium was used for murder. Dionysius, Tyrant of Syracuse, escaped the assassin's dagger in Schiller's ballad but not the drink laced with poppy juice offered to him a few years later by his son. The younger Dionysius felt that the old man had outstayed his welcome; but, when reproached by his mother, protested that the aged parent had died happy. A few centuries later Agrippina the Younger, last wife of the emperor Claudius, made use of the poppy to get rid of a succession of potential heirs and eventually of her husband, ensuring the succession of her son from an earlier marriage, the darkly handsome Nero.[7]

* * *

But these were celebrities and opium was to be the drug of the people. In the Hellenistic kingdoms which in the fourth and third centuries BC rose from Alexander's great conquests it was still largely an upper-class indulgence, popular in the officers' messes rather than in the taverns. To the rich it offered an escape from the pains of wealth, power and luxury, timeless causes of upper-class misery. In Athens it was fashionable among the city's golden youth. Xenophon put several endorsements in the mouth of his mentor, Socrates. One compares poppy juice to wine:

> Wine tempers the spirit while opium lulls the body to rest, revives our waning joys and is oil to the dying flames of life.[8]

But four centuries later in the Roman Empire opium was as widely consumed as wine. The reign of the Good Emperors – Nerva, Trajan, Hadrian, Antoninus Pius and Marcus Aurelius (AD 96–180) – marked a peak of civilisation whose decline Edward Gibbon would one day lament. Christianity was still underground but only sporadically persecuted. Yawning social inequalities remained; but in the home provinces like Italy and Rome itself ordinary folk were comfortable with themselves. Even the poor had their bread and circuses; and also their poppy cake. 'Cake' was in fact a misnomer. It referred to an unbaked malleable mass, not unlike today's marzipan, in which the sap of the poppy was mixed with sugar, syrup, honey, flour, fruit juice and egg yolk. (Only aficionados liked their opium raw.) In the big cities the delicacy was sold on street corners. The quality varied. Egypt continued to satisfy the luxury market and some brands sealed in Anatolia were ostentatiously expensive. But even the street-corner fare could pleasantly lift the spirit. Outside the big cities much of the drug (not generally considered a drug) was home grown. Prosperous villas and simple allotments would have a poppy bed; and guests would be offered a sip of the home brew. Children probably indulged surreptitiously as they would in ciggies a few thousand years later. A few drops sprinkled on domestic altars would appease grouchy ancestral spirits.

* * *

Latin poets like Virgil followed Homer in dropping pharmacological hints without going into pedantic detail. Anxious to visit his dead father, Aeneas descends into the Underworld. He and his guide, the Cumaean Sibyl, approach the gates of Hades – when, horrors! They are confronted by the slavering three-headed guard dog, Cerberus. But no problem.

> To Cerberus whose necks bristle with poisonous snakes, the seer [Sibyl] flings a bait, a mixture of honey and soothing drugs. Opening his triple throat in ravenous hunger, the beast catches it when thrown. At once he relaxes and sinks to the ground. As meek as a lamb, he closes his eyes in slumber and stretches himself over the cave.[9]

There is no question of the beast or the snakes being killed. In the fullness of time they will recover as would a patient after a routine appendicectomy under general anaesthesia today. But elsewhere Virgil refers to the poppy by name, recommending in his wonderful Georgics *spergens humida melle soperiferumque paparva* (giving dewy honey and soporific poppy)[10] and extolling the virtue of *Lethaeo perfusa papavero somno* (poppies soaked with the sleep of Lethe).[11] Even after becoming the darling of Roman society he remained a country boy at heart, at his happiest when celebrating rustic pleasures. (He was good at singing about arms and men too but heroic deeds never came as naturally to him as they did to Homer.) Yet his enthusiasm for the poppy was shared by his more urban fellow poets. Ovid in *Fasti* describes the goddess of sleep with 'her calm brow wreathed in poppies bringer of delectable sleep'.[12] Elsewhere he praises the poppy which 'induces deep slumber and steeps the vanquished suffering eyes in Lethean night'.[13] Nobody would surpass him in coining the memorable phrase. He also describes how Ceres puts her son, Triptolemus, to sleep with poppies before placing him over the fire to purge him of his mortality. Earlier, a few 'gentle sleep-bearing poppies' have dispelled her hunger.[14] Other poets extolled the juice both of the grape and of the poppy, each for the right occasion: the poppy for seduction and the grape for afterwards.

Even misinformation was available for those who delight in spotting the lapses of the all-knowing. Pliny got it wrong, claiming that poppy seeds were a powerful hypnotic while the latex was useful only in treating mild arthritis. Most citizens of the empire knew better. Yet Pliny erred in good company and his legacy survived. Seventeen hundred years later the Cathloic citizens of Paris were gripped by panic fearing that oil prepared from poppy seeds imported from the Protestant Netherlands led to impotence and sterility. The Medical Faculty of the University of Paris, charged by Louis XV to investigate the matter, took three years and substantial sweeteners from Dutch importers to come to the conclusion that it did not.[15]

* * *

In the first century the emperor Claudius' personal physician, Scribonius Largus, gave detailed instructions for how the pod of the poppy should be lanced, how the juice should be allowed to ooze out and dry and then carefully scraped off. He was a perfectionist, condemning those who allowed the scraping to be contaminated with thin shavings of the pod itself either through carelessness or deliberately to increase the weight. This was one form of adulteration difficult to detect since thin shavings did not alter the taste or the final consistence. He is sometimes credited with actually rediscovering a technique used by the Assyrians; but, like most successful authors of medical textbooks, he was a compiler of other people's wisdom rather than an original investigator. But his expertise was useful in dealings with the emperor and the murderously inclined imperial entourage.

What today would be called the pharmacodynamics of poppy juice was also investigated. Theophrastus, Aristotle's successor as director of Athens's famous School of Philosophy and later acclaimed as the Father of Botany,[16] gave a faultless account of the actions of the drug 'both blessedly beneficial and occasionally lethal'.[17] It was arguably the beginning of the scientific understanding of opium. He is particularly impressive when describing the hierarchy of its effects:

> The higher moral faculties are affected first . . . then the processes of logical thought . . . Base animal urges and essential but uncontrolled bodily functions are preserved longest.[18]

* * *

Beyond sober scientific probings opium was also acquiring a religious dimension. In Rome's multicultural pantheon some gods and goddesses developed a particular interest in the poppy. Sleep and Death, Hypnos and Thanatos, were usually represented as chubby twins clutching poppies as well as each other. On a famous statue Ceres, goddess of fertility, is shown holding a torch and poppy pods, said to relieve the pangs of labour pains. On a vandalised but still attractive wall painting in Pompeii she bends over a pregnant woman and pours poppy juice over her closed eyelids.

In the literary field opium found a welcome in the forerunners of air-travel blockbusters as well as in the more rarefied world of poetry. Lucian of Samosata (on the border of today's Syria and Turkey) wrote in Greek but was probably an Assyrian or one of the Jewish diaspora. He was funny and rude; and, judging by the number of copies of his works (or of works attributed to him) which survive, a best-seller. The title of his satirical *True Story* was clearly conceived with his tongue in his capacious cheek: by his own admission it is anything but true, a spoof rather on tall travellers' tales, as centuries later Don Quixote would be a spoof on knightly romances. In it he describes his arrival on the Isle of Dreams. The place is surrounded by a forest in which the trees are giant poppies and mandragoras. He and his companions gorge themselves and spend thirty blissful days in the happiest of pursuits – sleeping.[19]

* * *

But if priests and poets occasionally sang the praises of the poppy, doctors were becoming cautious. Aurelius Cornelius Celsus was a landed gentleman who practised medicine as a hobby; and his six-volume *De medicina*, completed in about AD 30, is that rarity, a literary as well as a medical masterpiece.[20] His contemporaries recognised his genius and dubbed him the Cicero of Medicine. (This was presumably meant as a compliment though the doctor's aphorisms were far removed from the rolling cadences of the famous orator.)[21] Celsus rightly urged that it is the causes of ailments that medical men should address, not the symptoms, and that to make patients deluded with poppy juice was a confidence trick. 'Pills and potions that for a time relieve symptoms are numerous; but they are often alien and harmful to the stomach and the digestive system.'

> Poppy juice has been used to calm tempers and induce pleasant dreams since the Trojan War and is still popular . . . but the sweeter the dreams, the rougher the awakening.[22]

Neither he nor any of his colleagues referred to the slippery slope of addiction. The Roman Empire and its eastern successor, Byzantium, had an elaborate legal and moral code; but nowhere is there a hint that the use of opium might lead to criminal or immoral behaviour. This contrasts with alcohol-induced misconduct which was well recognised, much written about and punishable with flogging and imprisonment and, under exceptional circumstances, even with death. Does the lacuna reflect innate moderation, a virtue much vaunted by Latin authors but not much in evidence in Roman history, or the relative impurity of the opium available? Perhaps both.

* * *

Unlike Celsus with his clever *mots*, his successor as the medical oracle of the empire, Galen, was a lugubrious professional. More than that: nobody has ever exercised so prolonged an influence on his own or probably on any other profession. Fourteen hundred years after his death the statement that 'Galenus dixit[Galen said so]' still settled disputes.[23] In the sixteenth century Paracelsus inaugurated modern medicine – or so he claimed – by publicly burning a few volumes of Galen's writings, a sacrilege almost as monstrous as trampling on the Bible. Today the force of Galen's authority is not easy to understand.

Born in 131 in Pergamum, then a flourishing commercial centre in Asia Minor, he enjoyed, by his own account, a pampered childhood and a privileged education. After settling in Rome he became the city's most expensive physician. But expense did not matter. The empire was riding on the crest of prosperity and the very rich were very rich indeed. Galen was a consummate name-dropper: 'when I was treating the emperor [or the emperor's brother, sister, cousin, aunt, uncle or poodle]' crop up regularly in his books. These were voluminous: it is indeed astonishing how so busy a practitioner had time for so much writing. Much reading, however, was not required: he tended to refer to other authorities as 'dim-witted', 'superstitious', 'asinine' and even 'murderous'. About his own surpassing wisdom, on the other hand, he had no doubt; and the sheer breadth of his oeuvre is impressive. He discoursed on anatomy, physiology, medical chemistry, surgical instruments, operations and the troubled mind. He himself summed up his achievements:

> I have done as much for medicine as our great and glorious emperor Trajan has done for the empire when he conquered lands and built innumerable roads and bridges. Hippocrates paved the way . . . but it is I and I alone who have revealed the true path of medicine.[24]

But like many men extravagantly successful in their prime, he became embittered in old age. The world's follies weighed heavily on his convoluted mind. Some elderly people become more tolerant of dissent or at least more hesitant about voicing it. Others move in the opposite direction. Galen belonged to the second category. To him in old age Hippocrates became too generalised, Celsus was a good stylist but a lightweight, Theophrastus was windy and most young doctors were fools. Galen himself had never favoured 'stupefying potions': now he expressed himself more forcefully. 'Above all,' he wrote, 'I abhor carotic drugs.[25] Some are merely useless but others are deeply harmful.' He was relatively kind to the poppy – as opposed to the abominable mandrake – as an antidote for older people against 'confusion'. 'Several eminent senators of mature years had benefited from the salve prepared with my own hands; but, to be safe, every detail of my practice has to be followed.'[26]

This would not have been easy, nor was it meant to be. Galen's pick-me-ups, mostly containing opium as their one effective ingredient, rarely contained fewer than twenty other constituents, most of them expensive and some, like extracts of the horns of young male rhinos, exotic. But he was only marginally more extravagant than other Roman prescribers. The daunting complexity of medical prescriptions in the Western world which lasted until the rise of patent medicines and the pharmaceutical industry in the early twentieth century was their legacy. But Galen was firm on one point. The accidental overdosing of children was not unknown and no respectable doctor should prescribe stupefying drugs for the very young.

Yet once again, about the dangers of addiction Galen had nothing to say. When discussing opium in his books, only once does he mention the need for a slowly but steadily increasing dose, and that only in passing. The drain on funds that can lead to crime seemed to worry him not at all. Like Celsus, he was at his most grimly minatory about constipation. The obsession had developed in the West after Rome's Eastern conquests, an importation like opium itself. Egyptian papyri are replete with warnings about the sluggishness of the bowels and ardent in their invocations to deities for painless and plentiful evacuations. In republican Rome such matters were rarely aired: what worried the elder Brutus was 'rage control', not 'bowel control'. But civilisation marches on; and by Galen's time colonic torpor rather than losing one's temper had become his clientele's chief worry.

There was one other concern. Opium taken regularly could affect sexual prowess. It could work both ways and neither extreme was a path to happiness. The possible dangers included not only diminished vigour but also impaired fertility and even diseases of the offspring. Galen himself, a Victorian before his time, tended to circumlocute. Ex-army doctor Dioscorides, the other person to stamp his personality on classical health and disease, had no such inhibitions.

CHAPTER 4

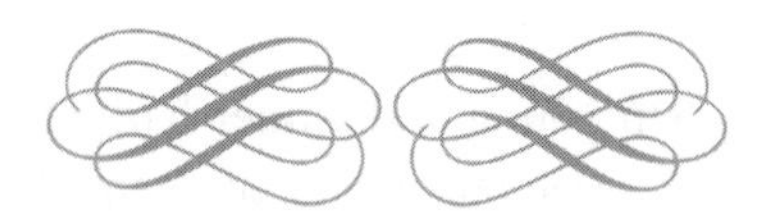

Rival brews

PEDANIUS DIOSCORIDES WAS BORN in Anazarba in Cilicia (today's south-east Turkey) and served for ten years as a surgeon in Nero's army. The peripatetic life let him indulge his passion for collecting plants, both exotic and ordinary, and his martial exploits earned him a plot of land in Macedonia. His profession also fostered a brisk literary style which is not without a certain parade-ground charm. In and after about AD 80 he compiled a six-volume pharmacopoeia in Greek – *De materia medica* in the more widely read Latin translation – which remained the prescriber's bible for a thousand years.[1] In this massive work he discussed the merits and drawbacks of 342 medicinal plants. The poppy was among them; but the space devoted to it was modest. The main competitors deserve a passing glance.

Dioscorides judged the calming effect of *Helleborus niger* or black hellebore, to be the most dependable. It was also a mild laxative. According to the somewhat scurrilous *Bibliotheca* of the writer Apollodorus of Athens, when the three daughters of Proteus, king of Argos, lost their reason and began to dance naked around the palace, the king sent for the handsome Arcadian shepherd Melampus, an expert in herbal medicine. His treatment with an extract of the plant restored the princesses' sanity or at least their sense of decorum. He was well rewarded by his patients and his potion continued to be treasured.

Hyoscyamus niger, a member of the great family of *Solanaceae*, provided the active base of a wide range of healing concoctions. They were particularly recommended by Dioscorides as being less likely than the poppy to blunt the higher faculties. Humans were not the only beneficiaries. The name 'hyoscyamus' implies that it invigorated hogs while the early English 'henbane' warned against its poisonous effect on fowl.[2]

The milk of the stalk of the mulberry tree (*Morus niger*) was a favourite for women in labour: it allegedly eased the pain without weakening the contractions of the womb. On the downside, it encouraged recklessness in man, a boon in battle but a nuisance in peace. King Midas's unwise wish to have everything he touched turn into gold was said to have been uttered under its influence.

The garden lettuce when young and tender had a high reputation as the mollifier of grief. In Ovid's *Metamorphoses*, after the death of Adonis Venus seeks solace by throwing herself on a bed of the plant. Lettuce is not specified in Shakespeare's reworking of the story but it is recognisable on Titian's great canvas now in the National Gallery in London.

The narcotic and soporific properties of the twining hop plant (*Humulus lupulus*) were appreciated independently of the brewing of beer. According to Theophrastus the action was discovered when workers breathing the scented air of hop houses started to spend much of their day in amorous dreams. The answer was ventilation; but hop pillows were still used as an antidote to the painful agitation of King George III during his spells of insanity.[3]

As today, different drugs had temporary vogues. Said to have been discovered by the mountain goats of Crete seeking relief from wounds inflicted on them by the arrows of hunters, the fashion for *dictamnus* (dittany), a plant still grown on the island, had something to do with imperial patronage: it was Caligula's confidence-boosting tipple before performing as a dancer in front of a petrified audience. He had no need to worry: his Teutonic bodyguard looking unfriendly ensured that his guests fell over each other marvelling at his genius.

Cannabis indica was more popular in the East than in Rome. Herodotus writing in the fifth century gave a memorable account of how in the form of a healing bath it was used by barbarians to treat their wounded.

> The Scythians are given the seed of this hemp, creep under rugs and then throw the seeds on red-hot stones. The seeds begin to smoulder and send forth so much scented steam that no Greek vapour bath could surpass it. The wounded Scythians howl with joy.[4]

In Indian lore the plant had names like 'leaf of delusion', 'exciter of desires', 'increaser of pleasure', 'cementer of friendship', 'mover of laughter', 'soother of pain' and 'causer of the reeling gait': most of those would later be applied to opium.

Hemlock (*Conium maculatum*) acquired an evil reputation after the execution of Socrates; but for long its prettily fronded green leaves were popular as inducers of sweet dreams. Even their scent was languorous and a cause of good cheer. But their tranquillising potency varied with the weather, the soil, the rainfall and the technique of extraction. After flowering they could kill.

Most famously, mandrake or *Atropa mandragor* was claimed by Dioscorides to have been a personal gift to him from the goddess of healing but was in fact far more ancient.[5] In Genesis, chapter 30, Reuben, Jacob's eldest son, finds it in the field and brings it to his mother, Leah, to 'open her womb' and let her bear children. Heavily encrusted with myth, legend and nonsense, as an inducer of bliss it surpassed poppy juice for centuries. Bishop Isidorus of Seville in the seventh century devoted a massive tome to it, adding in the Introduction that he himself could not have completed his labours without frequent recourse to it.[6] But mandrake could also arouse shrill hostility. The Elder Cato thundered against the 'young layabouts around the Forum befuddled by mandrake'. St Hildegarde in the twelfth century warned that the plant was formed from the soil from which Adam was created and 'will lull men into vain and unwholesome dreams'. The hair-raising risks attendant on its harvesting and the elaborate precautions to be taken filled many a cautionary text.[7]

* * *

Compared to such rivals, poppy cake when expertly prepared was generally regarded as benign though somewhat unpredictable. In the mid-fourth century AD Diocles of Carystos in Asia Minor recommended it as the best antidote against 'the impotent rage generated by toothache before the offending appendage can be removed by a skilled practitioner like myself'. But he also, in the stilted manner of his time, castigated it for failing to give up its secrets:

> There is much that is still unknown about you, wondrous flower. The potions prepared from your capsule will soothe some but cast others into melancholy. In some its effects will be immediate. In others it will be delayed for many hours. In some the medication will be tolerated. In others it will have unpleasant side effects. But even when you do not abolish the cause of pain and grief, the burdens will no longer prey as heavily on the victim's mind. How does it happen? It is time, poppy, to give up your secrets![8]

The chastisement was perhaps deserved but addressing homilies to plants, however droll, does not solve mysteries. The flower nevertheless continued to be held in affection. In the early empire it began to appear on coins and ornamental seals. In the catacombs it came to represent Christ's blood. During the last bloody anti-Christian campaign by the emperor Julian the Apostate in

AD 361 its representation was explicitly forbidden. The campaign collapsed, Julian was slain and the poppy re-emerged. As symbol of the faith it antedated the Cross. But solving its chemical secrets was not uppermost in people's minds. The Greeks were great poets and philosophers and the Romans superb administrators and builders of aqueducts, but neither had a feel for analytical chemistry. The chemical mechanism of but miracles exercised the minds of the early Christians not at all. The kind of pointless curiosity that animates much modern science and occasionally leads to useful discoveries was alien to these people, at best an eccentricity, at worst a waste of time. The answers to Diocles' whimsical questions would not emerge until the eruption of such an urge in Western Europe in the late eighteenth century. The consequences would then be mixed.

* * *

By the time Diocles penned his elegant stanzas his world was collapsing. Even before the final demise of the Western Empire barbarian hordes – Huns, Goths, Vandals, Visigoths, Ostrogoths and others now all but forgotten – were tearing it to pieces. The loss of poppy cultivation was part of the devastation. Barbarian chiefs and their flock had their own inebriants, above all fermented fruit juice: they had no use for complicated brews that required careful planting and lengthy processing. Opium disappeared from the Western world for almost a thousand years. That it was not entirely lost was due to a series of improbable events not in Europe or even in what remained of the Roman Empire but in the desert of Arabia.

CHAPTER 5

Affyon

THE RISE OF ISLAM remains without parallel. Within ten years of the flight of the Prophet from Mecca to Medina (AD 622 in the Christian calendar; year AH 1 in Islamic chronology) his followers had conquered the Arab Peninsula; and within a hundred years they were masters of a large part of the Byzantine Empire in Asia, Persia, Egypt, North Africa and most of today's Spain and Portugal. In the East under the Abbasid Caliph Harun al-Rashid, a contemporary of Charlemagne, Baghdad became the most populous and most beautiful city in the world.[1] It also became the centre of a unique cultural enterprise.

In Western Europe, slowly and painfully recovering from the barbarian migrations, the arrival of a few scholars from a faraway land was barely noticed. The explorers came in search of Greek and Roman manuscripts. Not many had survived – scrolls of paper are among the most perishable objects of any civilisation – and they were of no value to their illiterate new owners; but the mission was still not without risks.[2] Even to a simple-minded Vandal the fact that an object was wanted made it worth keeping. The search was nevertheless pursued with the dedication of collectors. The finds were carried back to Baghdad where a specially trained academy of translators rendered the texts into elegant Arabic.[3] The works most treasured were practical manuals on science, engineering and medicine; but an eccentric Nestorian Christian, Hunayn ibn Ishaq, became a familiar figure in the capital, roaming the streets and declaiming the *Iliad* in Arabic.

From this enterprise the sheer bulk of Galen's writings ensured that he would emerge as the commanding medical authority. His works were copied, analysed, quoted and abstracted. But during the golden age of Islamic medicine his views never went unchallenged. In contrast to the veneration accorded to his every utterance in the West a few centuries later, most Arabic texts of the tenth to twelfth centuries contain chapters correcting his mistakes. Even more determined was Arab resistance to the Galenic spirit of doctoring. Galen's disapproval of 'carotics' in particular found no echo in the Muslim world. The Prophet's prohibition on alcohol did not extend to painkillers and other beneficial medications. Poppy cultivation and the manufacture of opium or *affyon* (or *af-yum, ufian* or *asiun*) soon flourished in the Abbasid Caliphate, and a generation after the Arab invasion the poppy fields of Thebes were blooming again.

Cultivation was boosted by a new development. In the Greek and Roman world poppy cake was largely for home consumption or for the local market. Arab traders now carried the produce to India, Indonesia, China and along the North African coast to Moorish Spain. They travelled by dhow – they were skilled navigators – by donkey, by camel or, when necessary, on foot. Trading was in their blood. Their passion for a deal was mocked by the reserved Chinese and laid-back Indians, but what the travelling merchants were offering was good value: their reputation was more precious than gold. In some parts of the world opium was a novelty and aroused apprehension. But the traders persisted. They established staging posts along their routes, grand caravanserais in opium-friendly lands, safe houses and secret caves where their ware was officially banned. In these stations they rested, did their wheeling and dealing with local merchants and, most important, prayed. Though organised persecution was rare, Allah's help in hostile lands was essential. So was in many regions an armed escort. Most staging posts provided those, together with local guides. Travelling with treasure across bandit-infested deserts and mountains was dangerous and the routes chosen were often off the map. As they still are. For many of the movements of illicit drugs today follow the same paths; and the dugouts where Arab opium merchants sheltered a thousand years ago are the holes where home-made bombs and caches of heroin are discovered in Afghanistan today.

* * *

Happily, opium soon became welcome even where it had to overcome initial fears. Mostly in conjunction with traditional remedies, it was recognised as the best medication against an amazing variety of ailments. It stilled coughs, colics, headaches, rheumatic infirmities, obsessive fears, insomnia and rages. It quietened noisy children. It eased the last hours of the dying. In some Asiatic countries, notably in Java and other islands of the Indonesian archipelago and the far western provinces of China, poppy growing took root; but, fortunately for the international trade, Egypt and Mesopotamia remained market leaders. Rightly so, for the harvesting and processing of the plant was still an art requiring well-honed skills. In time high-quality poppy cake wrapped in trademark packaging of locally grown palm leaves became an international currency.[4] The material was compact, stable and its quality could be checked.[5] The last of course needed trained noses and palates. Tasting became a valued expertise, as it would be in Turkey and Europe a thousand years later.

* * *

Modern pharmacology is an Islamic creation. Even the best Roman pharmacopoeias had listed less than a thousand items. The pharmacopoeia of Abdullah ibn-al-Baytar who died in 1248 described more than 3,000, including 1,800 botanical species, 245 mineral preparations and 987 medicines of animal origin. They included plants never heard of in the Mediterranean world.[6] In famous academies Arab chemists – or al-chemists as they were soon to be known in the Christian West – developed the arcane techniques of filtration, crystallisation, sublimation and distillation. Even more innovative were the city pharmacies of Cairo, Damascus, Baghdad and other metropolitan centres. Far from the noise and bustle of the souks, men of taste and learning gathered to talk literature and philosophy. They would also sample the pharmacist's latest concoctions of *affyon*, a touchstone of his professional skills.

* * *

Two Islamic doctors would stamp their mark on Muslim medicine and posthumously on medical practice in Christian Europe. Abu Bakr Muhammad ibn Zakariya Razi, known in medieval Europe as Rhazes, was born in Tehran in 825 and spent his youth in the service of the great Persian ruler, Mansur ibn-Ishaq. When still a young man he wrote the first treatise on how to examine doctors before appointing them to positions of trust. 'Do not trust healers who know all the answers,' he advised. And 'the family of a conscientious doctor should always feel neglected'. The guide caught the attention of the Caliph in Baghdad who invited the author to preside over the medical showpiece of the Abbasids, the Al-Mu-tadidi hospital. The institution blossomed under Rhazes and his fame spread. He wrote on ethics, philosophy, music, theology, natural history and chemistry as well as producing a popular tome on home doctoring. A painful glaucoma blinded him in middle age; but, dictating to a 24-hour rotation of scribes, he kept a meticulous record of the effect of every kind of treatment. Purgation helped, as did the avoidance of anxiety; but the only medication which gave him true relief was an infusion of *affyon*. 'After this wonderful drug,' he recorded, 'the hurt remains but it hurts less.' In many parts of the Muslim world physical blindness was – as it still is – a gateway to spiritual enlightenment and the doctor died deeply revered.

But not by all. 'He spread himself too thin. He should have stuck to the examination of urine at which he excelled,' wrote his notoriously irreverent successor, Abu Aly al-Hussayn ibn'Abdallah ibn-Sina, known in the West as Avicenna. The heir was born in AD 980 in a village near Bokhara, the son of a tax collector. Bokhara today is the stamping ground of smooth oil men but in the tenth century it was a centre of Islamic learning, the arts and the crafts. Avicenna amazed his elders by reciting the Qur'an by the age of ten and practising as a healer by the age of sixteen; but he was no bookworm. In his autobiography he records that much of his later writing was done on horseback during military campaigns, in flight from his enemies, escaping creditors, in prison fearing execution, as vizier or chief minister to the Emir, and recovering from stupendous drinking bouts. The last may seem surprising for a Muslim sage; but, unlike Rhazes, Avicenna cultivated an image as one of the boys, boasting of his love of wine, women, song and the poppy. His death at the age of fifty-six was attributed to overindulgence in one, two, three or probably all four.[7]

But his self-portrait as a debauchee is not entirely convincing. Addicts in all ages have written pieces of inspired poetry, composed memorable music and created minor works of art; but none has compiled two monumental encyclopaedias, one of science and one of medicine. The latter, known in the West as *The Great Canon*, became, in the words of Ambroise Paré, surgeon to five French kings, 'the most famous medical book ever written'; and Dante honoured its author by placing him in the First Circle of the *Inferno*.[8] Unlike many works of high historical significance, it still overflows with wisdom. 'The unhappy eat either too much or too little. They should be made content rather than forced to eat or made to fast.' 'Patients seek your advice for two reasons. One is pain. The other is fear. Fear is the more powerful of the two. *Affyon* may relieve both but it is a remedy to be handled carefully.'

Carefully was a key word; and Avicenna became the Muslim world's unchallenged authority on opium. 'Used with circumspection, it may achieve by stealth and through the mind what more drastic medications of the body fail to do. It is one of Allah's signal gifts for which He should be thanked every day. But even divine gifts can be mishandled by bumptious men.'[9] He prescribed the drug in a variety of forms – as a 'cake', as an infusion, as a poultice, as a suspension, as a suppository, as a plaster and as an ointment. Whether or not he actually invented the technique, the *Canon* described light scarification of the skin to promote absorption from an otherwise non-absorbing surface. The method was later adopted for inoculations against smallpox. For cranky infants the recommendation was an equal mixture of poppy and dandelion juice. In exceptional cases – that is for patients to whom the taste of all forms was intolerable – fermented grape juice could be tried as a 'carrier'. The resulting linctus could not have been far removed from Sydenham's laudanum.

Well ahead of its time, the *Canon* also advocated precise units of dosage. The dose of *affyon* was usually specified in *wads*, lumps the approximate size of a pea. The *wad* – or something like it going by that name – was still used in South East Asia in the early twentieth century. How much activity a *wad* represented in Avicenna's time is uncertain; but the wide range of dosage recommended hints that many recipients may have already been habituated.[10] The *Canon* also suggests that in the Abbasid Caliphate the drug was not prohibitively expensive. Avicenna railed against medicines too costly for the poor; but 'kindly opium you can buy cheap in the market place'. He did, however, sound warnings: 'You must guard against too heavy adulteration: the trader should be known to you and be a reliable source.' There were words of wisdom for doctors too. 'The soothing effect of the drug is invaluable when patients are too dejected to listen to your advice; but the benefit is temporary and willing compliance is often followed by worse recalcitrance than before. Do not expect any thanks after the effects of the poppy have worn off . . . Collect your fee before you dose your patient with the poppy.'[11]

Two massive encyclopaedias might seem enough to fill a busy professional life, but Avicenna wrote a total of 270 works, including a slim volume of mnemonics 'for the hard-pressed practitioner in the market place without the time to ponder knotty clinical problems.'[12] His sarcasm directed at self-important colleagues must have made him many enemies; but his fame outlasted the hostility. Few other physicians would be remembered with veneration four hundred years after their death in a country they could never have heard of.[13] And some of his haunting non-medical poetry, for long negleted, stangely pre-echoed the thoughts and sighs of later literary addicts:

> How I wish I could know who I am
> What is it in this world that I seek?[14]

Did Baudelaire read this?

* * *

The Jewish contribution to the use of *affyon* was important. In the Iberian Peninsula the Moorish conquest in AD 711 was welcomed by the first diaspora and inaugurated a dazzling period in the arts and sciences. The Umayyad dynasty regarded the Jews as retarded cousins whose assimilation to Islam was only a matter of time; and the preparation of healing and calming potions was a field in which the people of Israel were recognised as experts. For the first time since the collapse of the Roman Empire fields of white poppies bloomed on the mountain slopes around Granada and the Caliph and his court in Cordoba demanded the best of their produce. But conflict between sophisticated rulers and fundamentalist tribes was already alive in the Muslim world; and peace and plenty rarely lasted for more than a few generations. Cordoba fell to fanatical Berber warriors crossing the Straits of Gibraltar in 1013 and thousands of Jews were massacred. Among those who escaped was the family of Moshe ben Maimon, better known today as Maimonides.[15] After much wandering they settled in Cairo under the stern but racially tolerant rule of Sultan Salah-ad-Din Yussuf ibn Ayyub, known in the West as Saladin.[16]

Maimonides soon became a celebrated healer. In the year 1185, when he was promoted to become physician to the Sultan, he was also invited through clandestine channels to become the personal physician of the crusading 'Frankish king', probably Richard the Lionheart. The second invitation he wisely declined. He was a doctor who inspired extraordinary trust: as a contemporary Arab commentator noted, 'Galen's medicine is for the body; but Maimonides' is both for the body and the soul.' One gets glimpses of him at work from his extensive correspondence.

> My new duties to the Sultan are becoming heavy . . . I visit him every morning if he feels ill or if one of his children or a favourite in the harem is sick; and I may have to spend most of the day at the palace. When I get back to Fustat [the poor outskirt of Cairo where he lived] I find the courtyard crowded with sick people . . . high and low, gentiles and Jews and of course Muslims, theologians and judges . . . I dismount, wash my hands and have my only meal of the day in a great hurry . . . Then I attend to patients sometimes till dawn. I talk to them sometimes lying on my back because I am too weak to get up. But I have to get up to visit those who cannot visit me because they are too enfeebled, in too much pain or because they are dying. To those in pain and to those dying I take affyon which some cannot afford and some do not know how to use. When they cannot drink it, they are helpless. I instruct friends and relations how to administer it to those who are too sick to swallow. It can be made palatable with honey and it can be given as an enema or as suppositories. It is a wonderful panacea. It is especially important to give it to patients

who will not recover. Prayers are helpful but are not enough. The dying must not be left to suffer. I hope one day to write a book about this.[17]

He never did write that particular book; but his approach to the doctor–patient relationship (a thousand years before that pompous phrase was invented) was summed up in the aphorism: 'May I never see in any of my patients anything but fellow sufferers.' The statement was in harmony with the spirit of Islamic medicine at the time but already differed from the evolving Western tradition. In the West the cultivation of a proper professional distance between patient and healer developed and practised by Roman doctors survived the fall of the empire and the advent of Christianity. It was deemed – as it still is – essential for the management of serious illnesses. To feel sympathy for patients was permissible. To expend thought and care on them was mandatory. But a feeling of fellowship was not calculated to improve the outcome.

Many differences flowed from this. Maimonides and his colleagues had no patience with silent suffering: 'The Lord gave us tears to shed . . . Do not try to stem their flow. When potions and vapours fail to ease your suffering, lamentations often help to relieve it.'[18] Such advice would have outraged Seneca or Marcus Aurelius. Nor would the great medieval doctors of the Church have approved. To them the purifying effect of suffering remained a core doctrine. Some of that suffering could be relieved but some would always remain beyond the art of the physician. And happily so, for some of the suffering would weigh in favour of the supplicant when facing the final Seat of Judgement. Islamic doctors, by contrast, were taught that Allah had created no painful ailment without creating a remedy for it, even if the remedy was still beyond the wit of humans.[19]

* * *

Moral endorsement of *affyon* was boosted by economics. Both the upkeep of an elegant court and holy war against the Christians were drains on the Sultan's treasury; and taxing poppy growing in the Nile Valley was a useful source of income. On occasion *affyon* could also be used in diplomatic bargaining: generous presents of the drug rather than threats of retaliation kept the turbulent tribes along the Upper Nile from military adventures. And it could be used as a chivalrous gesture. Muslim historians relate that when Richard the Lionheart lay in his tent before Acre prostrate with fever in the scorching heat, his foe sent him a caravan laden with ice from the Syrian mountains and a generous supply of Theban opium. True or not, the Lionheart was soon fit enough to slaughter many more Saracens.

* * *

Some branches of medicine developed under Islam centuries before they were recognised in the Christian West. Though medical injunctions in the Qur'an are few, one of them commands the faithful to treat the insane with compassion. 'Pagan superstitions' about the mad being possessed by the Devil were firmly rejected, leading to the establishment of the first specialist hospitals for the insane.[20] But even in specialist hospitals the best method of dealing with the violent and dangerous continued to be divisive. One school held that regular dosing with *affyon* was the best approach. Avenzoar, the Iberian doctor who died in 1173, listed twenty-three different mixtures each designed for a different kind of insanity. 'Some of these medications are costly and sustained dosing may lead to difficulty in feeding the patients. But this is preferable to keeping them shackled. Cages should be used only as a last resort, beating never.'

But more sombre perceptions were also gaining ground. Abulrayan al-Biruni noted in 1130 that travellers to the holy city of Mecca sometimes fatally overdosed themselves. The hot weather and the intensity of the spiritual exercises could impair practical judgement. And the doctor was increasingly concerned that those who had become addicted or even habituated sometimes became 'dangerously obsessed with their craving'. For the first time the prospect of lawlessness and crime loomed: '*Affyon* is a powerful and effective narcotic for all pains; but it can darken the sight, harden the spirit, weaken the mind, degenerate understanding and corrupt judgement.'[21] Abdullah ibn-al-Baytar enumerated 'the ten basic symptoms of an overdose: lethargy, lockjaw, uncontrollable itching, watering eyes, paralysis of the tongue, discoloured extremities and nails, profuse cold perspiration, painful but ineffective vomiting, convulsions and death'. Little could be added to the list today: it was obviously based on observing a not uncommon emergency. The best countermeasures were to keep patients talking and walking and to administer heroic purges; but even those could fail.

More frightening were the possible effects of the drug on the soul. It was in Islamic writings that the figure of the junkie (as he would be called a thousand years later) made his pitiful appearance. In addition to physical decrepitude, *affyon*, according to the judge and scholar Muhammed al-Zarkashi, could

> generate . . . laziness and sluggishness, kindle dangerous illusions, diminish the power of the mind and reduce all restraints . . . It makes the complexions yellow, blackens the teeth, riddles the liver with holes, inflames and rots the stomach, creates a bad odour in the mouth and adds a film and poor vision to the eye. It turns a lion into a beetle, makes a proud man a coward and a healthy man sick. If he eats, he cannot get enough. If he is spoken to, he does not listen. It makes a clever man dumb and a sensible man stupid. It takes

> away every manly virtue and puts an end to youthful generative prowess. It destroys the mind, stunts natural talent and blunts judgement. But it is the decline of the soul which is most grievious: nothing can stand in the way of acquiring more of the drug. There is no right and wrong, no do-s and don't-s. And he suffers. He is a glutton but cannot eat and he is sleepy but remains remote from slumber. He has been driven out of Paradise and threatened with Allah's curse unless he gnashes his teeth in repentance and puts his confidence in Allah.[22]

But the judge had no illusions. In reality the end was usually death and a swift journey to hell.

The uniquely two-faced character of opium had never before been so trenchantly exposed. The drug could be a beatific gift of Allah, Allah be praised. But it could also be the Devil's instrument leading to perdition. In sufferers from pain it could assuage the body and alleviate the mind. But it could also destroy valiant individuals and plunge their families into destitution. (That it could undermine the vigour of whole nations was not yet apparent.) How to act in the face of such an existential challenge?

No answer was forthcoming. By the end of the thirteenth century Arab medicine was in decline, as were the great Arabic-speaking caliphates. Cordoba fell to the Christians in 1236 and Baghdad was sacked by the Mongols in 1258. The Ottoman Turks, the dominant Muslim power for the next five centuries, were fighters, empire-builders, calligraphers and the makers of superb ceramics but had little interest in the healing arts. But in the wake of the Crusades opium was beginning to trickle back to Western Europe.

CHAPTER 6

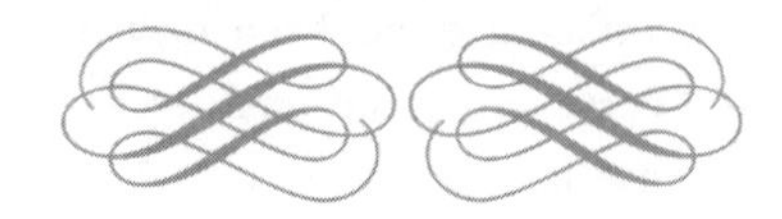

The sleepy sponge

WHILE ISLAM WAXED IN GLORY, Western civilisation floundered and almost sank. Even in the Neolithic Age poppy cultivation required a degree of social stability. For centuries in Europe there was none. Intoxication was of course common: for long periods in many parts of the continent drunkenness was probably the customary state of humanity. But news of a gentler and yet more exciting intoxicant was brought to the West – like so much else, good and bad – by the Crusaders.

The first of what were to be extraordinary migrations was triggered by the religious upsurge at the end of the first millennium. To the disappointment of many, Year 1000 did not bring the Second Coming; but daily expectation of it turned gazes to heaven – and also to the other place. Eternity in the latter would be a long time. In response to a papal call to conquer the land where Jesus had dwelt, taught and died in exchange for unprecedented pardons even for unspeakable crimes, tens of thousands flocked to the Cross. After a long and arduous campaign Jerusalem fell to the First Crusade on Sunday, 15 July 1099. It was a glorious moment (followed by the slaughter of the defenders) but the small Christian kingdoms in a sea of Islam could not survive. Other Crusades followed, not all of them propelled by pure faith. And fantastic stories of oriental elixirs inducing unimaginable delights and superhuman valour began to circulate in feudal castles and well-endowed monasteries at the other end of the world. Saracens primed on poppy juice lost all fear. Tartar horsemen and their mounts shed all fatigue. No wonder Christian campaigns sometimes ran into the sand.

The first samples of opium carried back to Western Europe must have been in a parlous state but they created a demand; and demand tends to conjure up providers. The real winners in the Crusades were arguably neither the

Christians nor the Saracens but a scattering of small settlements on islands at the northern end of the Adriatic. A century earlier these were poor fishing villages. Now the loot that accrued to them from ferrying crusading armies and their supplies to and from the Holy Land transformed them into the Most Serene Republic of Venice, a commercial power.[1] The Grand Council even blackmailed the Fourth Crusade to destroy the Republic's former trading partner but now its only rival in the eastern Mediterranean, Byzantium.[2] Yet keeping out of the actual fighting, Venetian traders maintained discreet but cordial relations with their Arab counterparts.

When, in the wake of hostilities, a clamour for Eastern goods arose in Western Europe, the city on the Lagoon was in an unrivalled position to respond. Seven days a week and in all seasons money men, middlemen, con men and the occasional honest trader crowding on the Rialto (their every blinking of an eyelid reported to the ruling Council of Ten) settled the price of the new commodities. The businessmen came from three continents, their skin colour ranging from the palest pink to ebony; but the combined sensitivity of their noses for what the markets would bear was infallible.[3] Almost invariably the mark-up on opium was among the steepest.[4] Marco Polo, one of the city's privileged but temporarily disgraced youths, travelling through Central Asia in 1275, met a caravan of a hundred camels laden with goods destined for his family's warehouses. The load included textiles, carpets, jewellery, glass, silverware, edged weapons, spices – and opium. In weight the last accounted for little but the profits from it would rival the profits on the rarest of jewels.

Where exactly did these profits came from? Mandrake still provides a negative marker. Throughout the Dark Ages an extract of the root of the mandragora plant remained the most popular painkiller. Illustrated Books of Hours depict the bizarre ritual of its collection: on extraction from the soil (ideally blood-soaked or soil surrounding the gallows) the plant was believed to emit a scream which could strike dead anybody unfortunate enough to hear it. The digging therefore had to be accompanied by a blast on a horn by a bystander.[5] But, despite the danger, the product was cheap. For a long time opium in the West was for kings, queens and princes of the Church. Mandrake was for everybody else.[6]

* * *

While still a precious commodity the processed brown mass of dried poppy juice was carefully weighed, parcelled up, tightly packed in waterproof cloth and carried from Venice mostly by Jewish merchants.[7] Later, as crusading zeal abated and it was safe for them to do so, they were joined by Arab doctors.

Both acquired a high reputation for skill in preparing and administering health-restoring medications. Holy synods thundered against letting such vermin near the bedside of about-to-die Christians; but fear and pain are timeless breakers of anathemas.

* * *

A second chink through which opium would start trickling back to the West was the foundation in about 1050 of the first new medical school since classical times. The legend of it being launched by a Latin scholar, a Jewish doctor, an Arab merchant and a Greek teacher was figuratively true. Protected by the ferocious but modernising Norman dukes of Sicily Salerno in southern Italy lay at the crossroads of civilisations. In 1063 Alphanus, a Benedictine monk from nearby Monte Cassino, travelled to Constantinople to study ancient Greek and Arabic texts and later taught at the medical school. His *Premnon Physicon* is the first surviving book to mention poppy juice in modern European literature.[8] His information was probably second-hand but the effects he described were the more enticing. A slightly later work, the *Liber isagogarum* (Isagogue meaning Introduction) by another dedicated monk and teacher, Constantinus Africanus, became a foundation text in the medical schools which sprang up in the wake of Salerno.[9] Bologna, Padua, Paris, Oxford, Cambridge, Prague and Vienna all possessed copies and learnt from them about the magic of the poppy. Outside academe a popular compilation of homely wisdom known as the *Regimen sanitatis salernitanum* (Salerno's Regimen of Health), a book of merry jingles composed some time in the early thirteenth century and eloquent about the joyous plant, became a best-seller. After centuries of circulation as manuscripts more than 200 editions were printed on the newly discovered printing presses in the sixteenth and seventeenth centuries in most European languages as well as in Latin, Hebrew and Persian.[10]

* * *

The reappearance of opium in the West was well timed to popularise the most important medical invention of the Dark Ages, the 'sleep-bringing sponge' or *spongia somnifera*. Primacy is never easy to establish in medicine; but the first detailed description of this ancestor of the anaesthetic mask seems to have been contained in the thirteenth-century *Chirurgia* by Theodoric, Bishop of Cervia, who in turn attributed the invention to his father Hugh de Luca. To ordinary folk *spongia somnifera* was too much of a mouthful and the invention soon became known as 'the sleepy sponge':

> Take the juice of the unripe mulberry, of hyoscyamus, of the juice of the hemlock, of the juice of the leaves of the mandragora, of the juice of the poppy capsule, of the juice of the wood ivy, of the juice of the forest mulberry, of the seeds of lettuce, of the seeds of the dock which has large round apples – each an ounce; mix all those in a brazen vessel and then place it in a fresh sea sponge. Let the whole boil until the sun lasts on the dog days so that the sponge consumes all the fluid.
>
> As often as there shall be need of it, place the sponge into hot water for an hour and let it be applied to the nostrils of him who is to be operated on, until he has fallen asleep. When the surgery is finished, in order to awaken him, apply another sponge, dipped in vinegar to the nose or throw the juice of the root of fenugreek into the nostrils. Shortly he will awaken.[11]

Or not. The danger of a permanent sleep is apparent from the frequent commands to make the patient inhale vinegar whenever 'the slumber appears too deep.'[12] Although the juice of the poppy capsule was only one constituent of a complicated cocktail, one wonders if on prolonged boiling with other juices some of it might have undergone a chemical transformation, known today as acetylation.[13] This could have generated traces of heroin, many times more potent than opium.[14]

* * *

The sponge was especially valuable to surgeons. In their bloody aprons they were still a lowly species compared to physicians in their billowing academic robes; but an elite was emerging even among the lower breed. Lanfranc was an Italian master surgeon who settled in Paris and there wrote his *Chirurgia magna* and the slightly abridged *Chirurgia parva*. Translated into many languages the smaller version became the surgical do-it-yourself manual of the age.[15] Despite its proclaimed practical aim, a relatively long section was devoted to what today would be called the basic sciences, including a copiously illustrated chapter on astrology. After the gruesome realism of operations the celestial sections are reassuring in their prognostic certainty; but the practical pages too contain pleasant surprises. One might have thought that in a brutish age the loss of male hair would be one of man's less pressing concerns. But one would be wrong. To cure a knight's baldness – both the total and the partial variety – the author recommended eighteen different ointments, depending on the age, social status and the colour preferences of the 'poor afflicted'. Poppy juice – 'a few drops will suffice' – was a regular constituent, especially when the application was expected to be painful. (What might have caused the pain is not clear.) Otherwise the drug was mentioned in connection with stemming the flow of diarrhoea but with the proviso that it was to be used only 'when a reliable preparation is available: otherwise dangerous'. It was still a costly remedy.

Lanfranc's pupil, Henri de Mondeville, was a grander personage. He graced the courts of Philip III and Philip IV of France, lectured at famous universities and had no difficulty in obtaining the 'essence of the best Egyptian poppy sold by reliable Jewish merchants'. Yet, though he recommended the extract for the mixture used for his *spongia somnifera*, he too had a restricted use for it outside surgery, mainly to combat intractable headaches.

The man who carried on de Mondeville's work, Guy de Chauliac, a native of Gascogny and a graduate of Bologna and Montpellier, was acknowledged for generations as the 'Model of the Christian Chirurgeon'. He himself stated in the Introduction to his *Chirurgia magna* that one of his purposes was to raise the status of his craft to that of a learned profession. The book is almost oppressively comprehensive, containing 3,299 references to other named works, including 890 quotations from Galen alone, foreshadowing the computer-generated monstrosities of today. He continued to advocate the soporific sponge generously infused with 'the best available opium', partly no doubt to save his patients from suffering but also to prevent them from 'thrashing about'. Violent reactions to pain were among the chief risks of medieval surgery both to patient and surgeon: even the best poppy juice was no guarantee against injury.[16] Like most of his contemporaries, the Gascogniard enlivened his teaching with anecdotes of his surpassing virtuosity, interspersed with bloodcurdling tales of the incompetence of his rivals. 'If you want to live,' he admonished his readers, 'make sure that your surgeon follows my instructions to the letter and that he is not inebriated.' The infamous Henri de Natteville, not otherwise commemorated, when under the influence of alcohol apparently killed six patients in a row by overdosing them with the sponge. By contrast, the brilliantly coloured illustrations of de Chauliac's original work show his patients, some wearing a crown or a mitre as their sole garment, grinning with delight as they have their perianal abscesses lanced by the master.[17]

Guy de Chauliac's English contemporary, John of Arderne, served under John of Gaunt in the Hundred Years War and then settled in Newark on Trent. Here, despite the distance from metropolitan centres, he was visited by the great, the good and the rich. He himself stated that he endeavoured to confine his practice to the high-born and wealthy who could afford his fees and were likely to benefit from his ministrations. He made an exception for genuine mendicant friars, former crusading knights with scars to show for their exploits, and his own children, grandchildren, nieces, nephews and their 'agreeable flock'. Not only were his charges exorbitant but he was also unsparing in the use of such expensive medicaments as extract of poppy capsules of the highest quality. He found it useful in ointment form as well as when consumed by mouth or inhaled. He also recommended a mixture of opium and henbane in wine for the worst kind of 'mental excitation'. The only risk attendant on this 'surpassingly excellent remedy' was the usual: instead of recovering their normal vile temper, his patients might become permanently becalmed. To prevent such a mishap Arderne recommended pulling the patient's beard every few minutes. He had no advice for the unbearded.

One cannot help being impressed by these majestic medieval texts, their mixture of the scientific and the extravagantly superstitious. Guy de Chauliac might recommend the urgent inhalation of a powerful opium mixture (always to be carried in readiness in a sealed container) for 'strangulation in the chest', a possibly life-saving measure;[18] and, on the next page, prescribe against epileptic fits that the patient write 'in his or her own blood on thrice-blessed parchment' the names of the Three Wise Men and recite thirty-three Ave Marias daily for three months. But carried out in the right spirit perhaps that too was life-saving.

* * *

These brave beginnings were cut short by the greatest medical disaster in the history of Europe, perhaps the world. The Black Death started in January 1384 in the Sicilian port of Messina where it was probably carried by three Genoese galleys. Genoa had fortified trading posts along the Mediterranean and the Black Sea and a few months earlier the crew had been besieged by a nomadic tribe in such a fort in the Crimea. An epidemic among the besiegers, almost certainly the plague, forced them to withdraw; but before withdrawing they catapulted their dead into the fort. It was the first known example of biological warfare and it was as effective as it would be today.[19] By the time the galleys arrived in Sicily 'disease clung to the very bones of the crew'. From the crew it spread. Crossing to the mainland it killed more than two-thirds of the population of Naples, including the archbishop and his entire clergy. It then progressed north. People died in their homes, in the churches, in the streets or on carts trying to flee. The illness then crossed the Alps and engulfed Germany and Western Europe. About a third of the population succumbed; and it was a painful death. 'The pestilence was so violent that after the first few days no man could succour another'. In Lyon the 'wailing and screaming could be heard day and night and the stench of bodies hovered around the countryside for many miles'. The previous centuries had not been times of ease in Europe; but this was a catastrophe beyond medieval comprehension. Statistics are often quoted and are horrifying; but they cannot convey the psychological impact of the calamity. The widespread madness of the flagellants and its consequences of torture, pogroms and murder were one side effect. But taking a more detached and perhaps inhuman view, the disaster also blew away the cobwebs of centuries. A desert of doubt, disbelief and desecration was the right soil for the growth of three revolutionary movements, the Renaissance, the Reformation and the cosmic and terrestrial explorations. They would transform everyday perceptions and provide the background to the career of the inventor of laudanum – or something like it.

CHAPTER 7

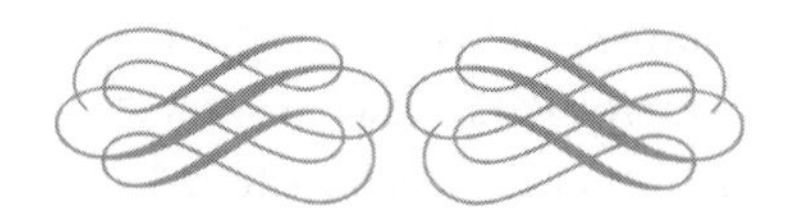

Greater than Celsus?

OPIUM HAS ALWAYS been divisive; and the man who launched its incarnation under the famous name *Laudanum* remains appropriately controversial.[1] To his devotees he is a maligned prophet. To his detractors his very name, Philippus Aureolus Theophrastus Bombastus von Hohenheim *sive* Paracelsus, spells buffoon. This particular jibe is unjust. He was the only son of Wilhelm von Hohenheim, a Swabian doctor of the House of Bombast, the word derived from the prosaic *Baumast* (building mast), not from bombastic; Theophrastus was in honour of Tyrtamus Theophrastus of Lesbos, a respectable follower of Aristotle; Aureolus was a pet name given to him by his mother because of his shock of blond hair; and Paracelsus, literally greater than or at least the equal of Celsus, expressed an aspiration rather than a boast.[2]

But he was certainly something of a fantasist as well as a genius. Born in the small Swiss pilgrimage town of Einsiedeln in 1493, he learnt the rudiments of doctoring from his father; and, after studying at the University of Vienna and perhaps at universities in Italy, he set out on his travels. These, according to his own account, took him from Portugal and Spain to France, England (where he claimed to have worked in both Oxford and Cambridge), the Netherlands, Sweden, Moscow, Constantinople, Rhodes, Crete, Alexandria, Sicily and back to Strasbourg. Such a journey taking less than two years in the Europe of his time suggests a sizable input of romancing, especially when punctuated by a spell as an army surgeon in Flanders and a course of lectures to the pope and his cardinals in Rome.

Both exceptional luck and mostly self-inflicted misfortune played a part in his subsequent career. After being expelled from various German cities where he tried to settle as a surgeon, he cured the gout of the famous printer Johannes Frobenius of Basle. Frobenius became a patron and in the Frobenius house he met some of the leading intellects of Europe. In a letter of thanks for his treatment Erasmus of Rotterdam wrote: 'I recognise the deep truth of your mysterious words [which always accompanied Paracelsus's healing ministrations] not by any knowledge of medicine which I have never studied but by simple feeling.'[3] It was not a negligible compliment from the man widely regarded as the greatest thinker of his day. Appointed city physician combined with a chair in the university Paracelsus at once aroused both scandal and enthusiasm. Many thought that he was bewitched and he was probably often high on drugs. He lectured in German instead of Latin; he wore a carpenter's apron instead of academic robes; and his ardour lent his oratory exultant power. But his utterances both by word and in writing became increasingly provocative: 'When I saw that nothing resulted from current medical practice but pain, killing and laming, I decided to abandon this miserable art and seek the truth elsewhere.'[4]

True perhaps but unflattering; and where exactly this 'elsewhere' was – and is – was never revealed.

On St John's Day he ceremoniously cast the works of Galen, Avicenna and other masters of the past on the bonfire.[5] Public burnings of papal bulls were popular in Protestant cities like Basle, just as conflagrations of the Reformers' works were the highlight of Catholic junketings in Koblenz, fifty miles further east, but Paracelsus was going over the top.

> Ye physicians, so-called, of Paris, ye of Montpellier, ye of Swabia, ye of Cologne, ye of Vienna, ye of Meissen [and many others] and those who dwell on the Danube and the Rhine, ye islands of the sea, thou of Italy, thou of Dalmatia, thou of Sarmatia, thou of Athens . . . ye Greeks, ye Arabs, ye Israelites, after me and not after you! Even in the remotest corner there will be none of you on whom the dogs will not piss. But I shall be monarch and lead! . . . So gird your loins and forget the past . . .[6]

Eventually even his students, at first enthusiastic, turned against him. A warrant for his arrest was issued. The charge was blasphemy, a reference presumably to his frequent invocation of the Almighty; and that carried the death penalty. He escaped from God-fearing Basle at night leaving his crucibles, alembics, books and manuscripts behind.

* * *

The highways of Europe would be his home for his remaining years. In some towns he was fêted but more often threatened by the 'pack that is large though their art is small'. He was a short, fragile-looking figure with a limp and an oversize head; but he must have been tough. He never ceased to write. He lived like a tramp; and yet his books were published by famous printers and discussed in academic circles everywhere in Europe. In 1531 he briefly returned to his place of birth where, in the little *Wahlfahrtskirche*, he underwent a spiritual experience. A mystical tract was the outcome of what he claimed was a friendly exchange of views with the Almighty. But was it God? Or the Devil? Or, most likely, the voice of opium? Many later addicts would adopt a strangely similar tone: instead of praying for divine help in their own troubles, it is they who seem to be reassuring their troubled Creator that all will be well in the end. The promise that 'I shall be Your witness' occurs more than once.

The fact that the tract was not overtly blasphemous only added to the disquiet it aroused. Better be on the safe side and burn the author. Others demurred. One never knew with miracle doctors: they might come in handy to stop a life-threatening fit engineered by the Devil. Eventually Paracelsus was left in peace but he professed to feel less and less at home in this world. In vain, he wrote, were his efforts to prove his purity of soul, wasted were his cures which he usually performed free of charge. Yet he never gave up and his short *Defensiones*, both a plea in his own defence and a summary of his experiences in a hostile world, are timelessly disturbing.

From his last three years only one letter survives, in which he declines to visit a patient some distance away 'because of my miserable weak body'.[7] But in 1540 he was summoned to Salzburg by the Prince Archbishop, Ernst von Wittelsbach, to attend to his nephew. The young man was going out of his mind (as Wittelsbachs were apt to do even in the sixteenth century). Arriving at the Inn of the White Horse on 22 September the doctor suffered a stroke. Summoning his last strength 'as is the duty of every doctor', he prepared for the archbishop a vial of his laudanum. The archbishop sent him a healing fingernail of St Isidore of Seville. The vial worked; the fingernail came too late. The man who aspired to be a soulmate of Celsus died two days later. He was forty-eight.

* * *

Modern editions of Paracelsus's works take up sixteen large volumes. Much of his writing is hermetic, but, whenever one despairs, a piece of insight pierces the fog. He preached that alchemists should 'make medicines, not gold'.[8] There was, he insisted, a chemical cure for most ailments other than those destined by the Creator to ease man's passing. (Nobody ever *died* in Paracelsus's books: to him the word – and perhaps the concept – did not exist.) Presciently he recommended zinc for skin complaints and iron for 'poor blood'; and he almost certainly used sulphuric ether – 'sweet oil of vitriol' – as a form of anaesthetic.[9] He persistently and assiduously studied the effects of opium both on himself and on others. To him the drug was essentially a purveyor of sleep, but sleep in a wider and deeper sense than the word commonly denotes: 'Such sleep does not apply to the sufferer only. It may apply to the disease too. An ideal medicine puts the disease to sleep without killing the patient.'[10]

Like many of his thoughts, the idea was taken up by later hypnotherapists and other practitioners of unorthodox medicine. Most important perhaps, opium helped to demonstrate his core idea, the interaction, even the unity, of body and soul. 'Opium does not cure the swelling in the abdomen but in the patient's mind it puts the swelling to rest . . . [and] a pain that does not hurt is no longer a pain.' The last suggestion has been pondered by many since.

It was a disciple who seems first to have used the word 'laudanum':

> The master had little black pills, not unlike mouse droppings which he called *laudanum* . . . He boasted that with those pills he could wake up the dead and expel the devil . . . and indeed he proved that patients who seemed to be dead suddenly arose and those possessed regained their good sense.[11]

His prescriptions for intractable pain in different parts of the body were complex but always contained the magical extract of the poppy:

> Take of Thebaic opium one ounce, of orange and lemon juice six ounces, of cinnamon and cariophilli each half an ounce. Pound those ingredients together, mix them well, and place them in a glass vessel with its covering. Let them be digested in the sun or in dung for a month. Then press out the mixture and place them in another vessel with the following: half a scruple of must and half a scruple each of corals and of a magistery of true pearls. Mix those, and after digesting all for a month, add a scruple and a half of the quintessence of gold.[12]

Did he himself believe in the contribution of pearls, corals, gold and indeed all ingredients other than the first? Probably. He was not a deliberate obfuscator. But, though after his death ardent Paracelsians sprang up everywhere in Europe and spread his teaching in France, Germany and England, the costly trimmings prevented his invention from becoming popular.[13]

But cost was not the only deterrent. To the medical profession his notion of pain relief without probing into the cause of the pain remained suspect – as it still is.[14] During the two centuries after Paracelsus the death throes of several kings were documented. In 1598 the dying Philip II of Spain refused (or was refused) pain-killing drugs lest they befuddle his mind and invalidate his declarations of repentance. In 1685 Charles II suffered from medical tortures for days after his stroke before the end came as a merciful relief. The gangrene which killed Louis XIV in 1714 must have been deeply distressing to him and those assembled around his bed. The last trials of saints like Teresa of Avila in 1589 have also been chronicled in pious detail. All were surrounded by the best medical talent of their day. None was offered opium to ease their suffering.

Nor was the drug generally regarded as yet as an easily marketable treasure. Columbus, Vasco da Gama, Cabot, Drake and the conquistadores all had poppy juice on their shopping list; but the item came a long way below gold, silver, precious stones, spices, amusing monsters and slaves. For at least a century after Paracelsus's death interest in his laudanum remained scientific rather than commercial.

* * *

Happily, though, 'scientific' in the second half of the seventeenth century in England signified something special. Amid revolution, civil war, the beheading of one king and the restoration of another, the foundations of modern science were being laid. It was only the hobby of a small coterie, many of them fellows of a new club not yet called the Royal Society, and some both exceptionally bright and irrepressibly curious. The discovery by Dr William Harvey of the circulation of the blood, followed by the revelation by Professore Marcello Malpighi of Bologna of a microscopic network connecting the venous and arterial sides, the missing link in Harvey's model, was of practical importance only to undertakers – the information made the injection of embalming fluids easier – but it puzzled these young men. Among them was Christopher Wren, recently ejected from Cambridge for his Royalist utterances and now trying his hand at astronomy as Savillian professor in Oxford; the Honourable Robert Boyle, a rich Irish aristocrat who maintained a private laboratory in the city; Robert Hooke, a prickly character but an experimental wizard; and Robert Lower, a Cornish doctor who, in his youth, worked as an assistant to the famous anatomist Thomas Willis. What they wanted to know in the light of Harvey's and Malpighi's discoveries was how drugs like opium worked.

On 17 October 1667 the group gathered in Boyle's laboratory over a chemist's shop on the High Street overlooking Brasenose College. They were to witness an experiment Wren was to conduct on a dog.[15] It was later described by Boyle.

The dog was tied to the four legs of a table and Wren made a

> small but opportune Incision over that part of one of a hind Leg where the larger vessels that carry the Blood are most easy to take hold. Then he made a Ligature on those Vessels and applied to them a certain small Plate of Brass . . . almost the shape and bigness of the Nail of a Man's Thumb . . . [with] four little Holes in the Sides, that, by a Thread passed through them, it might well be fastened to the Vessels. . . . This Plate being fastened on, he made a Slit along the Vein on the Heart side and dexterously inserted the slender Pipe of a Syringe; by which I had proposed to have injected a warm Solution of Opium in Sack. [This accomplished] the *Opium* was quickly carried by the circular Motion of the Blood to the Brain, so that we had scarce untied the Dog before the *Opium* began to disclose its *Narcotic* Quality, and almost as soon as he was on his Feet, he began to nod with his Head and faulter and reel in his Pace, and presently after appeared to be stupefied, that there were Wagers offered his Life could not be saved. But I, that was willing to preserve him to be whipped up and down a neighbouring Garden, whereby being kept awake, and in Motion, after some Time he began to come to himself again; and being led home, and carefully tended, he not only recovered but began to grow fat so manifestly that 'twas admired.[16]

It was the first demonstration of the effect of opium on the brain and the first intravenous anaesthetic. The latter aroused a ripple of interest among the cognoscenti but there was not to be another one for three hundred years.[17]

* * *

Desultory experimentation was carried out on the continent too. Inevitably the question of how Paracelsus's concoctions worked exercised the great mind of René Descartes. Mathematician, philosopher, anatomist and neurophysiologist, he had already located the human soul in the pineal gland, a tiny structure at the centre of the brain. He now speculated that invisible pores in its capsule might be how opium impinged. Perhaps he was right.[18] But his personal experience was less than happy. He tried a brew sold as *Laudanum Paracelsi* to help to lift his gloom after his arrival in Stockholm as the honoured guest of Queen Christina. Honoured but bullied: even in midwinter the queen, an eccentric bluestocking, insisted on starting her highbrow *conversazione* at four o'clock in the morning. *Laudanum Paracelsi* improved the philosopher's depression to the point of foolhardiness. He tried ice-cold baths on rising, caught a cold and died from it.

* * *

In Spain opium-laudanum had to contend not only with lay ignorance but also and more dangerously with the suspicious vigilance of the Holy Inquisition. The drug was regarded by many as a legacy of Moorish domination and therefore suspect. This did not entirely stifle experimentation, but it led to circumlocutions not easily interpreted by posterity. As might be expected from the son of a barber-surgeon and the grandson of a Cordovan physician, Miguel de Cervantes was exceptionally well informed on medical lore and his masterpiece, *Don Quixote*, published in 1605, contains details of numerous poultices, balsams and embrocations.[19] Even more intriguing is the possible interpretation of the madness – if mad he was – of his immortal creation in psychopharmacological terms. Would opium explain the Don's brave delusions? His mental state seemed to clear as his physical health declined. The drug is not mentioned – it would not be for fear of spies and busybodies – but the possibility has recently been debated by leading Spanish scholars.[20]

In Protestant lands the Inquisition was no danger but scientific pioneering was nowhere without risks. Johannes Siegmund Elsholz, physician-in-ordinary to the Elector of Brandenburg experimented with opium on animals – he may even have performed intravenous infusions like Wren – but he left incomplete and confusing notes. Perhaps he feared the displeasure of his animal-loving master.[21] Further east, in Transylvania, Prince Gabriel Bethlen experimented with the drug, a gift of 'my friend, the Sultan', on horses, his more docile courtiers and on himself. 'Its effect is truly amazing,' he noted, 'it lifts the spirit like prayer . . . It must never be allowed to fall into the hands of popish priests.'[22] That danger was slight: the juice was expensive and popish priests in Calvinist Transylvania were paupers.

* * *

The poppy also blossomed in literary circles. In England it became part of Sir Thomas Browne's scented imagery: 'The iniquity of oblivion blindly scattereth her poppy . . . but there is no antidote against the opium of time.'[23] Robert Burton, a chronic insomniac, recommended it to fellow sufferers who 'by reason of their continual cares, fears, sorrows and dry brains [are often] crucified by melancholy'.[24] Thomas Shadwell, dramatist, poet and Whig busybody, may even have been addicted: at least his devotion to the drug made him the butt of John Dryden's mockery:

> Tom writ, his readers still slept o'er his book,
> For Tom took opium, and the opiates took.[25]

But the fact that the habit was lambasted as an eccentricity suggests that it was still the distraction of a few. Shadwell made no secret of his attachment and it did not prevent him from supplanting Dryden as Poet Laureate and Historiographer Royal after the revolution of 1688. But a century would pass before, in a simpler guise, poppy juice would become the inspiration of a new and brilliant literary age.

CHAPTER 8

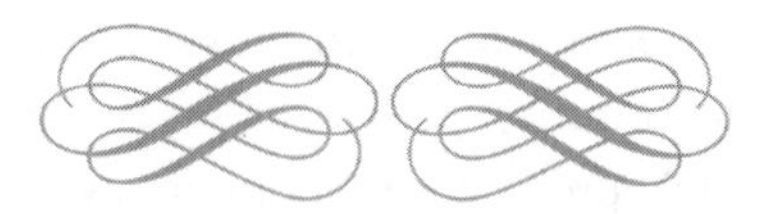

The tincture and the powder

In the last of the series of engravings entitled *Marriage A-la-Mode* Hogarth depicted the suicide of a foolish young countess. The date was 1745 and the story had begun six years earlier with her happy engagement to the scatter-brained but amiable heir to the Earl of Squander. By the time of the closing scene all the riches shown or hinted at in earlier images – the vast estate, the stately home, the family silver, the army of servants and parasitic hangers-on – had been lost. But worse: the young woman had just learnt that her lover, Counsellor Silvertongue, had been hanged for the murder of her husband. She is now dying in a battered armchair, her baby daughter held up to her for a farewell kiss.[1] The picture is full of haunting detail, like the compressed last 200 pages of a Victorian novel; but one small object draws attention to itself. An empty upturned bottle lies in the foreground. Labelled *Laudanum* it had obviously contained the poison which is killing the countess.

* * *

Opium had clearly come some way since the mysterious 'mouse droppings' touted by Paracelsus and his followers. The historic change came when the term 'laudanum' was resurrected by a London doctor, once eponymously famous, now little remembered.[2] Born in 1618 into a family of Somerset gentry, staunch Parliamentarians all in the Civil War, Thomas Sydenham was old enough to rise to captain of the horse in Cromwell's army.[3] After the Royalist surrender of Oxford he registered at Magdalen Hall but found the university comatose. He travelled to the continent, studied at some of the famous medical schools of the day – Leiden, Paris, Montpellier – and, after qualifying, settled in a house in Westminster. He married a Dorset girl, inherited an estate in Wynford Eagle, raised a family. His wide circle of friends and patients saved him from harassment after the Restoration.[4] He died aged sixty-five from the renal complications of gout of which he gave a masterly clinical description. The comparative paucity of other data led his first biographer to draw a parallel between 'our greatest doctor and our greatest dramatist, about both of whom little is known but both of whose name begins with S'.[5]

Greatest or not, Sydenham wrote well and a great deal and was deeply concerned with the relief of suffering. It was the last which led him to formulate a comparatively simple preparation of opium, capable of being standardised and given safely even to the young and enfeebled:

> The laudanum mixture to which I have referred is quite simple to be given in daily draughts and is easily prepared . . . one pint of sherry wine, two ounces of good quality Indian or Egyptian opium, one of saffron, a cinnamon stick and a clove, both powdered. Mix and simmer over a vapour bath for two or three days until the tincture has the proper slightly viscid but still easily poured consistence.[6]

The more he used the tincture, the more enthusiastic he became:

> Here I cannot but break out in praise of the great God, the giver of all good things, who hath granted to the human race, as a comfort in their afflictions a medicine of the value of opium, either in regard to the number of diseases it can control, or its efficiency in extirpating them . . . Medicine would be crippled without it; and whosoever understands it well will do more good with it alone than he could well hope to do with any other single medicine . . .[7]

His indications for it eventually ranged from cholera to bunions: his friends sometimes referred to him as Dr Opiophile. But the hint of mockery was affectionate: Sydenham was well liked and among the literati the purple liquid was a success. Its alcohol content promoted absorption and added an extra caress to the effect of the main ingredient.

* * *

Outside circumstances contributed to its popularity. One clue in Sydenham's writings is his reference to 'good quality' opium. The drug was suddenly available, an affordable luxury. The background was a momentous historical development wholly unrelated to the poppy.

The religious wars between Protestants and Catholics – with a sideshow from the Turkish Empire threatening the whole of Christendom – had convulsed Europe for nearly a century and a half. Many attempts at reconciliation had failed. Theological doctrines remained starkly opposed. Military forces were becoming evenly balanced. The conflict was cruel, fanatical and seemed never-ending. Except that in 1648 it did end. The Treaty of Westphalia embodied one of the most cynical compacts in history; but, because it signalled peace, it was deliriously received.[8] From one day to another the flames of yesteryear, literal and spiritual, became wisps of smoke.[9] Explanations for such cataclysmic changes always abound: none is ever wholly convincing.[10] But the effects – and side effects – are often spectacular. This was true of Europe in the mid-seventeenth century. The frivolous minor arts burst into bloom.[11] Among the more weighty consequences international trade revived. Transporting goods across the continent was once again safe. Maritime commerce too picked up. Opium was one of the first beneficiaries. As Arab traders had discovered centuries earlier, properly packaged the commodity was light, compact, durable and, weight for weight, as good as gold. In some respects better. In hard times gold was of no practical use. The demand for opium was, by contrast, compulsive. Enter Dr Sydenham with his splendid tincture. It made the drug safe and palatable. More precisely, it was the combination of the tincture and a best-selling powder introduced by one of the doctor's disciples. As often happens, the name of the disciple eventually became better known than that of the master.

* * *

Thomas Dover was born in Warwickshire in about 1660, all that is known about his background. In moments of exuberance he claimed to have obtained a medical degree in Cambridge; but if he did, he must have done so as a toddler. He emerges into documented history as an apprentice in the Sydenham household in Dorset. There he fell ill with smallpox and, as he never tired of relating, the doctor himself came from London to save his life.

> In the beginning I lost twenty-two ounces of blood. He gave me a vomit; but I found the experience of purging much better. I went abroad, by his direction till I was blind, and then took to my bed. I had no fire allowed in my room: my windows were constantly open, my bed clothes were order'd to be laid no higher then my waist [in January]. He made me take one ounce of his famous tincture every hour and two bottles of small beer, acidulated with spirit of vitriol every twenty-four hours. He came to feel my pulse every two hours day and night and to make sure that I had taken the tincture. The tincture saved my life.[12]

While recovering he was also, it seems, instructed in the rudiments of medicine. In 1696 he went to practise in Bristol, showed commendable courage during an epidemic of spotted fever (probably scarlet fever) and amassed sufficient capital to take the eighteenth century equivalent of a gap year. It was the time when plundering Spanish ships in the South Seas was the right of every Englishman and the pleasure of some of the best.[13] Apart from engaging in such exploits, Dr Dover (as he now styled himself) looked after his shipmates and earned their esteem and gratitude. On 2 February 1709 he also rescued from the island of Juan Fernandez off the coast of Chile a shipwrecked Scottish sailor by the name of Alexander Selkirk, the beginning of a literary legend.[14] He returned with a booty of £170,000 and set up in practice in London. Happily he retained many of his piratical ways. He scandalised the Royal College of Physicians by announcing that he did not intend to take examinations: he was going to cure people instead. His patients would be his diplomas. This he did – at least some of the time. Late in life, his youthful indiscretions forgotten, he published *The Ancient Physician's Legacy to his Country* in which he gave away the secrets of over a hundred unpublished prescriptions. Among them was the celebrated powder which would in time find a place in every pharmacopoeia of the civilised world:

> Take Opium one ounce, Salt-Petre and Tartar Vitriolated each four ounces, Ipecacuanha one ounce. Put the Salt-Petre and Tartar into a red hot mortar, stirring them with a spoon until they have done flaming. Then powder them very fine; after then slice in your Opium, grind them to a powder and then mix the other powders in with these. Dose from sixty or seventy grains in a glass of white wine Posset . . . covering up warm and drinking a quart or three pints of the Posset-Drink while sweating.[15]

The preparation first achieved fame as a 'diaphoretic' – a substance that would 'sweat away' any poison in the body – but was soon recognised as the remedy of choice against all forms of diarrhoea and abdominal discomfort. 'Drink the mixture before going to bed and all pain will disappear before the morning.' He hinted that occasionally – but not often – the patient's life would disappear with the pain.

> Some apothecaries desired their patients to make their Will and Testament before they ventured upon this remedy in the dose recommended; but nothing ventured, nothing gained.[16]

Soon both Sydenham's laudanum and Dover's powder were competing with other, sometimes cheaper opium-based nostrums. Venice Treacle, Mithridate, London Laudanum, Dr Bates' Pacific Pills would all achieve a degree of popularity; but drug loyalties are strong and none of the competitors would prove as durable as the two pioneering preparations.[17] Dover himself died in 1742, aged eighty-two, mourned by his patients, friends and several more or less legitimate families.

* * *

In Catholic countries the spread of opium tended to be a little slower. The reasons were social and economic rather than religious. In principle both Catholic and Protestant Churches were against inebriating drugs. They were also against sin. But Catholic countries were generally poorer; or, more to the point, in Catholic countries wealth was often distributed more unevenly. While in Protestant lands opium was gaining ground among the rising middle classes, in Catholic Europe it was still the indulgence of the upper crust. In the Hofburg in Vienna and much to his surprise the Emperor Leopold I found a 'Turkish concoction of the poppy' an excellent aid when composing his sacred motets; he even sent some of it to his beloved but bewitched cousin, Charles II of Spain.[18] The Venetian ambassador reported that the drug was 'madly in fashion' at the dissolute grand-ducal court in Florence. Gian Gastone, in particular, the last Medici grand duke, was mad, gay, a patron of the arts and a dedicated addict. In France Louis XIV's chief apothecary, Monsieur Pierre Pomet, took a lively interest in the new opium-based brews:

> Opium procures blessed rest by its viscous and sulphureous particles, which, being convey'd to the channels of the brain, agglutinates and slows down the animal spirits . . . good sleep ensues for the senses are, as it were, fettered by the viscous and agglutinating property of the drug.[19]

Among other excellent effects this was good for composing the 'Unseemly Hurry of the Spirit', useful in diseases of the breasts and lungs and supreme in preventing the spitting of blood, curing coughs, colds in the head and for 'stopping like magic' the looseness of bowels. It was also handy in 'hysterick cases'. His assessment of various brands was not unlike the ruminations of a connoisseur of wines:

> I have studiously compared the substances from Cairo or Thebes, the black and hard material from Aden, the softer gum from Turkey, the yellow and more crumbly sort from Cambodia and Decam in the East Indies and several other offerings by traders of uncertain origin. At this time I would regard only the Theban or Turkish as being of the first class. They are weighty and of good consistency, thick and more solid in feel than the Indian

> variety. The Theban opium is of a fresh reddish colour, almost like fresh aloes, of a strong poppy scent, of an acrid bitter taste, that will burn and flame; easy to cut and be dissolved in either fresh spring water, wine or spirit of wine, and pretty clean from dirt, excrement and filth. The Indian opium is softer, yellower, lighter, not so good of body and much fouler, in every respect inferior to the former.[20]

And more in the same unspeakable vein.

Disappointingly, Monsieur Pomet's royal master and his morganatic wife, Madame de Maintenon, disliked drugs generally, though she relented when opium was offered in the form of clysters.[21] But among courtiers the physicians Regis and Fede could not have functioned without the brew. As the irrepressible Madame de Sévigné reported,

> the poor Abbé Tetu is slowly going out of his mind and his overexcited state is painful to behold. He is only kept going with opium. The other day Regis and Fede gave the poor man ambergris instead by mistake and he nearly died . . . For some reason this kind of confusion seems quite common. I think the naughty doctors are too inclined to indulge themselves.[22]

At a less exalted level a peripatetic Capuchin monk, Etienne Rousseau, cooked up a commercially successful mixture, an oily liquid which became known in his own country as *La Brune* and in England as Lancaster Black or Quaker drops.[23] The original allegedly contained crushed pearls, together with nutmeg, saffron, crab apple and yeast as well as opium; but the crushed pearls were optional. The duc de Saint-Simon, a martyr to indigestion and his own crotchety temper, became an unexpected devotee of this plebeian brew. 'I could not suffer the vile manners of the young at Court nowadays without it,' he confided to his cousin. Thomas Jefferson took samples of *La Brune* back across the Atlantic and recommended it to his friends. It remained in the United States pharmacopeia till 1865.

* * *

Fads and fashions in medicine spawn books as they do in other fields. Mercifully, the majority are soon forgotten. The first modern monograph devoted to opium and published in 1700 was a cut above the rest. In his 450-page *Mysteries of Opium Revealed* John Jones proved himself an enthusiast. He could see no reason why readers should deprive themselves of an elixir which would not only dull pain but also induce 'Serenity, Promptitude, Alacrity, Expeditiousness in Dispatching Business, Assurance,

Ovation of Spirit, Courage, Contempt of Danger, Magnanimity, Euphory, Contentation and Equanimity.' What more could a man desire? If, after taking opium,

> He keeps himself in action and discourse of business, it seems like a most delicious and extraordinary *refreshment of the spirit* as upon hearing good news or any other great cause of *joy* . . . sometimes compared, not without good cause, to the gentle and yet intense *pleasure* which modesty forbids me to name.[24]

To the benefit of his readers the bashful Jones did not remain bashful for long. By page 258 he was comparing the 'Sensual Ovation, Magnanimity and Boldness' brought about by the drug to the 'Vitality of wild beasts in a rut' with 'increased *Seed production, Venery, Erections* etc' in men and breast development in women. All this was due to the stimulation of the *Venereal Membranes*. He also argued that this effect was due not (as his pitifully simple-minded colleagues believed) to fumes, vapours, auras and effluvia rising from the stomach to the brain but to a fatty stimulating salt already present inside the tissues. The wonderful effects of the drug therefore 'cannot be much wondered at, considering that [its] Active Principle, Sal-Volatile-Oleosum and Opium itself, are naturally carried in ourselves . . . being exalted by Nature to . . . stimulate us when necessary.'[25]

The idea that an opium-like substance was native to the human brain was not followed up for 300 years, but it would then revolutionise the neurosciences.

Surprisingly perhaps, Jones was also among the first in Western Europe to describe in clinical detail the agonies of withdrawal:

> Anguish and pain, followed by great, even intolerable Distress, Anxieties and Depression of Spirit, which, in a few days, commonly and in a most miserable Death, attended with strange Anxieties.[26]

But his final conclusions were upbeat:

> Since opium does not operate by causing a grievous sensation and there being no other way left by which it may operate, it must operate by causing a pleasant sensation. And what can cure pain and all its effects better than pleasure?[27]

* * *

Jones's doctrine was music to the ears of the famous Dr John Brown of Edinburgh – but not for long. Brown's general concept, that good health was the result of equilibrium between stimulation and relaxation and that all therapy should be directed towards restoring that balance, dominated medical thinking for a hundred years. But eventually 'Brunoism' had to face difficulties; and high on that list was the dilemma of whether opium should be classified as a stimulant or as a relaxant. Depending on dosage and, even more, on striking differences between individual responses, it was in fact – or could be – both or neither. This was an unsatisfactory state of affairs in a system which claimed to leave nothing un-pigeonholed or at least un-pigeonholable. What made the unpredictability even more vexing was that the question could no longer be brushed under the carpet. At the time when Jones published his pioneering book the drug was still the eccentricity of a few and might have been ignored. But as the century progressed, it was becoming available in a variety of new guises and not only in aristocratic, bohemian and criminal circles. Even in Brown's native Edinburgh, Athens of the North, opium soirées were becoming the rage among the fashionable young.

Nor could middle-aged models of common sense wholly ignore the trend. On Sunday, 23 March 1773 James Boswell wrote in his diary:

> I breakfasted with Dr [Samuel] Johnson whose heaviness of spirits of yesterday was much relieved having taken opium the night before . . . He protested that the remedy should be taken only with the utmost reluctance and caution. I mentioned how commonly it was used in Turkey and that it could therefore not be so pernicious. He grew warm and said that Turks taking opium were one thing, Christians taking the stuff were another.[28]

There may have been a grain of truth in the doctor's political incorrectness. Variations in racial susceptibility to opium have never been demonstrated and probably do not exist; but, as the use of the drug spread, the astonishing range of individual responses was becoming apparent. To many this was disturbing. A habit, they felt, should be either safe or unsafe, recommendable or forbidden, virtuous or evil, godly or ungodly. It should not depend on . . . what? Race? Sex? Whims of nature? Opinions clashed. Those who regarded addiction as inherently evil might forgive quivering junkies (not yet so named) to the point of offering them support on their way to moral regeneration or a merciful end; but they were offended by the notion that some people could take the concoction with impunity. Yet such individuals seemed to exist. Or did they? How far could the personal testimony of addicts be trusted? They were surely and by definition liars. And while the righteous were sometimes forgiving, addicts who were apparently unharmed by their addiction found the degradation of junkies an affront to their own innocent habit. Both sides had to face a difficulty. Selected cases could prove almost every shade of truth and its opposite.

In the past arguments about opium among the few who indulged in it tended to be good-natured, sometimes joshing. As the drug habit spread the disputes became intemperate, personal, even vicious. For those who had witnessed the despair, degradation and sometimes the death of a friend the stuff was poison. Those who had benefited from it, or thought they had, were fulsome in their praise. Scandal and gossip stoked the controversy. As a topic of pleasurable indignation drugs were beginning to trump royal infidelities and sinister pseudoreligious practices among the debauched rich. Many claimed to know the characters depicted by Hogarth intimately. (Did they exist? Never mind, however debauched, an earl was an earl.) Soon there would be live celebrities to feed the appetite for outrage. The new scandal sheets were cheap and sold in their thousands. They whetted the appetite of the young. This was, after all, the Age of Reason, much vaunted by all the clever *philosophes* in France and spouted about by trendy intellectuals on both sides of the Channel. In such an age everything was worth trying once. Or even twice. And as a result of happenings on the other side of the globe to do so was becoming almost patriotic.

CHAPTER 9

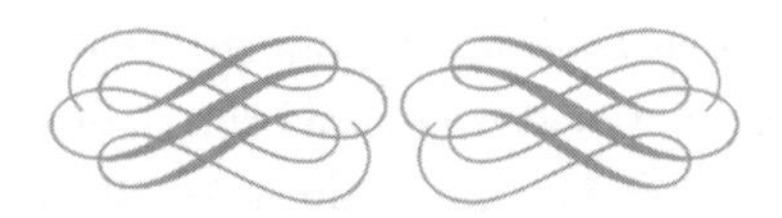

On the banks of the Ganges

THE RELATIONSHIP BETWEEN supply and demand is complex and in the case of opium has often been impossible to unravel. Was the drug suddenly available in Europe because the discoveries of men like Sydenham and Dover had created a demand? Or did the influx of affordable opium account for the success of their discoveries? Whatever the sequence, increasing opium consumption in the West coincided with the transformation of the pattern of poppy cultivation in far-flung corners of the world.

For centuries in India the processing of the home-grown poppy had been a domestic chore, usually left (like other domestic and most undomestic chores) to women. Most of the produce was for family consumption – the standard remedy for all common ailments as well as for the occasional outbreak of cholera – but some went to the local market and a choice selection as a tribute to the seat of the regional nawab, beg, sheikh, maharaja, nizam or in whatever title the emperor's territorial representative rejoiced. Or virtual representative. The Mughal emperors still held court in Delhi, successors to the mighty conquerors of India; but imperial power had largely fallen into the hands of regional potentates. Some of these ruled over a few dusty villages, others over rich lands several times the size of Britain. Their complicated disputes, sometimes erupting into savage little wars, as well as the infighting between branches of the ruling families, merged with a rivalry more fraught with consequences.

During much of the eighteenth century France and England were at war in Europe and, though news of declarations of hostilities and the conclusions of eternal peace could take a year to reach distant continents, the conflicts were duly enacted across the oceans. In India both sides tended to hire native troops and form shifting alliances with local rulers or would-be rulers. On 23

June 1757 a battle among the sodden mango groves of Plassey, a few miles outside Murshidabad, capital of the Mughal viceroys of Bengal, settled the fate of that important province.[1] It was a famous victory for Robert Clive, a twenty-seven-year-old academic dropout and failed apprentice accountant who had never had a day's formal military training. This did not stop the Elder Pitt, prime minister of the day, describing him as 'our great and heaven-born general'.[2] The Elder Pitt was right. Though this could not be foreseen, the battle laid the foundations of two hundred years of British rule in India. It also earned Clive the beginning of a vast fortune, a knighthood and a little later the barony of Plassey. It would also lead to the transformation of opium growing in the Ganges Valley.

For a century or more opium cultivation on the banks of the great river had been supervised in an easy-going way by the Mahratta Opium Council, a body formed by representatives of five semi-independent states of the sprawling Mahratta confederacy. The Council met at irregular intervals, exchanged tokens of esteem and gave formal permission for the poppy harvest to start. It also, in a somewhat desultory way, checked that the produce of different regions was not heavily adulterated. To nip in the bud any unseemly competition it then decreed the prices for the different grades of the opium gum. 'Under its benevolent guidance,' a European traveller observed in 1748, 'the rich land around Patna produced enough opium to satisfy most regions of India.'[3] But approval by a visiting European aristocrat was one thing; satisfying hard-nosed businessmen another.

After Plassey the Mahratta Opium Council dissolved and, for want of any other authority, its role was taken over by Clive's employers, the Honourable East India Company of England. Known in Britain as John Company and in India as Bahadur ('brave' in Hindustani), this alliance of businessmen and aristocrats had been conceived in the euphoria following the defeat of the Spanish Armada.[4] Dedicated to the pursuit of trade with the East Indies and beyond, the Company prospered and now possessed imposing headquarters in Leadenhall Street in the City of London. Overseeing the cultivation of the poppy on the other side of the world was not part of their original plan or brief; but in a growing empire (not yet so called) such extracurricular tasks had to be shouldered.

But success had come at a price. Greedy foreigners, above all the grasping French, had to be beaten off; and benighted natives were often slow to appreciate the benefits of belonging to a mighty global enterprise. Yet military campaigns were costly. Though the Company had rewarded Clive with a jewel-studded sword costing £10,000, some of the directors muttered about their paltry personal booty. The bribes paid to local princes whose well-timed treachery had ensured the successful outcome of the battle must be recouped. With the sudden growth in demand for opium at home, the drug promised to be the answer to their prayers.

But there were difficulties. The Company's inspectors soon described the 'shocking inefficiency and haphazardness' of the existing trade. One near-apoplectic emissary reported:

> Left to their own devices and their incomprehension of the meaning of 'efficiency' 'profit' and 'fair trade', it is surprising that the trade exists at all . . . Both the quality and the price are unregulated . . . When the harvest has been poor the local peasants and traders would not hesitate to *adulterate* their produce with water, thorn-apple pulp, molasses or worse, much worse. I have seen them add cow dung.[5]

Nor was it easy to make headway through the meshwork of caste:

> A third of the harvest might be lost through sheer negligence because the *gomastas* or low-caste supervisors would not talk directly to the even lower-caste *koeri* or cullers and show them how to cope with unwanted rain-fall and other hazards; and the *ryot* or plantation owner (or superior middle-man) would only worry about the end-product and his cut and would not talk to the *gomastas*.[6]

This could not be tolerated. Warren Hastings, Clive's successor, was the son of a poor clergyman, the first British representative to be styled governor general and the last to rise from the ranks. He knew the opium trade from the inside and hated it. 'The drug is not a necessity of life but a pernicious article of luxury which ought not to be permitted,' he declared to the directors on being appointed. 'I shall stamp it out.' So long as the profits flowed undisturbed the board were tolerant of such eccentricities. (They had to be. Some company men enlisted only to observe eclipses of the moon in distant parts of the world, others to collect snakes or butterflies.) But it took less than a year for the governor general's resolve to be eroded. Anticipating his grand opiophobe successors Hastings wanted to leave his mark on the subcontinent; and administrative reforms were badly needed. But reforms of all kinds in India always cost money. Faced with a tight-fisted company – later to be replaced by an equally parsimonious Treasury – opium was the only potential source of that commodity. But its harvesting and marketing had to be transformed root and branch.

Discipline, in Hastings's view, was the key to the changes needed. The obstacle in the way was the obstinacy of the opium growers. They were peasants. They cherished their ancient but deplorable ways. The threat and when necessary the execution of exemplary punishment for obstruction were increasingly called for. For more efficient control Hastings restricted opium growing to the states of Patna and Bihar. He commanded neighbouring provinces to provide the opium growers with the rice and fruit they needed to survive. The people between Patna and Benares soon grew nothing but poppy. This left them at the mercy of their neighbours. Neighbours in India were sometimes ancestral enemies. Government orders tended to be ignored. The people between Patna and Benares starved. Hastings decided that part of the population was redundant and should be moved. But even starving Bengalis clung to their miserable little plots in a fierce and unreasonable manner. Force sometimes had to be used. It worked but seemed to leave behind sullen resentment. But by the 1780s the area under profitable poppy cultivation was a quarter of a million acres.

* * *

Opium growing on this scale called for a change in the handling of the material. After a few hours' exposure to air the latex of the poppy becomes a sticky, dark brown, viscous gum, not unlike shoe polish. It is unstable and unrefined but moderately active; and, variously diluted, this was what the natives of India had consumed for centuries. No longer. The collecting pots of the scrapings were now emptied into progressively larger containers and eventually into shallow tilted brass dishes known as *thallees*. In those the gum was stirred and turned daily until most but not all the moisture had evaporated. It was then thrown into huge vats and kneaded to exactly the consistence required for moulding.

The moulding sheds of Patna were among the first factories of modern times. They employed children as young as seven and wizened old men. The workforce squatted behind tin vessels or *tagars*, big enough to hold gum for five 'balls'. Their basic equipment was a basin of clear water, a supply of carefully dried and preserved poppy petals, a cup containing *lewa* or inferior opium and a brass mould which could be opened. The moulds would be smeared with *lewa* and lined with poppy petals. Their delicate lattice would eventually form a complete covering and perfect insulation and became the trademark of genuine 'Patna'. The balls, the size of naval cannonballs weighing about 1.5 kilos, were lined up on racks in the drying rooms. Like lofty cathedrals, these were designed to let air currents flow through them; a gang of young boys turned the balls twice daily for some weeks, making sure than no thief or insects got to them. The latter were the greater menace. They came in dark clouds, trillions of them, fast, noisy and seemingly from nowhere, apparently attracted by the smell of the poppy. Allowed into the sheds they could destroy a year's harvest in less than an hour. Only a few minutes were available to seal the doors, windows and cracks. The menacing black clouds could then be dispersed with acrid chemicals. Thieving was risky. If caught (or suspected), the culprit was beheaded.

When dry, the product was weighed, valued and packaged into mango-wood chests known as 'catties'. Each catty had two trays, each tray holding twenty balls, each ball fitting into its individual compartment. Today these superbly crafted containers are collectors' pieces. At the time most were fed into open fires in faraway baronial halls where the astringent scent of burning mango, 'the smell of the jungle', to old India hands, was appreciated. When full, the chests were sealed with pitch and sewn into hides. Only a token gift now went to the regional lord. Most were transported by skiffs to warehouses in Calcutta to await the auction season. For this bidders came from all parts of the world. Most were riff-raff but some were grand personages with a retinue of servants or even armed escorts. The Chinese or other orientals – who could tell the difference? – often carried eye-watering sums of silver. The boats waiting in the harbour got steadily bigger. By the end of the century the show-stealers were the opium clippers built for the trade in Baltimore in the United States. Fast and sleek, under full sail they were a joy to behold. Soon the number of catties shipped from the Patna region every year exceeded 200,000. The net profit on every catty was reckoned to be £80. The total yearly revenue from opium for the London investors was said to exceed £2 million. All eminently satisfactory. But there was a gradual shift in the pattern of the trade. Increasingly the ships leaving Calcutta turned east rather than west. The Celestial Kingdom of China beckoned. It was close. It was huge. It was reputed to be inexhaustibly rich.

* * *

The personal fate of the man who had laid the foundations of all this riches was less happy. In addition to gaining a fortune, Robert Clive, now Lord Clive of Plassey, developed an intermittent abdominal complaint during his early years on the subcontinent. Periods of remission were followed by debilitating and increasingly painful episodes of colic, sickness and diarrhoea, suggesting regional ileitis, a chronic inflammatory disease, or perhaps a non-malignant lesion causing intermittent intestinal obstruction.[7] Whatever the cause, the only effective treatment was opium. Clive also maintained that the drug allowed him to accept his affliction with Christian fortitude. Unfortunately – though it did not strike him as particularly menacing – the amount required to cope with the symptoms slowly increased. His own personal supply was specially grown, processed and tested for him by trusted staff; and his Indian retinue followed him back to England on his visits home. By 1767 when he returned after his third and last spell as governor, he was taking doses which his son described as 'monumental'. On bad days he would shut himself into his study and communicate with no one but an old Indian retainer, guardian of his medicine chest.

Such a regime quickly killed some. Others, seemingly indestructible, became hyperactive. For years Clive seemed to belong to the second category. He built a Palladian mansion in Claremont in Surrey surrounded by a paradisical garden. He administered his scattered estates in England and Ireland. He had himself elected to the House of Commons – his was an Irish peerage[8] – and he vigorously pursued a policy aimed at making the government more directly responsible for Indian affairs. This was an admirable aim but entailed a guerrilla war with his former employers. He branded the Honourable Company greedy, dishonest, hypocritical and stingy. All this was probably true but it did not endear him to the directors. Piqued, they claimed to have been defrauded by him of millions. That too may have contained a kernel of truth. The dispute was complicated by the fact that it could take a year to obtain documents from India; and when they arrived they were often in tortured English nobody could understand. Or everybody understood differently. But most draining was the need for Clive to defend himself against his political enemies whose aim was to impeach him for 'scandalous extortion and corruption'. He was cross-examined in the House of Commons and in the Lords. Even those who appreciated his prodigious and often anonymous personal generosity sometimes felt uneasy. That he had accepted presents and had taken his share of booty was normal: it was the size of his fortune which made people gasp. India was beginning to be known in England for the contrast between the fabulously rich and the abjectly poor. The former were few, the latter existed in their millions. Hannoverian England was no welfare state; but the image was not appealing. English boys returning from service in India – unlike the forgotten majority who never did return – bedecked with the trappings of a maharaja aroused envy. Clive, though not a trained advocate, was, according to the Elder Pitt, one of the most effective speakers in Westminster when 'he was in the mood'. When he was not, he talked gibberish. His 'moods', many suspected, varied with his opium intake. Eventually, in 1774, he was vindicated – and was instantly offered the command of British troops in North America. He was a brilliant soldier and the offer remains one of history's tantalising 'if's.

On the day of his exoneration he invited his political enemies to a party to end all parties. He had his London house in Berkeley Square festooned with marvellous exotic decorations. Prodigious quantities of food and drink lay ready to be consumed, expensive presents to be distributed. But he snapped. Some biographies state that he took an overdose. In fact, though he may have dosed himself heavily beforehand, during the night before the party, on 22 November 1774, he stabbed himself with his penknife. It could not have been an accident. He was forty-nine.

* * *

By the time of Clive's suicide the bulk of opium from India was heading east, not west. Despite rumours of growing imperial displeasure the Chinese trade seemed to many the act of a benign Providence. Inevitably there were snags. Competition from the Portuguese, the Dutch, the French and even from upstart Americans was a nuisance but manageable. Pirates were a graver menace: it was their golden age. The thousand and one islands of Indonesia, like the Bahamas in the western hemisphere, offered them an unconquerable base. With every passing year their ships were getting faster and more nimble. Despite their semi-permanent state of drunkenness their seamanship was faultless and their gunmanship often awe-inspiring. (Many were deserters from the British and French Royal Navies.) Their bravery was born of desperation. Captured they could expect no mercy. And like mythical monsters they seemed to grow new heads every time they were decapitated. But with her growing naval power Britain would surely prevail. She had to. The China trade was too profitable to fail.

But all this entailed a change in the trade routes. Though Indian opium remained available in Britain – it was to be Lord Byron's favourite brand – Indian growers were now targeting China. It was nearer. The profits could be luminous. It was the poor and backward peasantry of the Anatolian highlands of Asian Turkey who were becoming the West's chief suppliers. Their produce, more potent than the Indian variety, seemed to be available in unlimited quantities.

CHAPTER 10

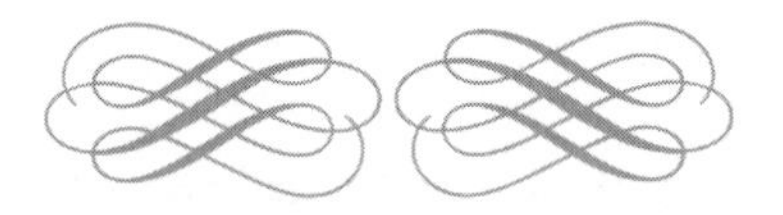

The Turkish connection

As in India, the peasants of the Anatolian Plateau had grown the poppy for home consumption for centuries. Who discovered the region's wider potential is uncertain; but given the growing demand for opium in the West, the development was inevitable. The soil and the dry temperate climate were ideal; and, no less auspiciously, the inhabitants were poor, hardy and ignorant. The area was neither too sparsely populated nor overcrowded. Around 1760 Greek, Turkish and Armenian traders began to make their way to the villages. To the local peasantry the sums they offered verged on the supernatural. Though this was the historic heartland of the Ottoman Empire, it had long been neglected. The Sublime Porte in their palaces on the Bosphorus were still gazing westward though Turkey's days of European conquest were over. Soon indeed the country would be known as 'the sick man of Europe.' But what happened in the Anatolian hinterland exercised the Sultans, the grand viziers, the imperial eunuchs and the ladies of the Seraglio not at all.

Yet it was to be a startling transformation. By the 1780s the poppy fields near Kara Chissar and around Magnesia were spreading for hundreds of miles. Unlike the Indian and Egyptian variety, the Turkish plants reached six to eight feet in height: workers in the field during harvest were invisible, submerged in a shimmering sea of green, gold and white. The fact that cultivation remained in the hands of peasants was both a strength and a weakness. Though it eliminated middle men, competition between families was ruthlessly exploited by the traders and drove the prices down. For the harvest all members of a kinship were pressed into service, though the hours in the field had to be limited. As a Western observer reported:

> the exhalations of these huge plants could be dangerous, especially early in the morning and after sunset. They were avoided by the farmers retiring in the evenings to their huts . . . They do not leave until sunrise. As soon as moisture in the air reaches a certain level, a strong narcotic smell develops. To anyone not used to it, this gives rise to a severe and dangerous headache after a quarter of an hour.[1]

Every part of the plant was used. The pressed seeds provided oil for cooking, lighting and, when exported, the manufacture of soap. Poorer households mixed poppy cake with their bread. What remained was fodder. Harvest took place in May to June and the dried opium packed in grey calico bags was sent by mule to Smyrna. By then dealers had purchased the crop, the money being advanced at an exorbitant rate of interest.

By the mid-eighteenth-century Smyrna, once known as the pearl of the Mediterranean, had become a 'poxy, plague-ridden and pestilential hell-hole'.[2] Ships' crews avoided disembarking, arranging for the brothels to visit on board. But the place was still geographically convenient and provided an excellent harbour; and it was there that a polyglot crowd of merchants and their retinues forgathered. The calico bags were opened in the warehouses by a public examiner, a well-paid official regarded as incorruptible.[3] He would scrutinise the content of each bag, judging the material by colour, weight, appearance and consistence. He could tell by experience if it was pure. If suspected of adulteration it was discarded. The final purity was expressed in carats, the traditional measure for gold and diamonds. Twenty-four carats was pure. Anything under twenty was burnt. So was merchandise that was still moist, even after drying in the warehouse. Damp opium was unstable and brought the firm selling it into disrepute. Bags which passed muster were repacked into zinc-lined, hermetically sealed but plain wooden cases for shipment. In contrast to the Indian merchandise, there was never anything fancy about Turkish opium.

With Britain virtually controlling India, it may seem odd that the Turkish growth quickly replaced the Indian; but Britain had had commercial links with the Ottoman Empire since the foundation of the Levant Company in 1581 and exported to Turkey a range of merchandise, especially cotton goods, for which opium could be exchanged. Transport from Turkey was cheaper and the mechanics of the trade were in place. Turkish opium was also more potent than the Indian brand: later analysis suggests that its morphine content may have been as high as 10–13 per cent in contrast to the 4–8 per cent of the Indian produce.

Not only the source but traditional routes too were changing. After a century of graceful and melancholy decline the Republic of Venice was snuffed out by General Bonaparte on 12 May 1797.[4] It was Marseilles which became the main reception centre on the continent for merchandise from Turkey and the Middle East and would remain so till the 1970s. Amsterdam and Rotterdam were also developing into the opium ports they still are. A small but steady trickle was smuggled into Italy along a coast line which was, as it still is, impossible to police. But most of the cargo was shipped directly from Smyrna to Britain in British vessels. Dover, Liverpool and Bristol were occasional destinations; but the hub of the trade was London.

* * *

The mechanics of the business evolved gradually. As the Levant Company declined – it folded in 1825 – aggressive wholesale importers were taking the place of the cosy price-fixing cartels. Purchasing opium was now by competitive deals. More than 90 per cent of the transactions were settled in Mincing Lane in the City, a significant proportion in Garroway's Coffee House near the Royal Exchange. The fortnightly auctions were conducted by a system known as 'bidding by the candle'. A small candle was lit at the start of the auctioning of each lot and the highest bid advanced before the wick burnt away was the winner. The ceremony was endearingly quaint and totally cut-throat. Mincing Lane dealers rarely worked under a 50 per cent profit margin. This could rise to 200 per cent but could also fall to nothing or end in a staggering loss. Risks were great. Profits could be greater.

The operation of market forces led to a steady fall in the retail price. Monthly publication of current price lists allowed chemists and wholesalers to shop around. Among the biggest buyers were Allen & Hanburys and the Apothecaries' Company. The latter displayed a notice every Saturday indicating the amount of opium they required and dealers who wished to sell sent samples by the following Tuesday. The buying committee was a group of 'highly qualified medical gentlemen' assisted by 'knowledgeable chemists'. They selected the best. Both sellers and buyers benefited from a stepwise reduction in import duty. From nine shillings per pound in the early 1820s it fell to four shillings in 1826 and to one shilling in 1836. It would be abolished altogether by the free trade agreement of 1860. But the most important reason for the cheapness was availability. When a drunken master ran the SS *Crimea* aground with the loss of several tons of cargo as well as most of the crew, the price in the City rose by barely 2 per cent and only for a few weeks. Like the much-vaunted hauls by Customs and Excise today, minor contretemps like wars and earthquakes caused no more than a blip in the price of the drug in the corner shops.

* * *

But the Turkish connection irked many a patriotic Briton. They felt that so precious a commodity should be cultivated within the empire or, even better, at home, not bought from the 'rascally Turk'. It was not an idle dream. After 1763 the Society of Arts vigorously promoted the study of medicinal plants and offered prizes for new discoveries. In 1785 Dr Alston, professor of materia medica in the University of Edinburgh, showed that opium could be produced from home-grown poppies; and in 1789 a 50-guinea prize was awarded to John Ball of Plymouth for the 'excellent quality' of his produce. A few years later Thomas Jones set five acres with white poppies in Enfield a few miles north of London and won a gold medal. But he insisted that it was his exceptional expertise in lancing the pod which accounted for his success; and that amateurs were wasting their time. Yet, after several not entirely successful efforts, Thomas Young, an Edinburgh surgeon, showed that opium could be cultivated even in the Caledonian cold: his venture earned him a diploma as well as a profit of £50–80 per acre yielding 56 pounds of opium, several hundred pounds of oil and a quantity of early potatoes prudently planted between the rows of poppies. But the most successful opium cultivators were Dr John Cowley and Mr Staines of Winslow in Buckinghamshire who, in 1823, received 30 guineas for 143 pounds of opium 'of excellent quality, collected from about eleven acres of land'.

What made these ventures ultimately uncompetitive was the cost of labour. Cultivation on any scale depended on hired hands; and, however low agricultural wages were in Britain, they could not compete with what families accepted as a princely wage in Anatolia. Distance mattered; but transport by sea was cheap. Various cost-cutting expedients were suggested. John Ball pointed out that with children making the incisions and scraping off the juice, requiring no great strength, 'the expence [*sic*] will be found to be exceedingly trifling'. Except for the occasional blinding headache, the work in the fresh air would be 'thoroughly sanitary'. (At least it was preferable to the mines or the mills.) His partners, Jones and Jeston, carried the proposal further, devising a pay structure with financial inducements. Employing seven or eight boys between the ages of six and twelve and an adult superintendent, a reasonable wage scale could be based on behaviour. 'To the youngest we give three-pence a day [ten working hours] and, if tractable and well-disposed, an additional penny for every additional year of age'. Others might favour a productivity deal with an average rate of eightpence a day and a penny for every extra half-pint of opium. Cowley and Staines commented that 'a general cultivation of opium would certainly benefit persons not fit for common agricultural labour like unemployed female lace-makers between hay-time and harvest'. In a later experiment they employed 'the more peaceable kind of Irish migrant', also paid a shilling for an eleven-hour day. Some wondered if growing poppies in Ireland might be the way to rejuvenate the Irish economy. But Turkish opium remained a quarter of the price of anything grown in Britain and probably more dependable in quality.

Home-grown opium was nevertheless welcomed by patriotic members of the medical profession, especially during the Napoleonic Wars. Thomas Jones's mixture was used by physicians at St Bartholomew's Hospital in London against rheumatic complaints, bowel disorders and, most successfully, in cases of 'hysteria'. Not perhaps quite so effective, it was better tolerated than the Turkish variety. Both differences could be explained by the lower morphine content of English opium but at the time they were attributed to unspecified impurities in the Turkish produce. Apothecaries at the Middlesex Hospital recommended a grain of English opium, more wholesome than the Turkish variety, against all disorders of the stomach. The suggestion was not welcomed by all. The redoubtable Mr Cuthbert Moores, Fellow of the Royal College of Surgeons, past ninety but still a dab hand with the scalpel, declared English opium to be 'totally useless: you might as well drink tap-water'. Dr Latham, a fashionable practitioner in Bedford Row, on the other hand, found that a one-grain dose of John Ball's brew relieved his own ticklish cough and he recommended it to all with a similar complaint. He even tried to grow the poppy in his own back garden. Experiments in the West Country continued for the best part of another century: the occasional white poppy blooming by the wayside and smelling delicious in parts of Somerset today is their legacy.

* * *

By 1805 about a quarter of opium imported from Turkey was re-exported, to the United States. The New World would eventually be the biggest market for the drug and its most determined foe; but the early years were welcoming.[5] *Blackwood's Magazine* confidently predicted that 'It is comforting that we are not to expect in Christian Europe or America to see the consumption of opium ever become so universal and perhaps even harmful as it is in some Mahometan countries were manly alcohol is forbidden.'[6] As in Britain, attempts were made to grow the poppy at home; and isolated experiments in Vermont were not entirely unsuccessful. But harvesting was never suited to slave labour, and non-slave labour was expensive.

* * *

The influx of cheap and potent opium inspired a rash of books. Not all were as confident as Jones's pioneering volume. Dr John Hill in his *Family Herbal* recommended caution and warned against the 'current belief' that opium could heal even the bite of a mad dog.[7] In his view Jones had 'merely told what he saw through an addict's eye', apart from being at times 'perfectly unintelligible'; a little harsh, but true. Dr George Young in his *Treatise on Opium* advised circumspection lest the habit 'developed a grip';[8] and Dr Samuel Crumpe in a searching and sensible *Inquiry into the Nature and Properties of Opium*, published in 1793, discussed at some length the grave consequences of deprivation and withdrawal: 'This can be the most painful experience imaginable, only physical weakness preventing the victims from destroying themselves . . . Nothing will relieve the suffering but more opium.'[9]

Without fully understanding it, both Young and Crumpe realised that Turkish opium represented a significant advance on the Indian variety, but the advance was double-edged. Compared to 'good old Patna' which was, at worst, ineffective, the action of the Turkish drug could be disturbing. Yet Crumpe confessed that he himself had taken it without any medical indication and found the ensuing euphoria 'exceedingly pleasant: indeed, my excellent spouse sometimes urges me to have a taster when my temper begins to fray'. And despite the perils, there was, at least for the first few decades of Turkish importation, little moral condemnation. Even sober medical commentators occasionally referred to the drug as 'The Hand of God' or 'The Anchor of Life'; and some quoted the great Robert Boyle, author of *The Skeptical Chymist* a century earlier, who suggested that poppy juice, through its 'occult secret' could be shown to affect beneficially 'animal spirits in the nervous system'.[10] Such an authority settled matters.

But not quite to everybody's satisfaction. Whether cautious or enthusiastic, the literature was beginning to note that opium did not just ease pain and induce a sense of well-being. It had – or could have – what were bunched together as 'side effects'. Once again, many attributed those to adulteration and hoped that the 'essence' of the drug would eventually be discovered and eliminate them. But that was not yet. In a unique way – at least in a way not shared by alcohol and other inebriants, not even by Indian opium – the drug could change both perception and responses to what was perceived. The flutterings of a candle could become the dance of fairies, a pleasant enough spectacle provided it did not lead to a conflagration. The noise of a pin dropped into a bowl could sound like a trumpet announcing the arrival of angels, an expectation apt to lead to disappointment. Soft breezes might whisper secret messages, some gentle but others inflammatory. The mundane smell of homely cooking could make previously placid characters fidgety, excited, dangerous. Passing fancies might grow into consuming passions. Dislikes might congeal into burning hatreds. Anything and everything could become intensified and exaggerated. Impulses could also become suicidal or murderous.

The occasional indulgence could be as treacherous as the established habit. It could, at critical moments, cloud the minds of previously level-headed citizens. The fashion – if fashion such an unwholesome craze could be called – alarmed those whose professional duty it was to keep a watchful eye on the common herd. Voices were raised from the pulpit. The Bishop of Rochester thought that in the eyes of God laudanum was 'not much better than fornication'. The Provost of Eton outlawed the drug among scholars. (Yet soon the Provost himself was observed wandering about improperly dressed but in a jovial mood late at night, suspiciously under the influence.) Doomsters in the respectable press feared that the plague might corrode the 'very fabric of society'. (New clichés always meet the needs of new emergencies.) Such fears were not groundless. Well-born youths under the influence were questioning Christian values. Their positively joyous response to deplorable intelligence from the continent, a revolution in France no less, was disquieting. Was civilisation tottering? Not quite. But, though nobody had yet given it a name, the Romantic Movement was creeping in; and it was creeping in wreathed in opium.

CHAPTER II

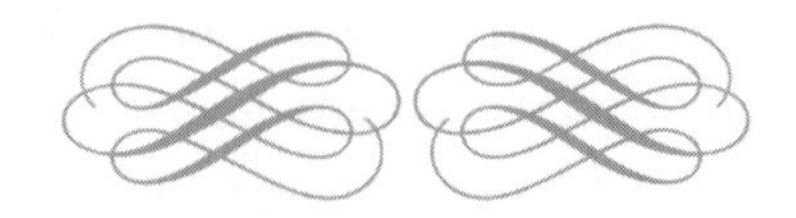

Romantic opium

It was a rebellion against both the past and the future. Romantic Man and Romantic Woman dismissed the old ideals of balance and moderation. They espoused freedom, fancy and fantasy. They exalted love, hatred and other irrational urges. They saw themselves as liberating the human soul, their own and mankind's. But they also recoiled from the ugliness rising around them. Of course ugliness had always existed and had to be accepted; but the new horrors – the belching chimneys, the teeming slums – were all man-made. Feudal lords in the Middle Ages lived more interesting but not significantly more comfortable lives than their serfs. Now the degradation of the poor could be contrasted to infinitely more gracious existence next door. Even the comparatively insensitive found the experience disturbing. The more timid tried to escape. Escape called for a sustainable dose of make-believe. Opium did not make the ugliness disappear but it could jog the imagination into more agreeable channels.

The drug had other effects too. It helped to overcome outdated inhibitions, rules of etiquette and even – or most of all – antiquated moral shibboleths. It prompted bold (if rarely kept) resolutions and brave (if often impractical) undertakings. It inspired words and music of great beauty. It helped the unimaginative to understand and sometimes even to respond to the whisperings of Nature. To sober and industrious folk the new cult was contrary to reason and irredeemably subversive. Opium-inspired rejection of material progress and the accumulation of wealth seemed to strike at the very roots of civilisation. Without the legitimate pursuit of profit social structures would surely collapse; chaos would ensue. But this was not the Romantic creed. And for a few decades, Romantic opium fuelled passion and pathos breathtakingly unrelated to money.

* * *

In 1805 Thomas Trotter, a Scottish naval surgeon, published *A View of the Nervous Temperament.*[1] In the book he portrayed the new generation, the youth who called themselves 'Romantic', as 'afflicted by the nervous disposition formerly the privilege of poets, artists and aristocrats'. And if not actually afflicted, acting as if they were. For Trotter believed that much of this was an affectation. But even posturing could be dangerous. He was concerned that ordinary people had begun to regard stimulants not as a luxury but as a necessity. The habit had started with tea and coffee. Now it was opium. It led to insomnia, hypochondria and hysteria. That called for more medication. And more medication led to more addiction.

But to children of the Romantic Movement addiction was no longer the predicament it had been to their parents. It was becoming almost fashionable. Men and women of sensibility, too fragile and vulnerable for this world, were admired and imitated. Plodding youths began to emulate those exquisite creatures who seemed to be endowed with a 'highly charged' temperament. The term 'highly charged' itself, borrowed from the new Romantic science of electricity, was mind-blowing. And the surest way to become thus electrified was to partake of the magical brew. One might disapprove of the Turkish massacres in Bulgaria and elsewhere – no Christian would condone such misbehaviour – but give the Turks their due: their opium was brilliant.

* * *

Like steam travel, Shakespeare and the laments of Ossian,[2] opium was conquering the continent too. France had been through the convulsions of a revolution and the Terror itself had ended in a bloodbath. As often happens after such upheavals, a manic mood of merrymaking gripped the survivors. Under the arcades of the Palais Royale in Paris but even in the unfashionable provinces the *incroyables* and *merveilleuses* wafted about in their diaphanous shifts on a cloud of opium. The revolution may have been a mixed blessing; but Danton's 'l'audace, de l'audace et encore de l'audace' still reverberated.[3] And though outside France Bonaparte would soon be turning into a monster, the spirit of change could not be squeezed back into the bottle. 'Audacity' took strange forms. On the continent as well as in Britain the elegantly morbid was 'in' and the robustly fit was 'out'. No less suddenly but with graver consequences, there was more to this than make-believe. The appearances of fragility and pallor were not always feigned. Tuberculosis – consumption to laymen and phthisis to doctors – was emerging as the most formidable and formative link between opium and the Romantic Age.

* * *

The illness was ancient. Some of the hunchbacks on Egyptian tomb paintings had probably suffered in childhood from Pott's disease or tuberculosis of the spine (as did the Hunchback of Notre Dame); and Arethaeus the Cappadocian gave a poignant description of the patient with the pulmonary variant of the illness:

> The youth with the croaking voice . . . the extreme wasting, the nails crooked and brittle, the eyes deeply sunk in their hollows but brilliant and glittering . . . the lassitude coupled with foolish gaiety . . . the shoulder blades like wings of the birds.[4]

But in the form in which it would cast a shadow on the lives of millions the disease rose up in England in the wake of what was soon to be called the Industrial Revolution, and, defying political frontiers, crossed the Channel and moved slowly across the continent. It then crossed the Atlantic.

It was, as to some extent it still is, surrounded by myths. It was often described as a painless disease and so it could be for months or even years. But sooner or later pain would come and it could be excruciating. In pulmonary tuberculosis, the commonest nineteenth- and twentieth-century form, the first manifestation was often a 'silent' haemoptysis or the coughing up of fresh blood. Keats, Schiller and a century later Chekhov were doctors who had been taught about the sign and recognised its significance at once; but the illness was soon so common that laypeople too understood its import. The pain would most often start with the spread of the disease to the throat. This made speech impossible and swallowing difficult. The only drug that relieved the suffering was opium. Mute, pale, wasted, but often with glittering alert eyes, it was recourse to the blessed tincture – sometimes by means of a straw and sometimes too weak to hold the glass unaided – that enabled sufferers like Chopin to die peacefully. A century or more later this would still be true of Aubrey Beardsley, Franz Kafka, Elizabeth Siddal, Sophie Munch, D.H. Lawrence, Katharine Mansfield, Alfredo Catalani, George Orwell and countless others no longer remembered by name.[5]

A few doctors disapproved of painkillers on principle or because they felt that pain was morally and even physically beneficial. Falling into the hands of such walking disasters caused much suffering. Dr James (later Sir James) Clark made the end of poor Keats in Rome a horror story.[6] Thankfully, most physicians were glad to prescribe laudanum in unlimited quantities: it made the patients' last days and the suffering of friends and families who kept vigil by the bedside bearable. If some of the friends and family later became addicts themselves, that could not be helped.

It is often said that tuberculosis transgressed social barriers and killed the rich, the titled and the powerful as well as the poor. This was both true and untrue. It did not mean that *Mycobacterium* was no respecter of class and wealth: few microorganisms were more obsequious. But for every 10,000 ill-fed working-class brats it did kill a Duke of Reichstadt, the cosseted son and grandson of emperors.[7] And one deeply mysterious attribute of the illness was true: its age incidence was unique. Tuberculosis struck down not infants or the elderly, generally the age groups most vulnerable to disease, but youths in their creative prime. John Keats had his first haemorrhage when he was twenty-two and Shelley had his earliest intimation of the disease, a painful attack of pleurisy, when he was twenty-four.[8] The illness would then smoulder with unpredictable flare-ups and remissions, the periods of low-grade intermittent fever often being anxious yet amazingly creative. But even in moments of high inspiration these youths lived in the shadow of death and had only two supports. One was that mysterious phenomenon, the *spes phthisica*, the hope of the tuberculous, an irrational and yet burning conviction that death would never come. The second was opium. It was the combination of the two which inspired their awesome courage and some of the greatest artistic and literary works of the age – or of any age.

* * *

After a night at the opera a medical know-all will usually explain that the death scenes just enacted by the likes of Violetta or Mimi are clinically impossible. The medical know-all will be wrong as well as a bore. Alexandre Dumas *fils* and Henri Murger knew what they were writing about.[9] Both had seen consumptive young women cast down in their ravishing youth. It is true that such deaths could be cruel and painful; but they could also be gentle and overflowing with love. What made the kinder alternative possible was laudanum. After seeing Jeanette de Cluseau in the first performance of the *Dame aux Camélias* Théophile Gautier wrote: 'Never have I seen a death so uplifting and full of hope, her pale face set off by the lace pillows, her soul shining through . . . grace, happiness and love to enchant the soul. Oh, the wonders of that blessed drink!'[10]

The scene needed little acting. Jeanette was dead six months later. And Virginia Poe, Edgar Allan's bride, continued to attend social functions in the evenings while coughing up tumblerfuls of bright red blood and being too exhausted to rise from her chaise-longue during the day; she enjoyed her last dance the night when she suffered her fatal haemorrhage. All this time she was on staggering doses of laudanum; and her husband kept her company. As often happened in such circumstances, he too became a user, almost certainly an addict. The inescapable question arises of whether the treatment in such cases shortened lives as well as relieved suffering. Contemporaries with greater experience than any expert today had no doubt that it did. But this did not stop them prescribing the drug. Quality of life was an invention of the Romantic Age propped up by opium.

* * *

Amid the endlessly mulled over marchings and counter-marchings of the Napoleonic Wars the *Mycobacterium tuberculosis* defied battle lines, blockades and counter-blockades. Of the three generals at Waterloo Wellington picked up the habit in the Peninsular War, probably after the Battle of Corunna: it was an opium-starved face which Goya sketched on the night of the carnage. The fiercely mustachioed Blücher was mocked by his subordinates because he carried not brandy but laudanum in his hip flask. The Bonapartes had been taught by *Mme Mère*, the frugal Letitia, to abstain; but in a crisis maternal homilies were forgotten. At least two of her sons, Joseph and Jerome, became users as their fortunes declined, and Napoleon could not face bidding farewell to the Imperial Guard (or what remained of it) without a fortifying draught. It could not have been his first, since laudanum appeared on his wants list on the island of St Helena.[11]

More surprisingly perhaps, the emperor's enemy and for some years prisoner, the saintly Pope Pius VII, recorded a gratifying experience. His efforts at conciliation with the emperor had failed and he was in despair. His spirits were restored by a medicine offered to him by his jailer's wife. He noted in his journal: 'I am told that this wonderful potion is prepared by ignorant and infidel Turks from a special kind of poppy. How inscrutable are the ways of the Almighty.'[12]

* * *

One of the pope's staunchest defenders (and one of literature's monumental hypochondriacs), the vicomte de Chateaubriand, assured his friends that he could not have finished *Le Génie du christianisme* without regular recourse to laudanum. The drug was often difficult to obtain during the Napoleonic Wars, but his needs were met by 'kindly and well-born Breton smugglers'. (To the vicomte the well-born was as important as the kindly.) After the Restoration convinced that his twinges of chest pain were manifestations of a fatal disease, he consulted France's most eminent physician, René-Théophile-Hyacinthe Laënnec. The inventor of the stethoscope was a fellow Breton and devout Catholic; and he spent three hours examining his illustrious patient. Eventually he announced that all the aristocratic organs were in splendid shape. The vicomte departed deeply dejected but continued to cherish his ailments and their only remedy, opium.

His doctor was not so lucky. After a bout of coughing he jokingly requested that a student apply the 'hearing tube' he had invented to his own chest.[13] But it was no joke: one of Laënnec's lungs was completely gone, the other partially so. He paid a farewell visit to his patient, the duchesse de Berry,[14] and informed her that she would need to find another doctor. He would be dead within six months. 'Never use euphemisms for dying,' he had often instructed his pupils. 'Death is part of life.' Enfeebled by loss of blood but with the pain kept in check by opium, he travelled to his beloved Brittany. The drug enabled him to visit the family chapel at Sainte-Croix. His wife later wrote that he returned to the *Manoir* fortified by 'l'alliance du Bon Dieu et l'opium'.

> Under the influence of this wonderful substance he suffered little during his last days. He dozed quietly most of the time. On 13 August 1826 he awoke, removed the rings from his fingers and jotted on the piece of paper on his bedside table: 'I am sorry. I almost forgot. This is such an unpleasant task for anybody else'. A few hours later he stopped breathing.[15]

* * *

Many physicians, who, like Laënnec, would spend their waking hours doling out opium to pale, coughing consumptives over the next century and a half were to tread a similar path. Both the illness and addiction were soon recognised as medical ailments. At least there was never any shortage of 'mild' or 'recovering' consumptive doctors and nurses to staff the sanatoria. Still unknown in Laënnec's day, after the mid-century these strange institutions began to spring up on mountain tops, along isolated stretches of seashore and in other remote regions. In these half prisons, half hospitals, the resident staff would often dispense the drug and satisfy their own need. Nobody minded: for once healers and the sick shared the same destiny. Why make the last journey more painful than necessary? Bless the poppy.

* * *

Inevitably it was not doctors or ordinary patients but afflicted poets, composers and artists who left the most eloquent testimonials. Keats wrote in April 1819:

> 'ere thy poppy throws
> Around my bed its lulling charities.[16]

'The Eve of St Agnes' is full of opium-inspired erotic images and the 'Ode to Indolence' is redolent of the drug:

Ripe was the drowsy hour;
The blissful cloud of summer-indolence
Benumb'd, my eyes, my pulse grew less and less;
Pain had no sting, and pleasure's wreath no flower.[17]

Sir Walter Scott disliked Laudanun and its effects but while he was writing *The Bride of Lammermoor* his doctor insisted on him taking six grains a day to counteract painful stomach cramps. When the novel was finished its author maintained that he could remember not a single incident, character or conversation from the chapters written under the influence; but he acknowledged that his description of Lucy Ashton's neurosis and final insanity, one of the 'blank chapters', was among his best. In the Brontë household the combination of opium and alcohol destroyed the family's pride and hope, Bramwell; and after his death the drug was unmentionable despite all three sisters dying of tuberculosis. Yet Charlotte included in *Villette*, her last novel, a haunting dream sequence which could have been opium induced and a vivid description of the relief afforded by a 'strong opiate drink' when she was deeply unwell and consumed with thirst. Instead of inducing drowsiness the concoctions made her 'alive with new thought ... I don't know whether Madame [Beck] had undercharged or overcharged the dose'. But opium in any dose could both soothe and conjure up strange and exhilarating new ideas.

In Germany Georg Philipp Friedrich Freiherr von Hardenberg, better known as Novalis, lost his beloved Sophie von Kühn to tuberculosis when she was fifteen; and he himself began to cough two years later. During his short life, most of it spent under the influence of laudanum, he invented that quintessential literary product of the drug, the poetic fragment. Obscure and teasing, these utterances were and are inexplicably memorable. He also created an opium-inspired religious poetry which enchanted some but disturbed many. His 'hymns' mixed evangelical piety with Marian mysticism and sexual allusions with spiritual frisson in a shocking but captivating way. Franz Schubert, usually unfailing in his empathy with words, found these outpourings difficult to set to music. At least until the final hymn. As the last vocal line, 'Ich fühle des Todes verjungende Flut' (I feel death's rejuvenating stream) rises above the stave, the notes convey the fluttering of a soul liberated from earthly suffering. The liberator was opium.

Schubert himself probably did not indulge even during his last painful illness; but his circle were soaked. Johann Mayrhofer with whom he shared a flat worked in the Imperial Censor's office, a deeply uncongenial occupation for a sensitive soul. What made it tolerable was escape into poetry and laudanum, the two inextricable. But in the end the burden proved too heavy and he jumped to his death from the window of their flat. The gentle and short-lived Ludwig Holty, a Protestant pastor, was deeply addicted, as was Johann Gabriel Seidl, poet of Schubert's last miraculous burst of happiness, *Die Taubenpost*. And the opium-soaked mediocre *Sturm-und-Drang* poet Johann Georg Jacobi 'sails to immortality having provided the words to one of the greatest songs ever composed'.[18] Some poets the composer admired only from afar. The Olympian Goethe liked his 'regular medicinal draught' when affairs of state and amorous intrigues at the court of Weimar threatened to overwhelm his great mind. Heinrich Heine imbibed during most of his exile in Paris: only the poppy, he wrote, could transform trite sentiments like love, despair, grief and mourning 'in our grating German language' into something approaching poetry.

Composers too were addicted. Despairing in love, Hector Berlioz conceived the 'March to the Scaffold' of his *Symphonie fantastique* in a laudanum-induced dream. All movements portray the fevered visions of a sensitive young artist, his fickle love appearing as a musical *idée fixe*. In a fit of jealousy the artist murders her. In the last movement the motif is cut short by the swish of the guillotine. About the same time across the Channel Carl Maria von Weber, kept alive by opium, conducted nine performances of his last opera, *Oberon*, at Covent Garden. He should have cancelled but his family back in Germany were desperately in need of funds. He was found dead in a pool of blood on the morning of what should have been the last performance. There was no money to repatriate the body. And in a darkened flat overlooking Place Vendôme, rented for the penniless composer by his Scottish patron, Jane Sterling, opium sipped through a straw made the passing of Frédéric Chopin a gracefully fading last Nocturne.

Opium sometimes bridged art forms. Adelbert von Chamisso, the scion of French aristocrats settled in Berlin, wrote the cycle of poems, *Frauenliebe und Leben* (A Woman's Love and Life) after marrying Antonie, an illiterate sixteen-year-old servant girl. He had already developed a cough partially stilled by laudanum. His young wife looked after him, concealing her own fast-advancing illness and increasing dependence on the drug. Eventually and contrary to the poetic sequence it was *der Mann* who was to mourn *die Frau*, not the other way round; but he survived her by only a few months. Twenty years later their happiness and grief inspired Robert Schumann to compose his greatest song cycle. The gigantic canvases of the heavily addicted John Martin, biblical and historical scenes dwarfed by wild and stormy landscapes, helped to shape the fantasies of Théophile Gautier, Victor Hugo and, above all, the somnabulistic travelogues of Gérard de Nerval.

And always it was the young. Thomas Girtin was twenty-four when he sketched his last, wonderfully tranquil images of London. Willy Turner, not a paragon of self-effacement, was heard murmuring years later: 'Poor, poor Tom, if he had lived I would be nowhere'. On his deathbed, Richard Parkes Bonington painted watercolours of Venice from memory in a Harley Street nursing home in a trance of laudanum. No other medicine would ease the stranglehold on his throat. When he could no longer hold the brush he continued to create shapes in the air with his empty hand until the moment of his quiet departure. He was twenty-six, in his friend Delacroix's words 'carried away like a young god by his own genius'.

* * *

On both sides of the Atlantic, the season of autumn was the addict poets' natural friend. On Lake Annecy in the Savoie Charles Hubert Millevoye wrote his last verse about the 'mournful colours of the dying season'. In Concorde, Maine, Henry Thoreau watched with fascination 'the hues of departure' glowing in the unnaturally vivid colours of laudanum. For Mikhail Lermontov in Russia the drug made September 'the month of unrequited love'. To Keats autumn was (as every English schoolchild knows) 'the season of mists and mellow fruitfulness'. In Italy – and in a miasma of opium – Count Giacomo Leopardi wept, watching the fluttering of the yellowed leaves in the park. In Hungary the last flowers of the summer still blooming in the valley inspired Alexander Petofi to pen the most beautiful lines in the language.[19]

* * *

To all these young people opium provided moments of happiness during the last months of their abbreviated lives. On others the drug imposed a long life of bondage.

CHAPTER 12

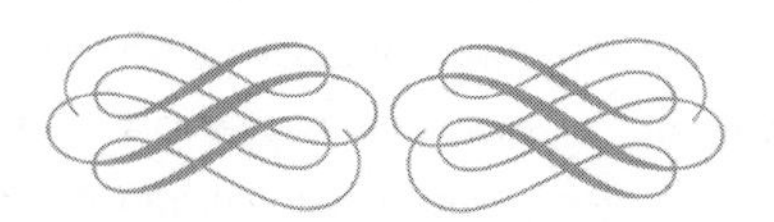

The pleasure dome of Xanadu

On 11 April 1816 Dr James Gillman, a general practitioner in the hilltop village of Highgate some ten miles north of London, received a letter from a colleague, Dr Joseph Adams of Hatton Garden.

> A very learned, but in one respect unfortunate gentleman, has applied to me on a singular occasion. He has been for several years in the habit of taking large quantities of opium. For some time he has been in vain endeavouring to break himself off it . . . His friends are not firm enough from a dread lest he should suffer by suddenly leaving it off; . . . but he has proposed to me to submit himself to any regime, however severe. With this in view he is desirous to fix himself in the home of any gentleman, who will have the courage to refuse him any laudanum, and under whose assistance, should he be the worse for it, he may be relieved . . . I could think of no-one so readily as yourself . . . Be so good as to inform me if such a proposal is absolutely inconsistent with your family arrangements. I should not have proposed it, but on account of the great importance of the character as a literary man. His communicative temper will make his society very interesting as well as useful . . .[1]

Dr Gillman had no intention of taking lodgers, however 'interesting'. Besides, he was well aware of the difficulties of weaning laudanum addicts off the drug. He had recently been told about the case of Mr William Wilberforce, the illustrious politician and anti-slavery campaigner, being treated by an eminent physician in Bath, the treatment ending in total failure. He had heard about many other fiascos. But he agreed to meet the patient without committing himself; and next evening Dr Adams drove Mr Samuel Taylor Coleridge to Highgate. The following day Coleridge came on his own, an occasion Dr Gillman would later recall:

> His manner, his appearance and, above all, his conversation were captivating. We listened with delight . . . and [when a friend who had been present departed] Coleridge informed me of the painful opinion which he had received concerning his case, especially from one medical man of celebrity. I was indignant. The tale was sad and the opinion given unprofessional and cruel, sufficient to have deterred most men so afflicted from making the attempt Coleridge was contemplating and in which his whole soul was so deeply engaged. In the course of our conversation he repeated some exquisite but desponding lines of his own. It was an evening of painful and pleasurable feeling which I can never forget . . . I felt indeed spellbound without the desire of release.[2]

Next day Coleridge sent a long letter to Dr Gillman:

> My ever-wakeful Reason and the keenness of my moral feelings will secure you from all unpleasant circumstance connected with me, save only one: *Evasion*, and the *cunning* of a specific madness. You will never *hear* anything but truth from me – Prior Habits render it out of my power to *tell* a falsehood – but, unless watched carefully I dare not promise that I should not with regard to this detested Poison be capable of acting a lie . . . Every friend I have (and thank God! in spite of this wretched vice I have many and warm ones who were friends of my youth and have never deserted me) will think of you with reverence.[3]

Years later Dr Gillman appended a footnote to the last sentence: ' "Vice" is too strong an expression. It was not idleness, nor sensual indulgence to contract this habit. No, it was latent disease.' On the evening of Monday, 15 April 1816, Coleridge arrived in Highgate to stay with the Gillmans for a month. He remained for eighteen years.

* * *

Many of the torments and tribulations of a brilliant mind in bondage to opium are revealed in Coleridge's notebook and correspondence; but the outward events of his life were unremarkable.

Born in Ottery St Mary, a small market town in Devon, on 21 October 1772, he was the tenth and youngest child of the Reverend John Coleridge. The father, the local vicar, was a saintly man much loved by his son; but he died when the boy was only nine. Samuel was then sent to Christ's Hospital, a boarding school founded in the City of London, known for its high academic standards and uncompromising moral rigour. He was rarely allowed home and his mother never visited him in nine years. He was often sick but dreaded the sickroom and put up with pain rather than endure incarceration. It was the kind of setting which nurtured many of those who left indelible marks on nineteenth-century English literature and letters.

At Jesus College, Cambridge, he won a gold medal for an ode on the slave trade in Greek but never gained a degree. Escaping from creditors he volunteered as a trooper in the Light Dragoons, an unwise move since he hated both drill and contact with horses. Fortunately, his mates volunteered to do the grooming for him in return for having their love letters composed by one of England's greatest lyric poets. Back in Cambridge, he made friends with Robert Southey, the two poets marrying two sisters, Sara and Edith Fricker. Coleridge came to hate his wife, by whom he had three sons and a daughter, and perhaps she was difficult. But to be married to an opium addict must be one of the most testing fates to try the love of a spouse.

In 1795 the poet met William Wordsworth and his sister Dorothy; and the friendship blossomed into a unique collaboration.[4] Their jointly published *Lyrical Ballads* (1798) is regarded by some as the starting point of the Romantic Movement in England. They toured the continent and Coleridge immersed himself in German philosophy and embarked on a series of translations and early exercises in plagiarism. After his return he settled in Keswick in the Lake District to be near the Wordsworths, probably a mistake. Soon Dorothy was expressing her horror at his puffed-up appearance and his drinking a quart of laudanum a week.[5] He hated being pitied; but he was indeed becoming a pitiful figure.

Between 1809 and 1817 a number of journalistic ventures ended in failure. A series of lectures in Bristol and London was plagued by periods of incapacity. Coleridge could still sparkle but his audiences became restless during his rambling digressions.[6] He was humiliated and soon he was destitute. Dr Gillman's house opened to him in the nick of time.[7]

* * *

But one has to look beyond surface events – or non-events – because they were shared by tens of thousands of less celebrated men and women in bondage to opium. Coleridge, like many addicts of his day, had suffered from a crippling and painful physical ailment since childhood. Rheumatic fever is virtually unknown in Western Europe today, though it is still common in parts of the developing world; but it was, for centuries, one of the most feared diseases of childhood. It began with a painful acute inflammation of a joint or several joints; but extra-articular complications soon followed. In particular, while the disease 'licked the joints, it bit the heart.'[8] Years later the victim would develop shortness of breath, swelling of the ankles, fluid in the abdomen, chest pain and a cough. Looking increasingly congested, exhausted and labouring for breath, the sufferer went into an undramatic but distressing decline; and the end was inescapable.

By the time Coleridge arrived at the Gillmans' house he was, at forty-four and in his own estimation, in 'all but the *Brain* an old man'. His heart was severely bitten and his chest pain and shortness of breath could be harrowing. In the absence of even salicylates, the salve to come for rheumatic fever and its sequels, opium was a godsend.[9] In Dr Gillman's care and on regular but relatively modest doses he survived. It was not a prognosis that most medical men would have pronounced at the time of his arrival. Until his move to Highgate Coleridge's increasing addiction – or slavery – to opium, its concealment and exploitation by others, dominated his life. To some extent he himself created the Romantic image of the opium-eating, opium-inspired and opium-enslaved artist; and he succeeded in generating a morbid fascination in the public which stoked his fame. But in 'The Pains of Sleep' he gives an almost clinical description of the junkie waking up from the sweet honeymoon with the drug to the nightmare of a lifelong liaison:

> . . . yester-night I prayed aloud
> In anguish and in agony,
> Up-starting from the fiendish crowd
> Of shapes and thoughts that tortured me:
> A lurid light, a trampling throng,
> Sense of intolerable wrong . . .
> Desire with loathing strangely mixed
> On wild or hateful objects fixed.
> Fantastic passions! Maddening brawl!
> And shame and terror over all![10]

As in most cases of addiction, it is difficult to date precisely the beginning and the end of the honeymoon. His own account is credible but – like all his reminiscences – probably at least half-imaginary.

> The habit into which I had been ignorantly deluded by the seeming magic effects of opium, in the sudden removal of a supposed rheumatic affection . . . by which I had been in pain and bedridden for nearly six month. . . . Unhappily, among my neighbours' and landlords' books were a large parcel

> of medical reviews and magazines . . . and in one of these reviews I met a case which I fancied very like my own in which a cure had been effected by Kendall Black Drops. In an evil hour I procured it . . . and it worked miracles . . . Alas! It is with a bitter smile, a laugh of gall that I recall that period of unsuspecting delusion . . .[11]

Whatever his poetic, religious and philosophical legacy, nobody before or since has documented with more insight the unfolding tragedy of an addiction. This is not to say that his documentation can be believed. His published accounts were almost never true, or not entirely true. But even deceptions can be revealing. In autobiographical fragments an addiction is never the addict's fault. He or she has been seduced by cruel circumstances, by the selfishness of society, by the indifference of the family, by a well-meaning but silly friend, by a malicious long-time addict or by that archetypal fiend, 'the pusher'. There is no indication who that person was in Coleridge's case; but around him clustered the usual cast of gapers and busybodies. Joseph Cottle, his publisher and subsequent editor, was titillated by the sordid drama and relished posing as a public benefactor by shedding tears over the evil in society. He would also busy himself collecting letters and other sensational tit-bits for later publication. Robert Southey would support Coleridge's family but his high-minded exhortations to his brother-in-law to 'pull himself together' and his refusal to help in any material way – even his foiling of any scheme of assistance proposed by others – is almost as unedifying as Cottle's lip-smacking. Most depressing to a doctor, almost all the medical men Coleridge consulted before Dr Gillman emerge from the encounters as ignorant, judgemental, useless, hypocritical and sometimes malevolent.[12]

But even the briefest picture of Coleridge – and of any addict – as a tormented innocent would be false. The first attack on him as a liar and a plagiarist was launched barely two months after his death by fellow addict Thomas de Quincey, and soon a chorus of academics rose to cast doubt on his life's work. The argument continued to rage – or at least simmer – until the publication in 1971 of Norman Fruman's meticulously documented and totally damning biography.[13] Some of the poet's lies were admittedly mere embroideries. Coleridge, like all Romantics, attached great importance to divine afflatus, the spontaneous bubbling up of words from nowhere; and in notes accompanying poems he often claimed that he had just 'fallen unwarily into composition' and that the lines enclosed were written 'in a hot fit of inspiration last night'. In fact the poems had almost always been painstakingly composed over a period of weeks or months and finished at some time in the past. On dissection even such popular favourites as 'Kubla Khan' were found to be ever so slightly 'borrowed'. Yet to label the poet – or any addict – a liar is an oversimplification.

'Truth' and 'lie' are among the most emotionally charged words in the language today and in public life the most morally weighted. But context is all. In the opium addict's universe fact, faction, fiction, fancy, fantasy and confabulation can become inextricably mixed. At one level Coleridge was a fervent seeker after the truth – or at least *a* truth. That truth could be to him more real than reality as it presented itself to his fellow men. He was also, for most of his life, a disputatious but fervent Christian. When in his letter to Dr Gillman he wrote that, 'true to his prior habit', he could not say an untruth, he almost certainly believed this himself. He was a damaged archangel (to borrow Fruman's subtitle) but he did not disseminate malicious lies; and in a world which thrived on gossip and innuendo he was an innocent. Yet, despite his occasional foray into the physical sciences, ostensibly based on factual observations, facts carried no moral imperative for him. It all depended on what use was made of them. This was – and still is – true of many devotees of the poppy.

The charge of plagiarism presents a similar conundrum. Sir Ernst Gombrich pointed out that even the greatest artists owe more to previous art than to life. This is usually well understood by artists but not always by critics. Coleridge borrowed on a grand scale or, to put no finer point on it, plagiarised without compunction. Many of his philosophical propositions were lifted from Kant, Schelling and Tennemann. But how many of the ideas of Kant, Schelling and Tennemann were lifted from earlier philosophers or even from drinking companions at the tavern last night? Even Coleridge's best-known poems emerge slightly mauled by Fruman; but the memorable images – the pleasure dome, the albatross – were Coleridge's.[14] Nevertheless, there was – and is – a gradual fading of conventional literary proprieties in the minds of many addicts. Do those proprieties matter? Perhaps.

CHAPTER 13

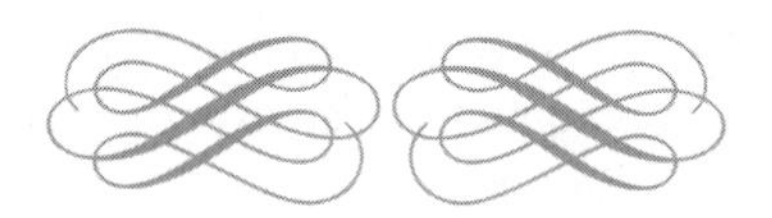

The opium eaters

COLERIDGE'S 'BONDAGE' WAS a private affair, known only to a small circle of friends. Thomas de Quincey's addiction was as public as he could make it. His *Confessions of an English Opium Eater* (1822) became a best-seller and made him famous, in the United States as well as in Europe, as *the* accredited face of Romantic opium dependence. The genre in which he chose to bare his soul (and most of his bodily functions) was ancient. To Christians St Augustine's chronicle of his conversion from profligate youth to searching Christian is the greatest autobiography ever written.[1] It was in his footsteps that many Puritans recorded their own spiritual journeys.[2] More recent in De Quincey's day were the posthumously published and shockingly frank *Confessions* of that odious genius, Jean-Jacques Rousseau. There can be little doubt that, despite the Englishman's explicit disclaimers, it was the Frenchman who provided the model.

The 'de' in De Quincey's name was a small conceit added by Thomas's mother to bolster her aristocratic pretensions. Her husband was plain Thomas Penson Quincey, a prosperous and kindly Manchester businessman, who died of tuberculosis when the younger Thomas was only seven. The family was left well provided for and moved to the fashionable spa town of Bath. Pupils and teachers at the grammar school there later remembered that young Thomas was short, that he had an extraordinary facility with languages and that he was a charmer.[3] Some of that charm would later give a special flavour to his letters and short literary pieces. He was seventeen in 1798 when he was bowled over by Wordsworth and Coleridge's *Lyrical Ballads* and became a convert to the Romantic sensibility. In defiance of his family's wishes he spent some time in solitary wanderings in Wales and then lived in London with a young waif,

Anne, who fended off their starvation by working as a prostitute in Oxford Street. One day she did not return.

Prolonged negotiations with guardians and money lenders eventually provided Thomas with an allowance sufficient to let him enrol as an undergraduate in Worcester College, Oxford. It was here, in 1804, that he developed the toothache for which, at the recommendation of a fellow undergraduate, he purchased his first dose of opium. It was an experience he never forgot:

> . . . in an hour, O heavens! What a revulsion! What a resurrection from the lower depth, of the inner spirit! What an apocalypse of the world within me. That my pains had vanished was now a trifle in my eyes; this negative effect was swallowed up in the abyss of divine enjoyment thus suddenly revealed. Here was a panacea for all human woes; here was the secret of happiness . . .[4]

Opium became his fate. As he often said, the drug did not 'create' anything but it enhanced beauties and intensified pleasures. He would travel to London and after a draught of laudanum attend the opera. Listening to the music became bliss, unspoilt by the chatter of fashionable pinheads in the audience. Time dissolved, space was transformed. After the opera he would walk the streets, reliving the experience without being aware of it. It was during this time that he wrote his famous – or to many infamous – eulogy:

> O just, subtle and all-conquering opium! That to the hearts of rich and poor alike, for the wounds that will never heal, and for the pangs of grief that 'tempt the spirit to rebel', bringest an assuaging balm – eloquent opium! . . . thou buildest upon the bosom of darkness, out of the fantastic imagery of the brain, cities and temples . . . beyond the splendours of Babylon and Hekatopylos; and from the anarchy of dreaming sleep, callest into sunny light the faces of long-buried beauties, and the blessed household countenances, cleansed from the dishonours of the grave. Thou only givest these gifts to man: and thou hast the keys of Paradise.[5]

The passage is remarkable since the lassitude that follows the high usually abolishes all desire to record the experience. On a more sober note De Quincey made four general observations. First, that his imagination could in part at least direct his opium-induced fantasies. Second, that his dreams of happiness were often accompanied by 'funereal melancholy such as is incommunicable in words'. Third, the normal senses of time and space were altered. And last:

The minutest incidents of childhood or forgotten scenes from youth, were often revived. I could not be said to recollect them . . . but placed as they were before me like intuitions and clothed in all the evanescent circumstances and accompanying feelings, I recognised them instantly.[6]

Opium, as one historian has commented,

was undoubtedly a Pandora's box of literary tools to the erudite and imaginative mind . . . To recollect otherwise lost experiences is, after all, the very stuff of the writer's art . . . Opium provided unique visual images, afforded a kind of mental time travel, opened up new ways of observing the mundane and acted as an *aide-mémoire*.[7]

Of course, even to an erudite and literary mind the tools could be treacherous. In cold daylight the products were often a jumble of ill-fitting parts, incomprehensible even to their creator. And, as addiction increased, marvellous visions tended to metamorphose into nightmares. De Quincey was not only an acute observer but also – a far less common gift among addicts – an acute recorder of what he had observed; and he saw the shadows as well as the highlights. An innocuous Malay trader whom he had met briefly became a terrifying incubus:

The Malay had become a fearful enemy. Every night for months, through his means, I have been transported . . . to China or Hindustan, stared at, hooted at, grinned at, chattered at by monkeys and parakeets, I ran into pagodas and was fixed for centuries at summits or in secret rooms from which there would never be an escape. I was an idol; a priest; I was worshipped and then cruelly sacrificed . . . Thousands of years I lived and was buried in stone coffins with mummies . . . I was kissed with cancerous kisses by crocodiles, and I was laid with unutterable abortions among reeds and Nilotic mud . . . Over every form of threat and punishment and dim sightless incarceration brooded a killing sense of eternity and infinity . . . The cursed crocodile became for me the object of more horror than all the rest . . . I was compelled to live with him and (as was always the case in my dreams) for centuries. When I sometimes escaped into Chinese houses all the feet of the tables and chairs became instinct with life and were soon transformed into abominable heads of crocodiles, their leering eyes looking at me multiplied into tens of thousands of repetitions.[8]

And yet De Quincey went on to live a modestly successful literary life, troubled only by periodic bankruptcies, narrow escapes from creditors, editorial feuds, vindictive reviews, intervals of drug-induced incapacity, death threats from outraged readers, academic snubs, ecclesiastical excoriations and other fixtures of a literary life then as now. His collected works have been published with reverent care and fill twenty volumes. They may not add up to a Dickensian oeuvre; but wherever one dips into them beguiling conceits, provocative insights and unexpected turns of phrase remind one that this was the golden age of the essay, the time of Lamb and Hazlitt as well as of poppy-heads like Coleridge and De Quincey. Some of the prose of the last is probably the most quirky and amusing. On the other hand, De Quincey never came near to fulfilling his boyhood dream of becoming a significant philosopher like Kant. In his day the *Confessions* were mutilated, pirated and translated into many languages. They were liked by Thackeray and extravagantly admired by Ralph Waldo Emerson and other American men of letters. Dostoevsky carried a copy in his pocket when he was led to his mock execution.[9] Baudelaire, like many Frenchmen an astute critic of English literature, declared that De Quincey had 'the most original mind of all English writers'. One wonders if Freud had read the passages relating to dreams. De Quincey did not use terms like the subconscious and Freud was more interested in cocaine than opium; but their interpretations are a similar mixture of insight and fantasy.

De Quincey's private life too was relatively contented, despite the financial drain of his addiction. Towards the end of his life he spent on the average £14 a week on 1,200 drops of laudanum, the approximate equivalent of £1,000 in today's money. His unbounded devotion to Wordsworth the man cooled but not his admiration for Wordsworth the poet. While living in Grasmere in the Lake District he married Margaret Simpson, the mother of his first child, William, born out of wedlock. The union proved those who predicted disgrace wrong. True, Margaret often despaired and financial worries probably contributed to her ill health and early death (for which De Quincey felt responsible); but the marriage was not without love and produced three more sons and three daughters. Four died in infancy; one son, Horace, was killed in the First Opium War. De Quincey himself died on 7 December 1859 at the age of seventy-four, a fair lifespan in the mid-nineteenth century.

* * *

As in earlier times – and to the perpetual consternation of the virtuous and campaigners against the evil of drugs – there were other opium eaters and rule-breaking survivors. George Crabbe pursued three careers, each with a measure of success. A poor boy from Aldeburgh, a Suffolk fishing village, he was apprenticed as a pharmacist and practised in his home county as a surgeon-apothecary for some years. He then went to London and became a writer and poet. Finally, he took holy orders and ended his life in 1832 at the age of seventy-eight as the rector of Trowbridge in Wiltshire. He emerges from his son's memoir as a kindly father and grandfather and a respected pastor of his flock.[10] But there was a second side to his life.

At forty-one he started to take laudanum to ward off attacks of migraine. These may have been brought on by stress. His much-loved third son died in a street accident and his wife began to show symptoms of a bipolar mental illness. He also developed a digestive ailment, perhaps a duodenal ulcer. From then until his death he continued to take slowly increasing doses and became dependent on the habit. He managed to keep this a secret from his parishioners and the addiction did not apparently interfere with his duties. But it affected his poetry, and for the better.

Crabbe's early literary efforts, a few classicising poems but mainly topographical sketches, were well composed but unmemorable. It was when his inspiration began to be suffused with opium that it took wings. Or so the chronology suggests. 'Peter Grimes,' one of a collection of poems *The Borough*, (1810), is set in the grim estuary creeks around Aldeburgh. It is the story of a young fisherman who murders his father and then abuses and by neglect kills two of his apprentices. He gets away with his crimes but becomes a social outcast, forbidden to employ another apprentice. Forced to live a solitary life amidst the mud banks, he is stung by guilt and goes mad. He rails against the spectres of his father and his apprentices which come to haunt him; and he tries to justify his crimes. At last, on his deathbed, in pervasively opium-inspired imagery he recounts his sufferings at the hand of his ghosts:

> And when they saw me fainting and oppress'd,
> He, with his hand, the old man, scoop'd the floor,
> And there came flame about him mix'd with blood;
> He bade me stoop, and look upon the place,
> Then flung the red-hot liquor in my face;
> Burning it blazed, and then I roar'd for pain,
> I thought the demons would have turn'd my brain.
> Still there they stood, and forced me to behold
> A place of horrors – they cannot be told –
> Where the flood open'd, there I heard the shriek
> Of tortured guilt – no earthly tongue can speak;
> 'All days alike! For ever!' did they say,
> 'And unremitting torments every day'.[11]

It is not only the guilt-ridden memories which make the poem opium-inspired. The impression of the endless mudflats too, the tides changing with terrible monotony, express a nightmarish world. Time could stop still in moments of delight or slow down wondrously by command of the poppy; but the juice could transform an unyielding rhythm into a torment. And Grimes's death is probably the most haunting description of an addict's end:

> And still he tried to speak, and look'd in dread
> Of frightened females gathering round his bed;
> Then dropp'd exhausted and appear'd at rest,
> Till the strong foe the vital powers possess'd
> Then with an inward, broken voice he cried,
> 'Again they come', and mutter'd as he died.[12]

* * *

Most of Crabbe's and De Quincey's literary contemporaries experimented with laudanum – and were influenced by it – without necessarily becoming addicted. Elizabeth Barrett was started on opium at the age of fourteen by her doctors for a disorder described by them with maddening Victorian opacity as a 'severe derangement of a highly important organ'. She continued to take the drug regularly throughout her life though she changed to morphine in her thirties. In 1840 she wrote to her husband, Robert Browning:

> Can I be as good for you as morphine is for me, I wonder, even at the cost of being as bad also? – Can you leave me off without risking your life, – nor go on with me without running all the hazards of poison?[13]

Robert took up this surely unique love correspondence using the drug as a metaphor of their devotion. 'May I call you my morphine?' And could she ever imagine that he 'might exist without taking my proper quantity?'[14] To her the drug was a medicine, not an addiction: gossip worried him but her not at all. For only a short period did she reduce (but not discontinue) her daily dose. After two miscarriages she began to fear that the 'poison' had destroyed her unborn children, and cut her intake by about a third. Her husband marvelled at her extraordinary willpower, 'equal to that of a thousand men'. It was as high a Victorian compliment as it could get. Her resolve paid off and she gave birth to a son at the age of forty-three. She was just fifty-four when she died in 1861, almost certainly from progressive pulmonary disease; but it is unlikely that her opium consumption significantly contributed to her passing.

Fanny Trollope, Anthony's mother, established a routine in her middle age of writing her book by night 'helped by laudanum and grey tea'. Harriet Martineau recalled that in the 1840s she had been told by a clergyman 'who knew the literary world exceedingly well' that 'there is no author or authoress who is free from the habit of taking some pernicious stimulant ... like laudanum'. But by then another social change was on its way, more shocking than the spread of opium to the literati. The drug was crossing the last social barrier and becoming part of the life of the poor, the ordinary and the destitute – in short, the people.

CHAPTER 14

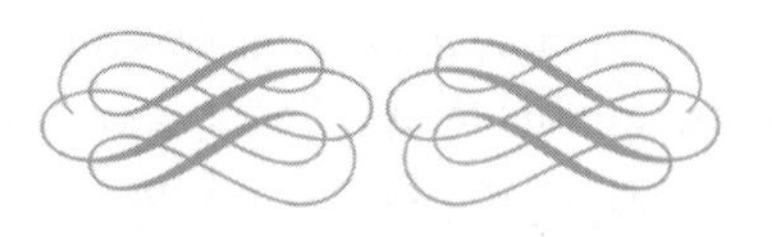

The people

During the early decades of the nineteenth century agricultural labourers, factory workers, miners, soldiers, sailors, pimps, prostitutes and their men- and women-folk still sought oblivion in gin, rum or home-distilled nameless spirits. By the mid-century opium had permeated all layers of society: it was consumed more widely than aspirin, paracetamol and all other over-the-counter analgesics put together are today. Between 1825 and 1850 imports to Britain rose from 23,300 kilos to 138,000 kilos a year. About a third of this was re-exported, mainly to the United States; but the annual increase in home consumption was still 4–8 per cent. In the first half of the century this was no more a matter of concern than cigarette smoking was in the 1950s. Indeed, opium was regarded by many as a boon, though with risks attached to it.[1] But what boon was without risks?

In most countries some regions became more deeply addicted than others. In Britain the Fens, the low-lying marshlands of Lincolnshire, Cambridgeshire, Huntingdonshire and Norfolk became widely known as 'The Kingdom of the Poppy'. In the cathedral city of Ely, in the heart of the region, there was no need to ask for a stick of opium or an ounce of laudanum in the shops. A coin laid on the counter meant only one thing. Dr John Hawkins described a King's Lynn farmer entering a chemist's shop, buying one and a half ounces of laudanum and drinking it there and then. He would then return twice on the same day for similar draughts and buy a half-pint to take home in the evening. A chemist in Spalding told Hawkins that he had sold more laudanum in his first year in the town than he had in twenty years in Surrey. A Holbeach grocer took about £800 a year for laudanum from the people of this poor working-class parish. Whittlesea with a population of 3,700 had five chemists dealing almost entirely in opium.[2] In Charles Kingsley's *Alton Locke*, published in 1850, a Fenman explains to a visitor:

> Oh! ho! ho – you goo into the druggist's shop o' market day, into Cambridge, and you'll see the little boxes, dozens and dozens, a'ready on the counter, and never a Fenman's wife go by but what calls in for her pennord o' elevation, to last her out a week. Oh! Ho! ho! Well, it keeps womanfolk quiet it do; and its mortal good against the pains.[3]

In many Fenland pubs opium was mixed with the beer. One enterprising Ely brewer added laudanum to his ale at source and, despite the slightly increased price, did well. The practice was probably illegal, but who cared?

Dr Henry Hunter who lectured in Cambridge and practised in the region wrote:

> In this part of the world a man may be seen in the field asleep, leaning on his hoe. He starts when approached and works vigorously for a while. Then he goes back to sleep, standing up like a horse . . . A man who is setting out on a hard job takes his pill of opium as a preliminary as a matter of course and many never take their beer without dropping a piece of opium into it.[4]

The writer Thomas Hood on a visit to Norfolk was 'much surprised' how widely pills of opium – or 'opic' as it was locally known – were used even among the poorest. Surprised but also not surprised. 'These Fen people living in the dreary, foggy, cloggy boggy wastes of Cambridge and Lincolnshire fly to it for a change of scenery. And who would blame them?' A local practitioner, Dr Rayleigh Vicars, also explained the popularity of the drug among his patients: 'their terrible lives could be temporarily brightened by the passing dreamland vision afforded by the poppy'.[5]

In some country areas laudanum was shared between man and beast. Many buying the drug stated that they wanted it for their pigs: 'they fat better when they're not crying.' Opium was regularly added to cattle feed. For a sheep 6 drachms of laudanum and for an unmanageable horse before taking it to the market two or three ounces were the regular dose. A child in Cambridgeshire died of opium poisoning in 1836 after accidentally drinking a potion intended for the calf. The accident was reported to the coroner, otherwise it might have passed unnoticed. The parents and owners of the calf were reproved. Not surprisingly, infant mortality in the Fenlands was 206 per 1,000 in 1850, higher than in Sheffield, the deadliest of the industrial cities at the time.

The reasons for such regional addictions were never entirely clear. In the Fens the damp climate promoted rheumatic aches and pains as it still does. 'Aigues', fevers and malaria remained common even after the advent of quinine at the end of the seventeenth century. Quinine cured the fevers but not the harshness of everyday existence. Opium did. But this could not be the whole answer. There were other marshy, low-lying unhealthy areas in Britain – the Thames Estuary or Romney Marsh had historic pockets of endemic malaria – where opium consumption was not particularly high. In the Auvergne in France, a province with some of the country's most luscious farmlands, as well as in the fertile Po Valley in Italy and in Andalucía in Spain opium consumption was entrenched.[6] In Transdanubia in Hungary poppy fields were the pride of Count György Festetich, the agricultural reformer. Fellow grandees visiting him admired his estate and for a time white poppy cultivation came to rival opera houses as an aristocratic fad. But home-grown opium could never satisfy the needs of the mushrooming industrial cities. That would soon overtake demand even in the most heavily addicted agricultural regions.

* * *

More than one circumstance made the drug part of life at all social levels. Falling cost was one: by the 1830s a lump of opium was cheaper than gin or all but the vilest home-distilled brew. But this was no more than a permissive cause. The drug was uniquely effective against cholera – or occasionally effective when nothing else was. The disease was a killer; and long before 'salt and water balance' was heard of, doctors realised that patients died because of massive loss of fluids. If the loss could be significantly reduced, there was a slim chance of survival. Opium sometimes seemed to achieve this. If not, at least it could make the end slightly less distressing. In Britain the epidemics of 1830–31, 1848–49 and 1853–54 were among the most effective promoters of the drug. But treatment had to start early and the doses had to be heroic: by the time diarrhoea was in full flow – sometimes within two or three hours of the onset – it was too late. A few babies and children who might have died from the milder 'summer diarrhoeas' and 'dysenteries' were also saved. It was an image-builder.

Another promotional point in Britain was the reluctance of the poor to visit the doctor. Even the cheapest consultation could cost a third of a skilled workman's weekly wage. The chemist's advice, on the other hand, was part of his sales pitch. By mid-century the consultation at the shop almost always ended with the dispensing of some form of opium. When the Sale of Poisons Bill was first discussed in 1857 John Brande of the Royal Institution submitted to

Parliament that the use of opium-based medicines was so general that any restriction on it was unenforceable. How could a chemist keep opium under lock and key when he dispensed it a hundred times a day?[7] On the continent doctors were cheaper and in poor regions they often accepted delayed payment or payment in kind; but poverty also bit more deeply. In south-western France beyond the lush vineyards of Bordeaux, including the region around Lourdes to be made famous later in the century, many departments were without a qualified doctor or pharmacist. Anodynes, mostly based of opium, were sold by hawkers or in a few general shops. In the United States doctor and chemist were often the same person.

* * *

In the 1840s the first popular patent opium medicines made their appearance. They were emblazoned with reassuring messages. For those who disliked asking for advice even from the chemist, self-medication became easy. In pubs Stott's Unique Fruit Cordial containing 3 per cent opium would keep the kids quiet while the grown-ups enjoyed their more potent – or was it? – refreshment. And how true the maker's claim proved to be! Barely had the kids consumed their first tumblerful than they were clamouring for another. Then some fell blissfully asleep. Among adults Sydenham's laudanum and Dover's powder were still popular; but they were competing with cheeky newcomers. Chlorodyne was market leader. This peerless concoction, basically opium made palatable with syrup and reinforced with a dash of an alcoholic tincture of cannabis, had been invented by an Indian army surgeon, Dr John Collis Browne, against cholera; and it worked wonders. To be pedantic, according to Collis Browne's restrained original report, it saved about one in 100 of those afflicted. But figures were soon forgotten. The inventor sold the formula for £25 to John Thistlewood Davenport, a commercial genius, who advertised the mixture as a remedy against every kind of diarrhoea, insomnia, neuralgia, migraine and all other forms of pain. The name could not be protected and 'Dr J Collis Browne's original Chlorodyne' spawned dozens of mixtures masquerading under the title but the original label was widely recognised. It also warned purchasers against crude imitations, 'as likely to kill as to cure'. Whatever the source (more or less), the drug saved a few lives in cholera epidemics. When cholera was not about, it was an all-round salve against pains and aches. But of course it did much more.

Like most opium-based medicines a spoonful of Chlory offered a temporary escape from the drudgery of the loom, the coalface or the plough. It had gin to contend with; but Chlory was kinder on the kids. It often led to incapacity but rarely to violence. Women took it more than men. It eased the hurt with which they struggled to raise a family and survive. And it silenced the scream which is heard in Europe today only in the concert hall: *Mutter, ach Mutter, es hungert mich. Gib mir Brot! Gib mir Brot!*[8] – so goes Mahler's lied. In England it was *bread*, in France *pain*, in Italy *pane*, in Hungary *kenyer*. The desperation was the same. Chlory did not stop the hunger but it reduced the cry to a whimper. And in 1850 it made the manufacturers an annual profit of £50,000, more than the income of a minor German principality.

Stilling hunger had always been a useful side effect of opium. It was a pity that the side effect had its own side effects. Regular dosing with the drug made children undernourished and sickly. By the age of four many had shrunk like wizened old men or, in Victor Hugo's memorable phrase, 'little yellow monkeys'. Few would ever benefit from free education, even when that became available. If and when they grew up, they would provide the next generation of the labour force, illiterate, poor, short-lived and dependent on opium. Or provide replenishment for the criminal underworld.

For criminals needed opium as much as law-abiding citizens. The underworld was not yet involved in distributing the drug – since it was not forbidden there was little profit in it – but its uses were many. Clients of dockland prostitutes might have to be rendered unconscious before being comprehensively robbed. A hard-earned gulp would counteract the aches and stiffness after a long night's toil. It would suppress for a time at least the symptoms of venereal disease. The greater the misery, the greater the need. Eventually the liquid might provide the last painless escape.

* * *

By the mid-nineteenth century the drug no longer had national or religious preferences. Catholic countries welcomed it as warmly as did Protestant ones and reactionary autocracies were as addicted as republics. Even after its deliriously debauched Congress of 1814–15 Vienna remained the tourist capital of Europe. London was foggy and dour. Paris was for milords and rich homosexuals. Rome was a den of thieves, many in episcopal purple. Berlin was irredeemably stodgy. The Kaiserstadt, by contrast, had the Dreitakt, *Kuchen mit Schlag*, the best opera in Europe, the delights of the Prater and *echt Wiener Charme*. What the guidebooks did not mention was that it also provided a livelihood for 15,000 registered prostitutes and accommodated 500 brothels ranging from the friendly family-run to the exotic *de grande luxe*. It was these as much as all the other delights that drew visitors of every age and nationality to the banks of the not-always-so-blue Danube; and their secret was not champagne (as operettas would have it) but opium. Madame's special mix prepared with 'virgin' poppy, not the heavily taxed prescription pills which, like tobacco and playing cards, were a state monopoly, could create a wonderfully relaxed atmosphere; and the same blessed brew helped the terminally exhausted hostesses after a night's revelries to reach oblivion in a never-never land.

CHAPTER 15

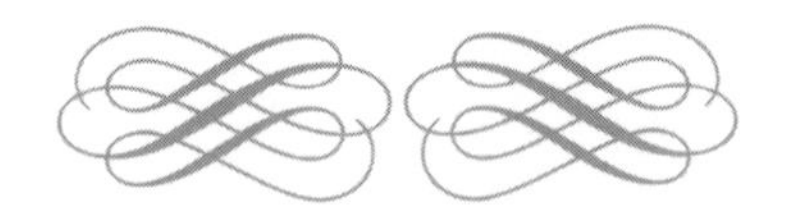

A salve for all ailments

DESPITE THE INCREASING recreational use of the drug, whether among exhausted prostitutes in the purlieus of Vienna's Stefansdom or working-class mums in the slums of Sheffield, in most countries poppy juice was still a medicine, indeed a salve for all ailments. What these were varied from country to country, from region to region and from class to class. In Britain coughs, colds and chestiness were the most common indications because they were the most common complaints. But gout and rheumatism remedies too contained opium as well as colchicum. In the 1840s Dr Rayner of Stockport prescribed opiate eye-drops for sleeplessness and against headaches. In 1854 the *Doctor*, a magazine written by 'eminent medical gentlemen' for 'the better class of sufferers and the indisposed', advised a correspondent complaining of lightning pains in the forehead to bathe her eyes five times a day with 'Dr Foresham's Occulin', a patent decoction of poppyheads. Patent opium-based remedies were recommended against toothache, earache, stomach cramps, flatulence, wind, nervous diseases, women's complaints and menopausal disorders. 'Hysterical symptoms and signs' were recognised if not always correctly diagnosed and usually responded with gratifying speed to opium lotions. Among more serious emergencies, opium eased the pain of patients with abdominal catastrophes like acute appendicitis and perforated ulcers. Before antisepsis, both tended to end in death. Ectopic pregnancies too were painful and usually fatal. Opium helped until the last journey. In moderation (because of the burden on charities or the public purse) the drug was used in lunatic asylums for those who could not otherwise be restrained, or before the arrival of important visitors.[1] In Bedlam, the famous institution in south London, those who came to be entertained by the antics of the inmates were expected to leave behind a gift of a bottle of good-quality sedative mixture. Good quality meant opium.

Whatever the experimental evidence of later times, the external use of the drug was also recommended. Buchan's *Domestic Medicine*, which ran into numerous editions, counselled the application of an 'anodyne plaster of opium and camphor' against all acute pains, 'especially of the nervous kind'. A 'liniment for the piles' was based on two ounces of emollient, half an ounce of laudanum mixed well with the yolk of one egg. 'Laudanum-enriched' healing balsams were recommended for bruises, sprains, chilblains, burns and recalcitrant leg ulcers. Opium was valued as a corrective against drunkenness and what was not yet called binge drinking. It was widely believed to cure delirium tremens. In 1846 a Manchester blacksmith took 6 drachms of laudanum while drunk to help him to sober up before visiting an important client. He passed out and then away. One sometimes led to the other. A Liverpool widow of intemperate habits laced her every drink with a few drops of laudanum and did so once too often.[2] Such cases were not news but made the coroner's court and therefore remain on record.

Accidents happened to the sober and virtuous too. In April 1848 Fanny Wilkinson, a sixteen-year-old household servant in Guisborough, sent for powdered rhubarb to cure her dysmenorrhoea. The general store had recently been taken over by a Mr Story, an industrious shopkeeper and a sound judge of preserves and draperies but ignorant of drugs. He also tended to get flustered when the shop became overrun with impatient customers. Saturday evenings could be hell. He had in fact tried to get rid of all medicaments when he bought the shop; but the previous owner had persuaded him not to as they were much sought after. The poor man confused a jar labelled '*Pulv. opii Turc. Opt*' with another labelled '*Pulv. rhub. Turc. Opt*'. Who might not? Fanny died from an overdose of opium and Mr Story had to stand trial for manslaughter. He was cautioned and told to refrain from dispensing dangerous drugs until he had had a chance to brush up his pharmacology.

Yet Mr Story was no different from hundreds of 'chemists and grocers' who picked up their dispensing skills wherever they could. It was for them that manuals like William Bateman's *Magnacopia* were intended. This massive tome described the preparation of an impressive array of opium-based potions, mixtures for astringent balls, gout remedies and corn plasters. They mingled happily with non-opium-based items like inks, grate-cleaning fluids, hair lotions, syrups for making jam and drops for exciting the passions. The book was advertised as a 'chemico-pharmacological library of profitable information for chemists, druggists, medical practitioners, surgeon-dentists, etc' and so it was.[3] Of course not everybody read their Bateman. Some of the apprentices who in the more prosperous establishments weighed and prepared the quarter-ounce, half-ounce and one-ounce loaves of children's op-cake were no experts in the written word. Ambitious pharmacists in London and the fast-growing cities in the Midlands mixed their own medications; and their 'specials' often commanded a faithful following.

Druggists and grocers were preferred to doctors by families whose prescriptions had been handed down through generations. A branch of the Quaker Fox family of Esher cherished a mixture against whooping cough and all manner of fevers bequeathed to an ancestor by a surgeon major in Cromwell's army. It was a concoction of equal quantities of chloroform, laudanum and an antimony solution laced with saffron and its effect was miraculous. To save on the cost they and many other families sent their bottles to the druggist to have them replenished from time to time. The practice was not without risk. Standing on the shelf the active ingredients tended to sediment and instructions to shake the bottle thoroughly before use had long rubbed off. The supernatant fluid was then tasty but ineffective. The last spoonful was lethal.

* * *

Recourse to the chemist and self-medication by the majority did not mean that doctors did not find opium an irreplaceable standby. Medical practice varied from one country to the other even more than it does today; but this was something on which the profession everywhere agreed. On 3 April 1844 in Vienna Ignac Semmelweis, future discoverer of the cause of childbed fever, was defending the main postulates of his MD thesis. This was an occasion of high drama which attracted the ghoulish and the idle as well as doctors and medical students. It was recorded virtually verbatim. Young Semmelweis's first assertion was:

> *Sine Opio et Mercurio nolle esse medicus.* (Without opium and mercury I would not want to be a doctor.)

Freiherr von Rokitansky, eminent pathologist and chairman of the examiners, had doubts about mercury. It was the only known remedy against syphilis (which was rampant) but its side effects could be severe. Would the candidate choose between his two 'essential drugs'? Semmelweis at once conceded that he might do without mercury but not without opium. The clinician Frantisek Skoda and the anatomist Hyrtl agreed: it would be difficult to recruit young people to the profession if opium did not exist. Certainly, nobody would choose obstetrics, the branch of medicine eventually chosen by Semmelweis, without it.[4] At that time childbed fever killed 10–20 per cent of healthy young mothers who had given birth at the University Maternity Hospital. It was a painful death mitigated only by large doses of opium.

* * *

Statistics were not yet an indispensable aid to illumination or obfuscation. Not surprisingly in view of the widespread self-medication, reliable information about the prevalence of the use of opium is still lacking. Professor Robert Christison of Edinburgh and G.R. Mart, a Soho surgeon, carried out a pioneering study in the 1830s, but their sample was small and unrepresentative. Otherwise patterns of use among different classes and age groups emerged fitfully, single cases shedding light on what was probably a widespread habit.

Martha Pierce, a sixty-year-old lace-maker and a widow with eight children, was incarcerated for stealing a ball of wool. She confided to her priest who doubled as the prison chaplain that, 'like most women in the neighbourhood', she had been using five pennyworth of laudanum a week for many years. It sometimes made her careless and forgetful but calmed her nerves. The chaplain claimed to know and love his flock and probably did; but honesty compelled him to record his surprise. It transpired that even some of his most assiduous flower arrangers were regular users. He found the information 'agitating'.

Doctors were as ignorant as everybody else. Medical consultations with the poor were brief: medical students were taught not to poke their noses into the private affairs and habits of their patients. The truth sometimes emerged accidentally. During the cotton famine in Lancashire in the 1860s the unemployed could not afford even a pennyworth of laudanum. As Dr Robert Harvey, later Inspector General of Civil Hospitals in Bengal and house surgeon at Stockport Infirmary at that time, recalled:

> Many applications were made to the Infirmary for lifeline supplies of opium by people too poor to buy it . . . I was much struck that the use of the drug was far more common than I had any idea of. Even among some of the very poor a habitual dose of ten to fifteen grains a day was not uncommon. They seemed none the worse for it until the supply was cut off. Then there was terrible misery. Some brought their screaming children along. The drug was for them, they claimed. I had no way of knowing.[5]

Dr Ainstie, editor of the *Practitioner*, reminisced about his time as a junior hospital doctor in the late 1850s:

> It frequently happened to me to find out to my surprise and by chance of patients being brought under my notice that they had been regular consumers of one drachm of laudanum, or even two or three, for years without any variation . . . They were often persons who would never think of narcotising themselves any more than they would of getting drunk but who simply desired a relief from the pains of fatigue endured by their ill-fed, ill-housed body and harassed mind.[6]

It was not just England. Pierre-Charles Louis, founder of scientific epidemiology, roamed the public hospitals of Paris in the 1830s collecting 'hard and reliable data' relating to the benefits of bloodletting. He was an obsessive investigator and his observations remain models of exactitude; but he did not record the use of sedative mixtures containing opium because the habit was 'virtually universal'.[7]

In the German-speaking world Prague enjoyed the reputation of a medical centre of excellence. One of its luminaries, Professor Friedrich Langhans wrote in his reminiscences about his experience in the early 1850s: 'In my first residency in the new *Frauenklinik* I was amazed. All the patients between sixteen and sixty demanded to be kept on their normal medication with Mist. Opii.'[8] He thought of writing a report about his discovery but was told not to waste his time: the fact was well known.

Not until the 1850s were reliable data wanted by anyone. Opium was not a 'national' or 'social' problem needing to be documented and discussed. To the few who thought differently, it was pointed out that many regular users, mainly of course among the wealthy, lived to a ripe old age. Some were in their eighties and nineties, their sallow complexion giving them away but otherwise hale and hearty. When in 1828 the reclusive 31st Earl of Mar died at the age of sixty-five, the inquest found that he had been consuming opium for fifty years, 49 grains of the solid and an ounce of laudanum a day. The insurers refused to honour his life policy, won their case in the lower court but were forced to pay up on appeal. Professor Christison's testimony that opium had 'undoubtedly shortened the Earl's life' was rejected. On the contrary, the learned judge ruled, for an individual suffering from such a painful condition as the late Earl's gout – and the learned judge himself was one of them – only opium made life bearable.

* * *

By the mid-century most people in Britain and Western Europe were introduced to opium as soon as they had left the womb, and many earlier. Up-to-date obstetricians liked to ease the tribulations of late pregnancy with a daily teaspoonful of a 'sedative'. At the first international conference of obstetrics in 1846 Professor Hector von Arneth lately of Vienna but now of St Petersburg claimed that routine opium in the last trimester bestowed significant benefits. A celebrated wit, he added that the practice made not only the patient's but also the obstetrician's life easier. Delegates from Stockholm and elsewhere promised to adopt his recommendation: others sounded notes of caution. Professor Dunoyer of Paris thought that the use of the drug should be restricted to the highly strung and that the dosage should be 'restrained'. Like many before him, he was concerned about constipation but did not think that uterine contractions were adversely affected. And 'one would have to deal with the occasional overdose'. This was a problem not confined to obstetrics. Popular guides to household practices like Mrs Beeton's indispensable volume in England or Mme Bonnefoix's Housewife's Handbook in France contained detailed instructions on how to deal with such an emergency. But even Mme Bonnefoix recognised that sometimes one could only pray to the Virgin or one of the more specialised interceding saints.

After a foretaste in the womb, the new 'baby calmers' emerged in the 1840s. The most popular was Godfrey's Cordial: a draught of the old 'Cordie' would stop the worst colic or at least the worst squealing. In 1851 a Nottingham chemist reported selling 600 pints a year. In Long Sutton in Lincolnshire a druggist would testify that he sold 2 gallons a month to a population of 6,000. It was popular on the continent too though not always under the same name. The English brands were highly esteemed: the Princess Metternich would not allow her children's nanny to use anything but Godfrey's. Later Mrs Winslow's Soothing Syrup was to become a close rival: the happy little angels on the labels lifted the heart. Street's Infant Quietener was near the top. Atkinson's Baby Preservative in its shiny wrapper aimed at upmarket customers: it would be the tincture of choice of the uniformed nannies promenading their preternaturally well behaved charges in Kensington Gardens.

But most preparations were targeted at the poor, not only at working mothers but also at a proliferating army of professional baby-minders. In some industrial cities these were in charge of dozens of infants and children during the long working day and were usually either disabled themselves and therefore home-bound or had a second home job. They charged about a fifth of the average wage and their aim was to keep their flock semi-comatose. Mums returning from a day's work and in desperate need of a few hours' sleep continued the dosing. Fatal overdosing at home and in the baby farms was common; but charges were rarely brought. When brought, they could rarely be proved. If they could be proved, juries were common folk who knew what was what. The loss of a sickly child was sad but could also be a merciful release. For some reason, as one observer put it, 'twins and illegitimate children almost always die'. Fathers were rarely in evidence. Mothers found guilty of carelessness were usually admonished. The horrific penalties for murder – infanticide was not yet a recognised offence – made any other verdict unthinkable.

* * *

In the United States non-specific patent brews based on opium were the sensation of the 1820s. Of course medicinal potions had long flourished in the New World: a brand of folksy quackery was part of the small-town American scene – or legend. (No mid-twentieth-century musical would be without the jovial vendor of fabulous remedies.) But opium brought about a change in style and probably effect. In the past the patent medicines were claimed to be effective against specific diseases. There had been drops against 'rheumatism', mixtures against 'consumption' and plasters against 'dropsy'. Now Perry Davis's opium-based Universal Pain Killer was invincible against all. It was not the only one. No ailment was a match for Samuel's Herculean Embrocation or Pond's Universal Pain Extractor. Makers of the new medicaments cultivated a vaguely anti-professional image. Because Pond's was of 'purely vegetable origin' (which of course included the poppy), it did not need the expensive ministrations of doctors. The best advice had already been sought; and endorsements by the most eminent, including Hippocrates, Paracelsus and William Harvey, were printed on the labels.

Another novel feature was the appearance of tiny tots on the packaging. The new panaceas were both 'uniquely universal' and 'incomparably delightful for your little ones'. In the past children may have been difficult to cajole to take foul-tasting medicines like cod-liver oil. No longer. Kids loved their gooey syrups and clamoured for more. And why not give it to them? These were truly the People's Remedies.

* * *

Concerned voices during the first half of the century were few. Or so it seemed to European travellers. More than anywhere in Europe the temperance movement was deeply embedded in the New World. Spokesmen and spokeswomen were sometimes gently mocked in private but never in public. They preached a core doctrine of the Founding Fathers and aspiring public men ignored them at their peril. But the movement's initial response to poppy juice was equivocal. To many opium called not addiction but pain relief. Few early nineteenth-century American physicians believed that it caused physical harm. Unlike the diabolical liquor it did not threaten the family and did not lead to violence. The mildly habituated often seemed a little distracted but were generally harmless. That virtuous man, the Reverend Walter Colton, chaplain to the United States Navy, positively enthused about the drug:

> If a man will take to stimulants, the juice of the poppy . . . has strong recommendations. It never makes a man foolish; it never casts a man into a ditch or under the table; it never deprives him of his wits or legs. It allows a man to be a gentleman; it makes him a visionary but his visions create no noise and no riots, they deal no blows, blacken no one's eye and threaten no one's peace. It is the most quiet and unoffending relief to which the despondent and distressed who have no higher resources can appeal.[9]

He even noted that under the influence of opium some of those whose religious faith had been fickle suddenly saw the light. Sadly, he also observed that though their attention could easily be caught, it soon tended to wander and their later recollection of their enlightenment was uncertain. And the *New York Times* of 1840 expressed the view that 'liquor generally arouses the animal, while opium subdues this completely. Indeed, in its place it awakens the diviner part of human nature and can bring into full activity all the nobler emotions of the human heart.'[10]

One effect of opium in particular could be contrasted to that of liquor. Opium doused rather than inflamed the 'baser passions'; or, as doctors put it, 'calmed the generative urge'.[11] Indeed, some practitioners advocated opium as a safe cure or at least an effective symptomatic remedy in acute alcoholic intoxication or, as commonly referred to, the 'DTs'.[12]

* * *

All this would change, and change over a comparatively short time. The immediate cause was to be a new development – or rather, the combination of two new developments. Both originated in Europe and both would take three or four years to cross the Atlantic. Even in Europe their significance was not immediately appreciated. One was the discovery of the long-sought-after 'principle' in poppy juice reputed to be more effective and virtually non-addictive. The other was a hollow needle which could be attached to a metal device and introduced almost painlessly under the skin. An aura of near-innocence would soon hover over old-fashioned tinctures like laudanum.

PART II

THE ESSENCE

CHAPTER 16

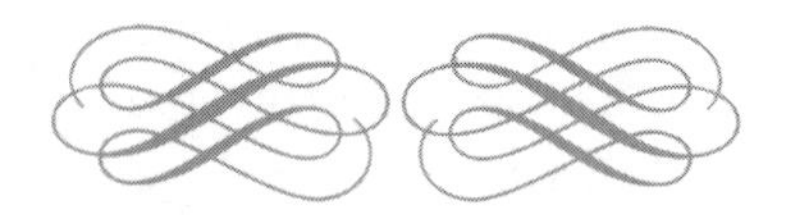

The shape of dreams

CHANGING TIMES HAVE their favourite cutting-edge sciences. Atomic physics attracted the scientific avant-garde in the 1930s. Microbiology was the dazzling field for new discoveries in the 1880s. Organic chemistry exercised the most original minds at the turn of the nineteenth century.[1] The discovery of the alkaloids of opium was one of their achievements.

The concept of an 'active principle' in opium was ancient but for centuries its isolation was regarded as a pipe dream. A few English Paracelsians tinkered with the material; none seriously tried to break the conceptual mould. The revolution in chemistry – Scheele in Sweden, Lavoisier in France, Priestley in England, Liebig in Germany and others both in Europe and in the United States – shattered it. In 1786 John Leigh, an amateur chemist of Virginia, published an *Experimental Enquiry into the Properties of Opium* in which he argued that getting rid of 70 or even 80 per cent of the raw gum would ensure that the effectiveness and reliability of the drug could be 'multiplied manifold'. Others followed. Seventeen years later Jean-Pierre Derosne in France prepared a salt which in animal experiments was about three times more powerful than the parent substance. What he had discovered was one of the minor opium alkaloids now known as *noscarpine* and formerly called *narcotine*. By then it was becoming obvious – and chastening – that there could be not one but as many as twenty *essences* in poppy juice, some with opium-like action but many without. In another age the realisation might have extinguished nascent enthusiasm. In 1804 it acted as a spur. Another French chemist, Armand Séguin, read a paper to the *Institut*, describing a stepwise process of separating all or most of the constituents of the dried juice. He may have been on the threshold of discovering morphine but was arrested for allegedly conniving in his other capacity as Controller of Drugs in the dilution of medical supplies to the army. The charges were probably false; but in the Napoleonic years he was lucky to escape with his life.

French authorities tend to emphasise the contribution of their countrymen (which was considerable) but morphine was eventually discovered by a 21-year-old German apprentice pharmacist in the small town of Paderborn in Westphalia. Friedrich Wilhelm Adam Serturner, born in the village of Neuhaus on 7 June 1783, had no university training and never acquired academic qualifications.[2] What he had was the chemist's equivalent of the gardener's green fingers. His laboratory equipment was self-made and his book knowledge of the 'basic principles of science', regarded in august circles as the bedrock on which all scientific advance must be based, rudimentary. His paid employment in the chemist shop in Paderborn left him little time but his curiosity was inexhaustible. Like many others, he had seen opium work wonders for a time in his tuberculous sister but then lose its effectiveness. Why this kind of thing happened was one of the questions that teased his mind and would continue to do till the end of his days. First in Paderborn and later in Einbeck in Hanover, still earning a modest living but in his own shop, he performed hundreds of experiments purifying and characterising extracts of opium and trying their effect on himself. In 1806 he published the results of a series of fifty-seven experiments in the *Journal der Pharmazie*. Twenty of those described the properties of meconic acid which he termed 'poppy acid'. He had discovered that poppy acid had a significantly different effect on various indicator pigments from the parent opium. He rightly deduced that the cause was something non-acidic but biologically active. This may strike the modern reader as deeply unremarkable; but it was either nonsense or revolutionary in Serturner's day. On evidence dating back to Descartes all biologically active organic compounds were believed to be acidic, their acidity being a key to their potency. Serturner sent his observations to Professor von Buon in Cologne. The professor commended Serturner for his industry but expressed the view that the pharmacist's results had to be mistaken.[3] Such mistakes were unavoidable in one lacking a systematic grounding in the principles of chemistry. It was not too late perhaps for him to immerse himself in the professor's own foundation text.

In private Serturner was a shy, solitary man but in chemical experimentation in his own estimation, *ein Ochs* – meaning totally stubborn. He continued his experiments, searching for the impossible non-acidic component. When he tried a comparatively simple chlorate precipitation method he hit gold – or better. He isolated a new compound which was not only alkaline but at least ten times more effective weight for weight than opium. He first called it *principium somniferum* but then succumbed to the lure of classical Greek recently unleashed on his countrymen by the great Johann Joachim Winckelmann. Serturner's chosen name *morphine*, was, contrary to what is often stated, derived not from Morpheus, a problematic Greek god of sleep, but from μορρη morphe, shape (as in morphology), short for the 'shape of dreams'. The official name in the *British Pharmacopoeia* introduced in 1885 was to be *morphina*.

For more than ten years Serturner and his discovery were ignored. He had no professional standing and his literary style was clumsy. Single-mindedly he continued to experiment on himself. He discovered most of the important effects of morphine, from early euphoria to late depression, from the dulling of pain to constipation. It was a remarkable achievement: in similar cases it would have been the combined effort of a dozen independent investigators. He almost certainly became an addict and was aware of the risks. It was these risks which impelled him to continue to press his discovery on the scientific community. In 1812 he wrote: 'I consider it my duty to attract attention to the terrible effects of this new substance I called morphium in order that calamity may be averted.'[4]

Five years later the warning attracted the attention of Joseph-Louis Gay-Lussac, professor of chemistry at the Ecole polytechnique in Paris.[5] Gay-Lussac was already famous. In 1802, at the age of twenty-four, he had ascended in a balloon to a record height of 6.4 kilometres to study the earth's magnetic field and the composition of atmospheric gases. More recently, in collaboration with Baron Alexander von Humboldt, he had established the structure of water. Like most good scientists he was not without a streak of paranoia and even accused his colleagues of keeping Serturner' discovery deliberately hidden because of the young pharmacist's German nationality and lack of academic standing. He was probably wrong; but without him Serturner might have remained unrecognised. Even with Gay-Lussac's championship another ten years elapsed before the *Institut* awarded the no-longer-so-young German a prize of 2,000 francs for 'having opened the way to important medical discoveries by his isolation of morphine and the exposition of its alkaline nature'.[6]

* * *

Serturner's work was revolutionary, extending beyond morphine, it triggered another burst of activity in organic chemistry. In 1832 Thiboumery and Pelletier isolated and named another opium derivative, thebaine after the ancient Egyptian home of high-quality poppy; and a few months later Robiquet identified the second most useful alkaloid in opium, codeine.[7] 'The spirit of Lavoisier is alive again in France,' Jean Marchais exulted. So it was, but no more than in England where Humphry Davy, a weedy-looking and patchily educated country boy from Penzance, was making his friends giggle with his newly discovered laughing gas. Two French chemists, Caventou and Pelletier, developed a comparatively cheap and simple process for isolating morphine of nearly 90 per cent purity; and François Magendie, leader of a cohort of bright young doctors who deplored the traditional hit-and-miss approach to treatment and claimed to be a new breed of *medicin-savants*, described the first case where the substance was used for deliberate palliation.

His patient was a married woman of thirty-eight with an aortic aneurysm who, for some years, had been seeking relief from the gnawing pain in her chest.[8] On several occasions she had attempted suicide; and to save her soul even if he could not save her body her husband, a devout Catholic, now kept her under constant surveillance. They had consulted most eminent doctors in Paris as well as pharmacists, herbalists, magnetisers and a selection of out-and-out charlatans. Since she claimed that opium in every form had made her violently sick, Magendie gave her morphine. The effect was, in Magendie's words, 'miraculous … there is no other word for it'. Of course her condition was incurable and the pain was still there; but she became 'calm and composed and almost happy'. She died in her sleep nine months later reconciled to her Maker. Magendie went on to demonstrate the value of the new drug in the palliative care of two other terminally ill patients. These were historic moments.

But for a few years morphine, still manufactured in small private laboratories, was probably used for suicide and homicide as often as for palliation. A sensational case of multiple murders in Lyon – a whole family exterminated by a greedy heir who happened to be a competent chemist – inspired Balzac to introduce the drug into literature. In his *Comédie du diable* the Devil boasts that this new invention called *la morpheine* has caused a gratifying increase in the population of Hell. But by the 1830s the drug was available commercially both in France and in Germany and not excessively priced: it was the best-selling medicine at the Pharmacie de la Cloche on the Left Bank in Paris, a meeting place of bohemians. Magendie published his *Formulaire pour la préparation et l'emploi de plusieurs medicaments* in 1826, one of the seminal books in the history of medicine with a lengthy section on morphine. Within two years it was translated into English and went into several pirated editions in the United States. For the first time doctors could prescribe accurate doses of a painkiller that was at least ten times more potent than the best Turkish opium.

Opium did not of course disappear. Jonathan Pereira's authoritative British *The Elements of Materia Medica and Therapeutics* published in 1839 still described 'good-quality laudanum' as the most valuable remedy available against pain without mentioning morphine. His generally upbeat comments probably reflected current orthodoxy:

> Some doubt has recently begun to be entertained as to the alleged injurious effect of opium eating on health and its tendency to shorten life. It must be confessed that in several known cases which have occurred in this country, there had been severe ill effects. But ... we should be ... careful not to assume that because opium in large quantities, when taken by the mouth unsupervised, is a powerful poison, and when smoked to excess injurious, that therefore the moderate employment of it is necessarily detrimental.[9]

But Pereira's insistence on *good-quality* opium was significant. Adulteration was a theme that was cropping up more and more. In F. Accum's sensational *Treatise on the Adulteration of Food and Culinary Products* published in 1820 opium was mentioned as the most frequently adulterated drug, while the author also warned about the drug being surreptitiously used to 'fortify' other expensive medicaments.[10] A range of substances could be used as adulterating agents, but the most common was flour. Even more blood-curdling was Accum's *Deadly Adulteration and Slow Poisoning Unmasked* ten years later. In the first issue of the journal of the newly established Pharmaceutical Society in 1840 John Bell chose illegal 'cutting' as the most worthy cause for campaigning.

How serious in fact the problem was is no longer easy to tell. Contrary to what was suggested at the time, adulteration in Turkey was probably rare: the penalty for being caught was death. But in Britain trends were changing. Acts of Parliament protecting the purity of food and drugs had existed since the reign of George III; but they had rarely been invoked. The bulk of what people ate, drank or took as medicine was home grown or originated in their village or surrounding region. Itinerant traders and stallholders at the local farmers' market were known to them.[11] The change was one of the side effects of the migration of country folk to the towns. It meant the alienation as well as the physical distancing of workers from the cultivators of their food and medicines. To mothers in the slums saving a farthing could also be a powerful inducement to buy the less expensive but also less trustworthy product.

For those who could afford it, morphine, usually dispatched in sealed glass ampoules, was safer. But to make the drug the landmark it potentially was – that is to take advantage of its purity, strength and the precision possible in its dosage – a convenient and effective way of administering it was wanted. Clysters had been tried with opium; now morphine suppositories coated with wax or animal fats enjoyed passing popularity. Direct inhalation, as distinct from smoking through a pipe, was messy and tended to produce nausea. Opium pipes were still an exotic device in the West. Skin patches were tried as well as application to lightly scarified surfaces; but morphine, even more than laudanum, sometimes caused painful blistering. Something radically different was needed and something radically different materialised.

It was the coming together of two strands of ingenuity, one ancient, the other recent. The invention of the syringe has been attributed to Hero of Alexandria, the Thomas Edison of Antiquity.[12] The device was certainly old; and it had been adapted over the centuries to such medical procedures as enemas and early and gruesome attempts at animal-to-animal and animal-to-man blood transfusions. But what was recognisably a hypodermic syringe with a sharp, pointed hollow needle was invented in 1841 by a surgeon in Lyon, previously known only for his charitable work among the city's poor. Charles-Gabriel Pravaz's purpose was to inject not morphine but coagulating solutions into painful varicose veins, an operation at which he excelled and which is still occasionally performed. His one and only syringe and needle were made to his specification by M. Charrière, a well-known local instrument maker. The syringe was of silver, the needle of platinum. For safety Dr Pravaz carried the precious object not in his medical bag but in a specially constructed pocket in the silk lining of his top hat. Since he was unaware of any wider application of his device he would be forgotten today but for the friendship of a former fellow medical student. Louis-Jules Behier had migrated from Lyon to Paris and had become a social lion; but he kept in touch with his former friends. He was 'charmed' by Charles-Gabriel's clever invention and immediately adopted it, referring to it as his *Pravaz*; but used it not to inject coagulant into working-class varicosities but morphine into aristocratic buttocks. His medication for a vast range of ailments was, by his own account, a *succès fou*. He later for a short time styled himself 'morphiniste impériale', the only known bearer of such a title. After that the resonant but perhaps slightly sinister sounding '*Pravaz*' became the name on the continent not only of the hypodermic syringe but also of morphine addiction. At least it was a code.

Other pioneers are now unjustly forgotten. As early as 1836 Dr Hyacinthe Lafarge of the small wine-growing town of Saint-Emilion suggested a way of introducing morphine under the skin with a specially designed pointed lancet. The manoeuvre required a certain amount of skill and never caught on outside a small circle of Parisian doctors. Dr Isaac Taylor of New York started to insert the drug through a small incision and a blunt-nozzled syringe. The method had a brief vogue but was painful and occasionally caused prolonged bleeding. More successfully, Francis Rynd of Dublin developed an ingenious syringe equipped with a trocar for giving drugs subcutaneously and sometimes, more or less accidentally, intravenously. In 1844 he became physician to the Meath Hospital, Dublin's first free hospital for the poor; and there he treated what he described as neuralgias, excruciating pains arising in the nerves, with an assortment of sedatives, including morphine but also with mixtures which would raise eyebrows today. He later wrote in the *Dublin Medical Press*:

> The subcutaneous introduction of fluids for the relief of neuralgia was introduced by me in this country at the Meath Hospital in May 1844 . . . Since then I have treated very many cases, and used many kinds of fluids and solutions with variable success. The fluid I found most beneficial was morphia in creosote, ten grains of the former to one drachm of the latter.[13]

Transparent glass was substituted for silver by Jean Lenoir of the Necker Hospital in Paris in 1853; and the method of turning the piston in a spiral groove was replaced by simple pressure and a tighter fit by the famous instrument makers, Luer's. Their name remained attached to the subcutaneous syringe in English-speaking countries as well as in France till the mid-twentieth century. But the man who is credited with the introduction of the method in Britain was Dr Alexander Wood of Edinburgh. A fellow of Edinburgh's Royal Society as well as a much sought-after physician, he had travelled extensively on the continent and had seen syringes used in France. He was impressed; and in 1851 he ordered a couple to be made to his specification from Messrs Ferguson's of South End. He showed the device at a medical meeting in London, where it aroused a mixed response. His listeners' main concern was accidental injection into a vein, artery or nerve, still a rare mishap. 'Always withdraw before injecting,' Wood pronounced, meaning that one should ensure that one has not (or has, as the case may be) hit a vessel. The advice is still regularly intoned but accidents still happen. A year later Wood's design was improved by Dr Charles Hunter of St George's Hospital, London, by the addition of a locking mechanism for the needle. This avoided the messy detachment of the needle from the syringe under pressure while injecting, another contretemps. In 1856 Dr Fordyce Barker of New York was probably the first to give a hypodermic injection in the United States.

All was now ready for great and beatific advances. Both Wood and Hunter were enthusiasts who came to share with each other and their acolytes a disastrous misconception. Not only did they believe – rightly – that morphine by the hypodermic route acted more quickly than by mouth but they also persuaded themselves that it would abolish 'the considerable inconvenience of an appetite developing'. The old dream in a new garb was to prove as treacherous as ever. The term 'appetite' hints at the belief widely held at the time that craving for morphine was not significantly different from hunger, and that hunger was an essentially oesophago-gastric function. 'Remove the act of swallowing,' Wood wrote, 'and the desire for more morphine would be assuaged.' Firm in this delusion he and his followers started to use hypodermic injections indiscriminately in a range of trivial as well as major complaints, from 'rheumatic aches' to inflamed eyes, from dysmenorrhoea to headaches and from hangovers to full-blown delirium tremens. Some doctors taught their patients to inject themselves, a practice later frowned upon by the profession. But who was to stop them? Victorian ladies were rarely the helpless ethereal creatures portrayed in romantic fiction; and many became experts at wielding the *pravaz*. It could be done unobtrusively under the dining-room table while consuming a formal meal, in a box at the opera while listening to the soprano's trills of terminal despair or, as observed by Flaubert, in church pews while apparently absorbed in prayer. Small 'ladies' models', some with delicately engraved silver plungers, were advertised in the press and a 'unique' set in gold in a finely worked ivory *étui* could be purchased as a '*cadeau d'amour*' from a famous Bond Street jeweller still in business.

The drug itself was now sold in a range of vials, a single-dose phial being capable of being broken daintily in a lady's handkerchief. For gentlemen the entire gear could be accommodated in the ivory handle of an elegant walking stick. Some of the elegant walkers became addicts. In 1854 the first fatality was recorded. Tragically and ironically it was a Mrs Wood, a relation of Dr Alexander Wood (but not his wife as stated in most books on morphine). She had suffered from migrainous headaches for many years and had taken what was referred to at the time as an accidental overdose. A popular hostess and worker in good causes, her death came as a shock to many and tempered the enthusiasm of Dr Wood and his circle. But the first death remained a passing shadow: subcutaneous morphine was here to stay.

From London, Paris and New York 'morphinism' – a new word – spread across Europe and the United States. Like laudanum in its early days the drug was for a time a luxury. The mid-century bloodlettings in the Crimea and across the temporarily disunited United States would gain it a wide and classless following. But by then another war, 'the most wicked ever conducted under the flag of this country' and the only one named after a drug, had been fought – and to all appearances won.

CHAPTER 17

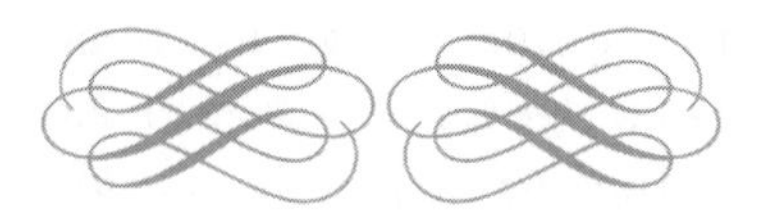

The most wicked of wars

I THE CELESTIAL KINGDOM

Few have seriously questioned Mr Gladstone's judgement of 1840 that

> a war more wicked in its origin, a war more calculated in its progress to cover this country with a permanent disgrace, I do not know and have not read of … I am in dread of the judgement of God upon England for our national iniquity towards China.[1]

He was talking of the opium wars between England and China – it could not have been any other – and he knew what he was talking about. He had just recovered from a nervous breakdown after a futile trip to Italy with his much loved twenty-four-year-old sister Helen who had become addicted to laudanum during a painful illness.

Nor have many questioned a more recent judgement that 'Britain has earned vast revenues from the opium trade by poisoning a substantial proportion of the Chinese people.'[2] The war, or rather wars, were the most profitable commercial enterprise in Britain's history as a colonial power, not excluding the bonanza of the slave trade.

Yet leading sinologists have criticised the 'mythology of the emaciated Chinese addict', a victim of European greed, at least in isolation:

> We should also remember the peasants carrying their lumps of poppy juice to market, a boon to the economy, the boatmen wrapped in their blankets passing round an opium pipe in the twilight, and the Chinese gentleman smoking peaceably at home with cultivated friends.[3]

Such idyllic caricatures are rarely convincing; but two mitigating facts can be recalled.

First, Britain did not introduce opium to China. The revered Chinese surgeon, Hua To, who lived in the period of the Three Kingdoms in the third century AD, probably used opium as an anaesthetic when performing major operations. In 987 the emperor ordered the compilation of an imperial herbarium in which opium, *ying-tsu-su*, was mentioned, *su* being the name of a pod; and the poet Su Tung-p'a praised the provider of sweet dreams. But poppy juice remained a medicinal concoction and a luxury until, towards the end of the seventeenth century, an entirely new vice was introduced by Western sailors.

The habit of tobacco smoking was acquired earlier in the century by Dutch and Portuguese sailors from the native people of North America to whom it was an act of deep religious significance. It then sped around the world like (in Chinese parlance) the East Wind. Dutch sailors in Java began to add a pinch of opium and arsenic to their pipe mixture, partly to give it an extra kick but partly also because it was supposed to prevent malaria. (Perhaps it did, at least by keeping mosquitoes away.) Within a few years and despite thunderous anathemas the innovation reached Formosa, today's Taiwan, and then mainland China. Eliciting a prophetically counterproductive response, it became sufficiently fashionable at court in Peking to be banned by the Emperor Tsung Cheng; and, though the edict did not remain in force for long, the damage was done. Since tobacco was not available in China, the smoking was of opium from the start. The slow ceremonial of preparing, filling and lighting a pipe seemed to suit the oriental temperament (insofar as such geographical temperaments exist) and high-class Chinese rejoiced in the refined craftsmanship of the necessary implements. Smoking was also a remarkably efficient way of introducing opium into the body: the effect was comparatively mild but almost instantaneous. Yet even in China until the start of opium importation from India the habit was confined to the leisured classes, in a vast country a tiny minority.

Critics of the opium wars may also overlook the fact that by the beginning of the nineteenth century the Celestial Kingdom was in one of its phases of decline. This was not immediately apparent. The empire was still the largest in the world. It had a fast-growing population of about 500 million, more than that of the whole of contemporary Europe. Its borders enclosed an area twice that of what was then the United States. To most Chinese what lay beyond these borders was a matter of complete indifference. The country had a continuous history stretching back at least five thousand years, and, despite occasional popular uprisings and foreign invasions leading to a change of dynasty, its social and spiritual fabric had remained largely unchanged. The notion of personal liberty under the law had never been part of either. Social morality was based on obligations as taught by Confucius, the Great Teacher, a duty to the family, to the kinship, to the village, to the province and eventually to God's representative on earth, the emperor. All land belonged to him; and he had the right to tax both the soil and its chief product, rice. The mandarinate, scholars who had passed a series of rigorous examinations, administered the realm on his behalf according to Confucian principles – at least in theory. But the emperor's power was absolute.

In 1651 the Manchu dynasty – known as the Dynasty of Purity or Tsing – originally the chieftains of a nomadic tribe, defeated and replaced the last great native line of emperors, the Dynasty of Light or Ming; and, though to strict traditionalists the Manchus would always remain barbarians, they became, like past conquerors, more Chinese than the Chinese. Three successful and long-lived emperors further enlarged their already huge domain, secured its frontiers, redistributed the land (which led to a sudden and potentially dangerous increase in population) and built a fabulous complex of palaces in Peking at about the same time as Louis XIV of France built Versailles. But grandeur, isolation and the worship of tradition carried the seeds of destruction. Religious wars and dynastic jealousies in Europe had long preserved China from intrusion by the European powers; but these circumstances were not under Chinese control and were perilously susceptible to change. In a historical perspective the episode which seems like a watershed was the embassy sent by the Honourable East India Company of London with the blessing of His Majesty's Government in 1793 to the last great emperor of the Celestial Kingdom, Ch'ien Lung.

* * *

The mission was led by Lord Macartney, a charming Irish lawyer who had already proved his manifold talents as envoy extraordinary to that ample and amorous lady, Catherine the Great of Russia. The expedition had cost the East India Company the huge sum of £80,000 and had brought to Peking a distinguished delegation of academics as well as traders and a splendid assortment of presents.[4] It foundered on the kowtow. The bowing, kneeling and placing of the forehead on the floor nine times was expected of all those approaching the Son of Heaven and the mandarins saw no reason for making an exception. For the sake of commercial gains Macartney was prepared to knock his sturdy head on the floor any number of times if one of the mandarins would do the same in front of a supposed likeness in oils of George III in full coronation rig. But even the lowliest Chinese official refused to do anything so ridiculous; and, thoroughly miffed, Macartney dug his heels in and refused too. After a wholly informal as distinct from an official audience Macartney recorded that 'the Empire is like an old, crazy though formerly first-rate man-of-war which a fortunate succession of able and vigilant officers had managed to keep afloat but which, when left without such guardians will drift until dashed to pieces on the shore.'[5]

He was right. The eighty-three-year-old emperor regarded the British mission as a charade. He insisted that the British continue to funnel all their trade through the already overcrowded port of Canton at the mouth of the Pearl River and that there would be no official British embassy in Peking.[6] Not now, not tomorrow, not ever. The offer of controlling the opium trade which Macartney considered to be his trump card was not mentioned, a foretaste of the mountainous documentation of the opium wars in which the drug itself would never be spelt out by name.

The emperor also addressed a personal letter to his distant but trusty servant, King George III of Great Britain:

> Our ways have no resemblance to yours, and even if your envoy were competent to acquire rudiments of them, he could not transplant them to your barbarous land. Strange and costly objects do not interest me. As your ambassador could see for himself, we possess all things and of the highest quality. I set no value on strange and useless objects and have no use for your country's manufacture.[7]

As an English traveller summed up the situation,

> The Chinese have the best food in the world, rice, the best drink, tea, the best clothing, cotton, silk and furs, the best metal implements, the best ceramic artefacts, the best houses and the best means to travel. They do not need a penny's worth from us or anyone else.

And yet, there was in China's attitude an element of hubris which Sophocles would have recognised as calling for divine retribution. And the immediate obstacle to maintaining friendly but distant relations soon became apparent. China may not have wanted anything the West had to offer; but the reverse was no longer true. The West wanted tea.

* * *

The English love affair with tea began in 1664 when King Charles II received a gift of two pounds of strange-smelling wrinkled black leaves on the occasion of his nuptials with Catherine of Braganza of Portugal. He thought the concoction prepared from the foliage foul but for once his loyal subjects disagreed. Within a few decades they consumed 12 million pounds a year of the dried leaves. What accounted for such an unprecedented acquisition of taste was a mystery at the time and to many remains so. Inevitably, there was talk of cultivating the shrub in some accessible part of the world – Ireland was in the frame – but such projects would take time. (Plans did not come to fruition for another hundred years.) In the meantime the only source of the indispensable brew was the Celestial Kingdom.[8]

Demand was not the only problem. The core difficulty was reciprocity, or rather the lack of it. Although China did buy some manufactured goods from England – Wedgwood's pottery had a temporary vogue in the land of superlative porcelain as quaint folksy artefacts – the trade imbalance by 1750 had reached a yearly £6 million. This was unsustainable. Then another complication arose. The only form of payment the Chinese would accept was Spanish silver, superbly minted massive coins from the South American mines, and after Spain allied herself with the rebellious colonies of America supply of it was running out. Yet Britons were clamouring more than ever for their daily – or thrice-daily – fix. Something had to give; and that something was the Chinese trade embargo on opium.

* * *

The Chinese government issued their first edict forbidding the importation of opium in 1779 and the original enactment was reinforced in 1799. As the document proclaimed:

> foreigners obviously derive the most solid profits and advantages; but that our countrymen should pursue this destructive and ensnaring vice is odious and deplorable.[9]

This was strong language but still aimed as much at Portuguese smugglers and Dutch, American and French freebooters as at Britain; and initially the directors of the East India Company, solid respectable citizens all, had every intention of complying. They were also anxious not to arouse Chinese displeasure lest it interfere with the Company's small but legitimate trade in other goods. Their governor general in Bengal, Warren Hastings, at first agreed but then changed his mind. Both he and his successor, Lord Cornwallis, were opiophobes and held the trade in contempt; but they needed funds to implement their reforms and their only source of money was the poppy. No subsidies could be expected from London: however benevolent, the Company was in business to make profit, not to distribute charity. A very British compromise was eventually reached: for several decades the smuggling of Indian opium into China was forbidden in theory but promoted in practice.

Unlike Britain, Canton, a picturesque port with gilded temples and beautifully laid out public gardens, overflowed with silver. Howqua, the chief of the local syndicate of merchants allowed to trade with barbarians, was reputed to be the richest man on earth. The main, indeed the only saleable British merchandise was referred to as 'the gentle and soothing garden balm'. Woollens were also listed but sales in them were paltry. (The Chinese disliked their rough texture.) Few if any were deceived but appearances were preserved. Howqua subsidised a host of Christian charities in lands as far away as the Bahamas. From Canton caravans of donkeys and camels carried the contraband into the interior.

Huge deals between the Chinese representatives and Indian and British traders were settled on board opium cutters according to time-honoured custom. Following carefully choreographed exchanges of courtesies over cups of tea, concerned enquiries were made after the health of a long list of local businessmen and officials. Their degree of well-being and residual ailments indicated the bribes expected and judged to be appropriate by their spokesman. Interspersed with banquets, theatrical entertainments and the handing over of gifts, the ensuing bargaining was described by one new arrival as resembling a Chinese opera. Nothing happened, everything evolved. Gentle intakes of breath and ghosts of a smile indicated varying degrees of assent. Barely visible flips of a finger represented well-nigh uncontrollable rage. But the volume of the trade was rising; and, though profits rose gratifyingly with it, perceptive observers felt uneasy. The very success of the enterprise spelt danger. In 1782 the first cargo of finest Patna opium found no buyers and had to be dumped. By 1830 the officially forbidden trade was reckoned to be the largest international commerce in any single commodity anywhere in the world. Only the highly evolved trading pantomimes and the sizable sweeteners meticulously apportioned and distributed up the hierarchy to reach the Imperial Treasury guaranteed the peace. And that not for ever.

* * *

In 1833 a reformist British Parliament dedicated to the doctrine of free trade abolished the East India Company's monopoly of trading with the Far East and self-styled merchant-adventurers began to flock to Canton like bees to the honeypot. There, lost in a strange world, they readily accepted the leadership of two Scotsmen, both old China hands. William Jardine, forty-six at the time, was a medical graduate of Edinburgh University and a former naval surgeon who had learnt all there was to know about the opium trade as a ship's doctor in the East India Company Marine Medical Service. James Matheson, the son of an impoverished Highland laird, had gone east and had started to trade in opium at the age of eighteen. Through family connections he had secured for himself the appointment of Danish Consul in Canton, not an arduous but a useful office. (The number of Danish subjects in the Celestial Kingdom was estimated at three.) The two men met while sampling the delights of a Macao brothel and immediately recognised like minds. A partnership was concluded. Jardine was the worker, Matheson contributed style. In their first and last moment of patriotic afflatus they chose a white St Andrew's Cross on a blue ground as their ensign. After that, sentiment was to play no part in their affairs. Both men were committed to making a fortune as quickly as possible, the future credo of the opium trade generally. When asked by a House of Commons committee fifteen years later if they were ever troubled by doubts about the morality of their enterprise, Jardine answered:

> When the East India Company was growing and selling the stuff there was a formal declaration of the House of Lords and Commons with all the bench of bishops in attendance that it would be inexpedient and indeed foolish to throw it [the opium trade] away. It could indeed promote the spread of the Gospels which missionaries in our own employ have in fact accomplished. I think our moral scruples need not have been so very great.[10]

* * *

But the Chinese Canton oligarchy were confounded by the brashness of the new arrivals. Peking too was becoming restless. Until now, whatever the creeping erosion of the rules, the Imperial Treasury had always squeezed a satisfactory cut from the exchanges of opium for tea. Now the opium trade was spreading unprofitably for Peking along the coast – Jardine's brilliant innovation – and the usual mechanisms for transmitting tokens of regard were beginning to malfunction. Coincidentally or not, the realisation dawned that 'foreign dirt' was poisoning the lives of thousands, perhaps hundreds of thousands, of ordinary Chinese. It was a danger foreseen by wise emperors of the past but not always acted on by corrupt officials. Peking was not unaware that the *Chinese Courier*, Jardine and Matheson's house organ, had recently informed its readers: 'Perhaps nothing could contribute more to the final reduction of the Chinese to reasonable terms in their dealings with ourselves than this … sapping of their moral energy.'[11]

Other developments too caused concern. Jardine and Matheson were revolutionising sales techniques. No longer did they rely on humble local dealers.

They had secured the services of Chinese-speaking Europeans, including that of the Reverend Dr Karl Gutzlaff, a Prussian-born Protestant missionary and linguist. The doctor thought well of opium and was happy to act as the firm's spokesman, interpreter and salesman provided his own demands were met. Apart from a comparatively modest commission he wished to be allowed to distribute with his packages of opium selected chapters of the Scriptures in his own translation. What could be more fitting than the 'wholesome coupling of bodily wellbeing and spiritual uplift?' The biblical extracts were handsomely printed in the firm's own printing shop and carried the firm's imprint. They also carried interleaved promotional literature for opium, 'the best way to achieve the spiritual enlightenment necessary to grasp the message of the Gospels'. But news of this aroused misgivings in Peking. Both religious and commercial proselytising had long been forbidden. The two activities combined did not cancel out the mischief associated with either. Resentment continued to simmer and would one day inevitably erupt.

* * *

This was dimly perceived by local British officials in Canton. They were lowly functionaries; but Chinese complaints, threats and warnings were duly forwarded to London and a certain 'recently observed restiveness' among the natives was reported. Sensing that all was not well on the Pearl River, His Majesty's Government decided on pre-emptive action. A Royal Superintendent of Trade, a nobleman of weight and seniority, inevitably flanked by two assistant superintendents of lesser but still gentlemanly status, would be appointed to sort out the misunderstandings. He would enforce a code of conduct on rowdy European traders. He would chastise Chinese trouble-makers. A meeting with the Chinese viceroy in Canton would take place. The two grandees would speak the same language though in different tongues. There was a hidden agenda too. The superintendent, it was hoped, would not only reside in Canton but also open a back door to the appointment of a fully fledged British ambassador to Peking.

The plans were grievously misconceived. No Red Barbarian – not to mention their big-footed womenfolk – had ever been allowed to reside within the walls of a Chinese city. In Canton foreign traders were confined to comfortable quarters called the Factories on the south side of the river opposite the walled port itself.[12] Their needs were met by skilled Chinese servants and by daily deliveries of food and other necessities. Chinese vendors of cloths and luxuries visited and did a brisk trade. Doctors too checked on the health of the inmates and dispensed Chinese remedies. Unlike most contemporary European medicaments, these tasted delicious and contained generous helpings of euphoriants. (Like many obsessionally hard workers, Jardine himself was a full-blooded hypochondriac.) The residents' post was collected and, after cursory censorship, transferred to boats destined for Europe. The sportingly inclined could ride into the countryside to shoot exotic birds provided they returned by nightfall. Traditionally mindful of children as well as the old, the Chinese had equipped schoolrooms and playgrounds. Chinese nannies were provided. Christian services in improvised places of worship were permitted. But regarding residence within the Chinese city, there was no reason to expect a change of heart. Nor was there to be one.

William John, 9th Baron Napier of Merchistoun, Lord Palmerston's appointee as Royal Trade Superintendent, was a Scottish aristocrat, a retired naval officer, a breeder of prize-winning sheep and, like his famous ancestor, an accomplished mathematician.[13] Everything about his selection and mission foreshadowed future practices in negotiating with the Chinese: the episode would otherwise be barely worth notice. As a devout Presbyterian Lord Napier disapproved of all inebriating brews (other than the locally distilled 'Water of life'); but Adam Smith was one of the elect, almost on a level with Euclid and John Knox; and the doctrine of free trade had to be defended. The alleged embargo on a harmless medicament for which there was an unquestionable demand was an outrage. Otherwise about China and the Chinese opium trade he knew nothing. Nor was he keen on foreign adventure. He accepted Lord Palmerston's invitation because, like most of his kin, he was deeply in debt. Even by Scottish baronial standards his ancestral castle leaked. He had two marriageable daughters to provide for. Duty and circumstances thus combined to settle the matter; and on the last day of 1833 he and a domestic retinue of twenty-four embarked for the East.

* * *

The Royal Superintendent and his troupe arrived in Canton in the scorching heat of July. All felt seasick and exhausted but the bustle of a thousand crafts of every kind and size in the splendid bay lifted the spirits. Yet something was missing. Advance messages had been sent to the Chinese viceroy; and no doubt it was felt that that august personage would wish to welcome His Majesty's envoy in person. But no august personage showed. At last a gang of impertinent Chinese customs officials came aboard, queried His Lordship's credentials, ogled his daughters and refused the party permission to disembark outside the Factories. The notion of being allowed to communicate with His Excellency Li K'un, let alone meet him in person, aroused ribald hilarity.

This was the more vexing since Lord Napier's instructions in London had been ambiguous. It had been envisaged that in case of a dispute his martial experience at Trafalgar as a seventeen-year-old midshipman thirty years earlier might prove invaluable. On the other hand, he was 'under no circumstances to appeal to our naval forces which were [as ever], overstretched'. How, in case of trouble, the two courses were to be reconciled remained unspecified; but Lord Napier had been given the honorary rank of admiral, and admirals were expected to cope with ambiguities.

A succession of humiliating events ensued. Reluctantly Napier took up temporary residence in the Factories and issued a proclamation in Chinese. In this he appealed to 'the thousands of industrious natives who would be victimised by the obstinacy of their Government and face ruin if they did not listen to the envoy of His Majesty'. This was unwise. Blusterings were usually ignored by the viceroy but a blatant incitement to rebellion elicited what he discribed in his dispatch to Peking as a measured reply:

> A lawless foreign monkey has issued a notice. We do not know how such a barbarian dog can have the audacity to call himself an Eye [that is an official] ... though a savage from beyond the pale would have restrained him from such an outrage. It is a capital offence to incite the people against their rulers and we would be justified in obtaining a mandate for his immediate decapitation ... While I consider this ... I formally close all trade until he goes.[14]

In his only dispatch to reach the Prime Minister, Lord Grey, in London, Lord Napier was moved to characterise the Chinese government as being 'in the extreme degree in a state of mental imbecility and moral degradation, dreaming themselves to be the only people on earth and being entirely ignorant of the theory and practice of international law.'[15] In a private letter to Lord Palmerston, probably inspired by Jardine, the Trade Superintendent expressed the view that though the situation was fraught, there was nothing which 'three or four frigates and brigs with a few steady British troops, *not sepoys*, could not settle ... And nothing else, in my opinion, will.' In the long term he was to be proved right.

But before that could happen His Lordship was declared by his personal physician to be under severe stress and in urgent need of rest. Few regretted his departure. The only useful function of the cutter *Louisa*, summoned by Napier to intimidate the Chinese, was to convey him ailing to Macao. There, after a few days' harassment by the constant pealing of popish bells he died. To the last he refused any brew made of the diabolical plant, the poppy, whose sale to the Chinese he had been sent to promote.

* * *

Lord Napier's mission seemed a humiliating but trivial episode; but it gave China false confidence. In Peking it nurtured the belief that, as in the past, peremptory refusal to deal with the Red Barbarians would keep them at bay. But the past was the past. With Lord Palmerston bestriding the political stage in Westminster, his fellow peer's humiliation would not go unpunished. And the opium trade was far too profitable to go away.

Only a few months after Lord Napier's departure an untrumpeted but portentous event took place in Canton. On the quayside the citizenry were watching with puzzlement the arrival of an entirely new kind of vessel. It was a 'cartwheel ship that put axles in motion by means of fire'. The *Jardine*, a 58-ton steamer, had been built for Jardine and Matheson in Aberdeen ostensibly to carry the mail between Macao and Canton but, as a slightly rusty cannon on deck proclaimed, was capable of defending herself. The ship was forbidden by the viceroy to steam up the Pearl River and her master had no wish to start a war; but her presence was awesome enough. Mount a gun on a steamer and for the first time in history one had a man-o'-war independent of wind and tide and able to beat the monsoon. Against such a beast Chinese junks and sampans would be useless. Ushered in by the opium trade, gunboat diplomacy had arrived.

II THE FIRST OPIUM WAR

Napier's successor, Captain Charles Elliot, RN, was the thirty-four-year-old sprig of another Scottish noble house – Lord Minto was his uncle – brave and honourable. None of his other accomplishments – he was a gifted watercolourist and possessed a pleasing tenor voice – were relevant to his mission. He abominated the opium trade and held in contempt all those who made a living from it. William Jardine in particular he detested, describing him as 'a shifty fellow and, though a doctor and an Edinburgh graduate, *not* a gentleman'. But as a naval officer he was resolved to do his duty by his country. His difficulty was that after Lord Napier's fizzle (as the episode became known) his country's government was far from resolved what his duty should be and how, whatever it should be, it should be carried out.

There was one exception. Lord Palmerston, foreign secretary, was not in doubt. After much manoeuvring – neither he nor William Jardine wanted to appear to be the initiators – the two men met and discovered that, if not exactly soulmates, they were of one mind about Chinese opium. Both wanted to teach China a friendly lesson in international cooperation, at the point of a gun if necessary. Both wanted to enforce the principle of free trade which Lord Palmerston had absorbed as a student in Edinburgh. Palmerston also wanted to establish Britain's place in the world as heir to the Roman Empire. Jardine's aims were more modest: he wanted to multiply his millions. Palmerston explained his position in a letter to his friend, Lord Auckland, Governor General of India:

> The rivalry of European manufactures is fast excluding our productions from the markets of Europe. And we must therefore unremittingly endeavour to find in other parts of the world new vents for our industry… If we succeed in our China expedition, Abyssinia, Arabia, the countries of the Indus and the Pacific will at no distant period give us important extensions to our commerce.[16]

This was a blueprint for Victorian empire building and summed up the justification for the opium wars. Morals, sovereignty, Christian virtue and national rights were lofty concepts to be prayed for but they should not be allowed to impinge on practicalities. Not in the short term. Eventually of course British rule would be blessed by all.

Jardine, in turn, perhaps a little carried away, offered all the Jardine–Matheson opium clippers to serve as ancillary craft to the Royal Navy, each opium vessel under the command of a naval lieutenant. Even to Palmerston's most faithful followers the idea of the Royal Navy providing officers for a private opium-smuggling operation was going a step too far; and Sir Robert Peel, new leader of the Tory opposition, thought that his moment had come. On 12 March 1840 he introduced a motion of censure in terms that, he hoped, would enable critics on both sides of the House to vote for it. It led to a memorable debate which probably reflected literate opinions in the country as a whole.

Sir James Graham, opening for the opposition, reminded the House that a sixth of the public revenue of Britain and India was now derived from the tea tax and the profits from opium. This was intolerable. Captain Elliot had repeatedly asked for clear instructions to regularise or abolish the traffic but none so far had been issued. The opium trade should be squashed and Captain Elliot's instructions clarified.

In reply and at his oratorical best (or worst) Thomas Babington Macaulay, at thirty-nine secretary of state for war, made a pulse-quickening appeal for a 'patriotic policy'. His verbal mists congealed around Captain Elliot's glorious action in running up the Union Jack the moment he arrived at the beleaguered Factories in Canton, 'reminding our countrymen that they belonged to a people unaccustomed to defeat, to submission and to shame'.[17] Sidney Herbert for the opposition tried to redress the balance by pointing out that 'we are about to be engaged in a war not only without a just cause but indeed any cause other than the pursuit of profit ... Opium is a poison and we should not be party to poisoning a whole country.'[18]

But the most rousing speech was delivered by William Ewart Gladstone, at thirty-one still the rising hope of unbending Toryism.[19]

> Does the Prime Minister know that the opium smuggled into China which irrevocably corrupts its people comes exclusively from British ports or ports that we control and in British vessels? We require no preventive service to put down this iniquitous traffic . . . We have only to stop the sailing of the smuggling vessels . . . The Chinese gave us notice to abandon our contraband trade. When they found that we would not, they had every right to drive this infamous and atrocious traffic off their coasts . . . Our flag must not become a pirate flag protecting a godless and deeply sinful enterprise . . .[20]

Gladstone comfortably occupied the moral high ground; but Palmerston, though no orator, had the commercial aces up his sleeve:

> I wonder what the House would have said to me if I had presented it with a large naval estimate for a number of cruisers for the purpose of preserving the morals of the Chinese people who were disposed to buy what other people were disposed to sell them . . . Ending our opium trade would simply increase the Turkish and Persian crop available to willing traders from France, Russia and the United States.[21]

Trade was trade. The House divided and defeated Peel's motion by 271 to 262 votes. The legitimacy of the opium wars was established.

* * *

In the mean time the exchange of insults in Canton continued; but a new personality had emerged on the Chinese side too. It followed the realisation in Peking that the country's silver reserves were draining away and that the number of opium addicts, mostly men aged between twenty and fifty, could now be reckoned in millions. The drug's happy but drowsy victims included the Son of Heaven's three sons, a serious blow to the succession. No Chinese emperor would ever again emulate a long line of heroic ancestors. On the last day of 1838 the emperor made the rare appointment of a High Commissioner 'to investigate port affairs in Canton' – in effect to stamp out the drug trade.[22] The man chosen was fifty-three-year-old Lin Tse-hsu, the son of a village teacher and one of the last to have risen by merit alone in the mandarinate. In his former province as a viceroy he had acquired the reputation of being just, firm and merciful, old Confucian virtues, and he had handled the illicit opium trade with vigour.[23] He was no mere administrator. For his uprightness he was known as Lin the Clear Sky, one of the few officials impossible to bribe with either money or opium. Plump and unprepossessing, with a straggling black moustache and a thin beard, he was also a considerable poet and an interpreter of the stars.[24] Most important, despite his traditional attitude to government, he was one of the modern Confucians who had made strenuous efforts to understand economics and who wanted to interpret the work of the Great Teacher to justify the reforms that were obviously needed.

On arrival in Canton he issued several edicts. They banned opium and threatened importers and dealers with death but they also offered addicts wishing to be cured facilities for rehabilitation. The last was a bold innovation not appreciated at the time and not followed anywhere for nearly a century. He also, some time in April 1839, penned the first of a series of personal letters to 'Our Much Loved Sister, the Queen Victoria of England'.

> Let me enlighten your Majesty . . . There is a class of evil foreigner that makes opium, brings it for sale, tempting fools to destroy themselves merely in order to reap profit. Formerly the number of opium smugglers was small; but now the vice has spread far and wide, and the poison has penetrated deeper . . . We have decided to inflict the most severe penalties on opium smokers and dealers . . . This poisonous substance is manufactured by certain devilish persons subject to your rule . . . It is of course neither made nor sold at your bidding. I am told that in your own country opium smoking is forbidden under severe penalties . . . It would be better to forbid the sale of it here too or, even better, to forbid its production . . . When this is done not only will we be rid of this evil but your people will also be safe . . .
>
> Suppose there were people from another country who carried opium for sale in England and seduced your Majesty's people into buying and smoking it; certainly you would deeply hate it and be bitterly aroused.[25]

Lin would have been shocked incredulous to learn that opium was at that time imported to England by a company operating under the auspices of the Crown, that it was freely sold in corner shops and street markets, and that it was consumed by both sexes and all ages. Of the powers of the Queen of England, then a twenty-one-year-old opinionated but inexperienced young woman, he had also been misinformed. And yet, beyond the misunderstandings and the circumlocutions, he was surprisingly prescient. The long-term interests of both Britain and China would have been best served by suppressing the opium traffic; and this was a decision which could be taken only by a strong British government. It was possibly the last time when such a course might have been effective. It might conceivably have influenced the drug trade – and history – for at least half a century. But Lin was ignored.

* * *

By May 1839, 1,600 Cantonese violators of the Imperial Edict had been arrested, 42,741 opium pipes had been confiscated – Lin understood the fetishist attachment of addicts to their personal paraphernalia – and 41,845 catties of opium had been impounded. But this huge haul was only a fraction – probably less than a tenth – of what was accumulating on rocky Lintin Island offshore and in the warehouses of the Factories; and Lin was determined to destroy it all. If the foreign merchants would not deliver their hoard voluntarily, he would force them to do so. No bloodshed would be necessary. He would merely withdraw all Chinese servants from the Factories and cut off their supplies. The second measure was ineffective: at the right price Cantonese merchants risked sneaking food and other necessities in at night. But being without cooks, butlers, valets, barbers, maids and nannies was a serious deprivation. Secretly, moreover, Elliot was in sympathy with Lin. He saw the evils of the opium trade and admired the viceroy's firmness. After eight weeks and acting on his own initiative, he decided that the merchants should comply. More than a little cavalierly he promised on behalf of the Crown that they would be indemnified for their loss. Like most of his class, he was without any grasp of practical economics and had no conception of what he was promising. In fact he was committing Her Majesty's Treasury to dispense the staggering sum of £2 million to compensate merchants for a contraband which they had no right to possess and had long given up any hope of selling. (Fully honoured, which it never was, it would have precipitated a grave financial crisis.) Well pleased, the traders surrendered their opium and it was destroyed by Lin's troops. Pilferers were shot. The stench could be smelt for many miles upriver. It was the high point of Lin's career. British warships were on their way; and he was beginning to overreach himself.

* * *

What provided the spark for open war was a rowdy but inherently minor escapade. A group of British sailors rowed ashore, visited a tavern, became drunk, demolishing a Buddhist shrine, and got involved in a brawl. This was par for the course; but, whether or not by accident, they also killed a Chinese reveller. Judicial executions in China were commonplace but private murder was a serious matter. Lin demanded that a particular British sailor nominated as the killer should be handed over to be tried by a Chinese court. Sympathetic though he was to all matters Chinese, it was inconceivable to Elliot that an Englishman, murderer or not, should be so tried. He compensated the dead man's family from Treasury funds and tried the brawlers himself. He meted out moderate punishment to several of the accused for being drunk and disorderly; but since he could not identify the slayer – witnesses were too drunk at the time to remember – the culprit went free. This was English but not Chinese justice. After a lengthy exchange of testy notes, Lin marched with a military force on Macao. It was holiday time and most British traders and their families were enjoying a spell of freedom from the constraints of Canton. The enclave was undefended and indefensible; but it was not even a British colony. It was ruled – more or less with Peking's blessing – by a Portuguese governor and it was the see of a Roman Catholic archbishop. Elliot kept his head and ordered British residents to board the merchant ships lying off Hong Kong. Lin, without a naval force and careful not to antagonise the Portuguese, ordered any Briton found ashore to be shot. None was; but, as William Jardine noted with satisfaction in a letter to Matheson, this was war at last. 'With a bit of luck the price of opium will hit the sky. We're in business.'

* * *

Jardine was rarely wrong. Matters duly escalated. In the first week of September a British merchant fleet anchored in what is today Hong Kong's Western Harbour was blockaded by Chinese war junks which Lin had eventually assembled. An ultimatum from Elliot was ignored. The British opened fire. Within ten minutes the Chinese armada was sunk. Some of the crew were saved; most drowned. The skirmish known as the Battle of Kowloon was the first action in what *The Times* of 25 April 1840 solemnly promoted to the status of a war. Shocked, Lin withdrew to defend Canton. On 3 November two newly arrived British frigates routed and sank another Chinese fleet. In January 1841 Elliot sent two warships to attack the Chinese forts guarding the harbour. The forts were captured after a token resistance. Despite his strategic advantage Elliot did not attempt to capture the city. A glance at his terrestrial globe persuaded him that occupying China was beyond his means. But the war spluttered on. Sailing north along the coast, he besieged a succession of unfortified towns. Some surrendered without a fight, others were overrun with considerable bloodshed. The British suffered losses from malaria and dysentery but none in battle. In many spheres China was far ahead of the West. They did not include the manufacture of weapons of destruction. Chinese cannon were beautifully crafted but, mounted on immovable stands, they were for celebrating birthdays, not for killing people. The Tartar cavalry, the only Chinese force to hold opium in contempt and who were therefore fit to fight, exhibited spectacular feats of bravery but with the armoury of Genghis Khan. For the past fifty years engineers in the West had been refining the steam engine. Their Chinese counterparts were taking pride in well-crafted wheelbarrows.

The court in Peking was a hive of intrigue. Lin's advice to carry the war into the interior (as Russia successfully did in 1812) instead of trying to defend insignificant coastal ports was ignored. But the decisive factor was opium. Many of the Chinese officers were addicts incapable of commanding. Their regular troops were often too dopey to fight. In August 1842 the Royal Navy occupied Shanghai and started to sail up the Yangtze River. The emperor sued for peace. By then the ships had reached Nanking. There, on board HMS *Cornwallis,* a treaty known to Chinese historians as 'The Most Unequal Treaty' was signed. Canton, Amoy, Foochow, Ningpo and Shanghai were declared 'Treaty Ports' and were to open to foreign trade. 'Reparations' of 21 million silver dollars were imposed on China to cover the cost of the indemnity for the opium destroyed by Lin. As an afterthought Hong Kong, beginning to blossom in a cloud of opium, was ceded to Britain.

* * *

Both main protagonists of the First Opium War were dismissed in disgrace. Lin was summoned to Peking, tried by a court presided over by the emperor and sentenced to death. The sentence was commuted to exile in the 'cold provinces', the unfriendly north-easterly region of the empire.'You have been no better,' the Son of Heaven declared, 'than a wooden doll.' Elliot too was recalled and reprimanded for trying to bankrupt the Treasury. Queen Victoria, descended of thrifty German housewives, noted in her diary that 'Captain Elliot *completely* disobeyed instructions and almost gave away much of my Treasury'. But this left the captain's character unblemished and he was still Lord Minto's nephew. He was appointed British chargé d'affaires to the newly constituted Republic of Texas, one of the Foreign Office's cold provinces at the time. Five years later, let bygones be bygones, he was promoted to the governorship of Trinidad and eventually to that of St Helena. Following Napoleon's sojourn there thirty years earlier the last posting carried with it a knighthood and the rank of admiral. In his memoirs he remembered with pleasure his only meeting with Lin. 'He was an honest Chinese of the old school, the poor devil, and a great hater of opium. We fought but in fact we agreed about most things. I wish there had been more of his sort.'[26]

* * *

Otherwise the treaty was well received in London, *The Times* bestowing its blessing. The exceptions were a few notoriously high-minded old Catholics and Low-Church Evangelicals. Lord Ashley, the future Lord Shaftesbury, thought commerce in opium no better than the slave trade. A few bishops and headmasters also expressed dissatisfaction Dr Thomas Arnold of Rugby declaring:

> We have been waging a war for the introduction of a demoralising, evil drug which the government of China wished to keep out, and which we, for lucre of gain, still want to introduce by force; and, in this quarrel we have burnt and slain in the pride of our supposed superiority. I fear God's punishment for this wicked iniquity.[27]

He was in a minority. In any case, a more decisive phase of the war had already begun.

III THE SECOND OPIUM WAR

Relations in Canton remained tense; but before reaching boiling point there erupted the bloodiest civil war in history. At least twenty but probably nearer thirty million people perished in the Taiping Rebellion, slain in battle, exterminated in massacres or judicially beheaded. The leader of the uprising was twenty-three-year-old Hong Ziuquan, the son of a peasant family in Guandong province in south-western China, perhaps a genius, perhaps a madman, perhaps both. After repeatedly failing state examinations he passed through periods of despair, catatonic trances and the study of the Bible. He finally declared himself to be the direct descendant of Christ, Muhammad and the Buddha. He had met God, an old man with a golden beard, who gave him a sword with which 'to kill the Demons who torture my people'. Hong's actual teaching emerged as an amalgam of Christian Puritanism, the social ideals of the French Revolution, monastic sexual segregation (which did not apply to the founder) and a ban on any substance of abuse including opium. Inspired like St Francis of Assisi by the Gospel of St Matthew – 'if thou wilt be perfect go and sell what thou hast, and give to the poor and thou shalt have treasure in heaven' – he also abolished private property. All plunder was to be pooled in a common treasury.

Starting to preach in the rugged area known as the Thistle Mountain in the extreme south-west of the empire Hong attracted his first converts from local tribes outside mainstream society. But then, following three years of drought and famine, his movement took wing. Within another year he had recruited over 20,000 converts soon to be known as God-worshippers. Most astonishingly in the eyes of contemporaries, those who joined him were able to rid themselves of their addiction to opium. This was something no imperial edict, promise or threat had ever achieved. As had happened before and would happen again, voluntary abnegation created a fanatical force. The expropriation of land gained the rebellion the loyalty of millions of peasants. On 1 January 1851 Hong proclaimed a new dynasty, the T'ai-p'ing T'ien Kuo, meaning Heavenly Kingdom of Great Peace. Deeply alarmed, the emperor ordered 'the extermination of the vermin' and recalled Lin from exile to carry out his command. Lin died on the way; and without him the opium-soaked government troops, 'formerly feared as tigers but now regarded as sheep', were repulsed by an army wielding scythes and pikes.

Contact between the Taiping and the British authorities in Hong Kong was established through missionaries; and there was a moment when the fate of opium in China hung in the balance. To many Europeans the Christian overtones of the rebellion were inspiring; and so was the rebels' prohibition on opium. Could they be the future for which the faithful had been praying and many had suffered martyrdom? The Bible in Chinese, censored and adulterated but still the Bible, was circulating in Taiping-occupied territories. Illiterate warriors were known to carry pages from the Gospels close to their hearts. Crucifixes were worn by their women.[28] The Jesuit Provincial in Macao sent Father Lecueur, a Mandarin-speaking member of the Order, on a fact-finding mission. The Father travelled hundreds of miles. Then, one evening, in a clearing he came across a crowd apparently engaged in some form of ecstatic communal worship. The ritual seemed both alien and familiar. It was clearly a happy affirmation; but of what? Then he recognised two words. One was 'halleluya' with the accent properly on the last syllable. The second was unmistakably 'Jay-sus'. He wept.

On his return Lecueur recommended further contacts and explorations and his superiors in the Order were not unsympathetic. But wiser – or more practical – counsels prevailed. In London Queen Victoria expressed her 'abhorrence' of the 'massacres perpetrated by this peasant revolt in China'. Those nearer the scene were increasingly alarmed by the Robin-Hood-ish undertones of the rebellion. No doubt entered the minds of Jardine and Matheson. The God-worshippers with their irrational ban on opium were to be destroyed. It was against this background that the Second Opium War was to be fought. China had to be defeated and forced into the comity of poppy-loving nations but the suitably compliant Manchu dynasty rather than the quasi-Christian Taiping were to be preserved. Weak emperors, not powerful prophets, were the ultimate guarantors of the free flow of opium.[29]

The Taiping army eventually marched two thousand miles and on 28 February 1853 occupied Nanking. About 100,000 Manchu defenders, state officials and the rich with their children and women were herded into temples and set alight. The rebels were to hold the city for ten years.

* * *

Back in Canton the trigger for the resumption of hostilities, known as the *Arrow* incident, was as absurd as such triggers often are.[30] The crew of a vessel called the *Arrow* which had a long and murky past but which may have been flying a British ensign was boarded in the harbour by Chinese customs officers probably by mistake. The mostly Chinese crew were arrested and taken ashore but released after a protesting letter from the acting British Consul, Sir Harry Parkes. The subsequent exchange of accusations, goadings, threats and finally armed action revolved around (insofar as it is possible to disentangle what they *did* revolve around) a public apology. Finally Parkes ordered Rear Admiral Sir Michael Seymour to occupy both Canton Bay and the city itself. This was accomplished with no great difficulty; but to what purpose? No occupying force was available and without it the port would be impossible to hold. But public opinion back home was reassured by Sir Michael's order that British guns should remain silent on the Sabbath.

But events in Canton led to another debate in Parliament and eventually to the calling of the 1857 general election. The Crimean War had recently ended in 'total triumph',[31] and the voting classes were ready for new excitement in some remote part of the world. Palmerston and his party won with a landslide. During the campaign, while in public staunchly defending Parkes, Seymour and their crew, Palmerston concluded that they were in a dangerously hysterical mood. He approached James Bruce, 8th Earl of Elgin, direct descendant of Robert the Bruce and son of the famous (or infamous) saviour (or vandaliser) of the Parthenon. The earl accepted the post of overall commander with misgivings. Like Lord Napier before him he had inherited from his ancestors but especially from his art-loving father mountainous debts. He needed a job.[32] But there were several delays. The Indian Rebellion, known at the time as the Mutiny, had diverted troops from other bases, including China. Elgin himself stayed in Calcutta as guest of the governor general. He witnessed the reprisals meted out to the mutineers (and those labelled mutineers) and they horrified him:

> I have seldom, from man or woman since I came to the East [he wrote to his wife], heard utterances which were less reconcilable with the hypothesis that Christianity had come into this world. Detestation, contempt, ferocity, brutality

> and vengeance is all I have seen from my countrymen since my arrival. Oh Lord, deliver me from such evil whether directed towards Indians or Chinamen.[33]

He might have been even more horrified had he known that within a few years he would be the victim of the same poison.

* * *

As if the world's great powers had sensed that the final act in the Chinese drama was about to be unveiled, their emissaries and appropriate military contingents began to congregate along the China coast. Nobody professed to want a war but nobody wanted to be left out of the distribution of the spoils or contracts to be signed. French public opinion had also been inflamed by the murder in central China of a saintly missionary, the abbé Auguste Chapdelaine. He had probably been mistaken for a follower of the Taiping but explanations never achieve much against popular fury. Baron Jean-Baptiste Gros, France's plenipotentiary, arrived on the new battleship *L'Audacieuse* to investigate what had happened, to exact justice and to stand by.[34] Elgin, at first suspicious of French intentions, gradually established friendly relations with the baron. Their intimate bond transcended nationality: it was their shared dislike of Sir John Bowring, Britain's new consul in Hong Kong, a bore, a bully and a bigot.[35] Then William Reed, a friend of President Buchanan of the United States and professor of history at the University of Pennsylvania, arrived in a behemoth of a warship, the *Minnesota*. Though intimidating, she was useless since her hull could not navigate the shallow rivers of China. Like the President, Reed disliked the opium trade and was determined not be sucked into a war to promote it; but he wanted to be on hand as a mediator (and just possibly a beneficiary). Last to line up was Russia's Count Euphemius Putiatin, his country's hero of the Crimean War, aboard the oddly named paddle-boat *Amerika*. The vessel was equipped with only six ancient guns but the count was confident that the Son of Heaven could not refuse to receive a personal representative of his fellow autocrat, the Tsar. He was to convey to Peking that if China would cede Manchuria to Russia without a fuss, the Tsar would help his imperial brother to root out the Taiping. In both expectations Putiatin was to be disappointed. The emperor would not receive anybody not related by blood to the Tsar and the defeat of the Taiping, the count was told, was imminent. This was fantasy but could not be contradicted. Manchuria, moreover, the homeland of the dynasty, was not for sale. In a huff the Russian joined his American, French and British allies.

* * *

In March 1858 Elgin with a British fleet and Baron Gros with two newly arrived French gunboats sailed north and arrived at the mouth of the Bei He River, the gateway to Peking. Gros was anxious to avoid unnecessary bloodshed until rebuked by the Quai d'Orsay. The only blood likely to be shed was Chinese and therefore none of his concern. The emperor and his court departed into the interior for what was described in the official bulletin as a hunting trip. In Tientsin, a mere twenty miles from Peking, the conquerors were rapturously received under the misapprehension that they were bringing opium. More auspiciously, the emperor at last sent peace emissaries.

* * *

On paper the Treaty of Tientsin gave the Western powers everything they wanted. Most importantly, Western travellers, including traders and missionaries, would be allowed to travel unhindered within the empire and a permanent British embassy would be established in Peking. Reparations would be paid. The Chinese solemnly promised not to refer to Britons as Red Barbarians in future. In his curiously ambivalent way Elgin again refused to mention the opium trade by name. Yet basically, as Baron Gros pointed out, opium was what the war had been about. But while trade figured prominently in Elgin's report, the baron meticulously observed French proprieties.

> Je suis heureux de pouvoir annoncer aujourd-hui a Votre Excellence que la Chine s'ouvre enfin au Christianisme, source réelle de toute civilisation, et au commerce … occidentale. [I am happy to announce today to Your Excellency that China is at last open to Christianity, the real source of all civilisation, and to commerce.][36]

Double talk ruled. Reed on behalf of the United States formally suggested that Britain should ban opium exportation to China. As if to underline his charge that 'half the country had already been poisoned' the transcription of the treaty into Chinese took longer than its negotiation since most of the scribes were semi-comatose. In response Elgin pointed out that the United States consuls in Canton and in Shanghai, both Reed's personal friends, were partners in the largest merchant houses trading in the drug. Reed then expressed the hope that high taxation would make the imported poison unaffordable to all but the very rich. Russia got nothing and Putiatin departed uttering imprecations.

In fact, none of the signatories had any intention of honouring the treaty. The more humiliating the conditions, the more the 'patriotic party' in Peking seethed with resentment. The higher the expectations on the Western side, the more those on the spot (though not the powers in London, Paris,

Washington or St Petersburg) realised that nothing short of the crushing military defeat of the enemy could ensure their realisation. But, complicating matters, it was now the agreed policy of the West to preserve the Manchu dynasty in some form or another. A takeover by the opiophobe God-worshippers who now ruled over a third of the country and still threatened the rest, and what Gros described as 'la balcanisation de la Chine' had to be prevented. These often self-contradictory utterances dominated the war over the next few years.[37]

* * *

For a time Lord Elgin returned to London and joined Lord Derby's cabinet as Postmaster General. After China this was a rest cure. To keep the opium war in the family he was replaced in China by his younger brother, the Honourable Frederick Bruce. Bruce, an amiable youth, was soon out of his depth. As he himself recognised, his interest in Scottish ecclesiastical history was of little use when in charge of a multilingual armada bent on enforcing unenforceable conditions on an unfathomable enemy. Fortunately the troops were now commanded by two hard-bitten generals, Cousin de Montauban on the French side and the taciturn General Sir James Hope Grant on the British. Yet, to everybody's consternation, when hostilities were resumed the Chinese won their first and only victory.

Only forty miles from the capital the advancing Western battleships were stopped by three bamboo booms, each three feet thick, laid across the Bei Ho River. The device had first been tried under Kubla Khan and it still worked. The boats ground to a halt and soon got entangled in the booms. The bombardment from the shores was also of unprecedented effectiveness. Hope Grant, wearing a profusion of gold braid, chose to command his troops from the deck of his flagship. He was hit by a sharpshooter and forced to turn over command to his deputy, the brave but reckless Captain Shadwell. By the end of the day all the ships had been immobilised and one British gunboat had exploded. Although the Americans were still officially neutral, Commodore Josiah Tattnall, hearing of Hope Grant's plight, ignored diplomatic niceties and in his steamship *Toeywhan*, the most modern in the Western fleet, hastened to the rescue. 'I'll be damned if I stand by and see white men butchered by the Chinks before my eyes,' he bellowed, sailing past the British flagship. Of butchering there was in fact no danger. The Chinese forces under the command of Prince Gong, the emperor's younger brother, remained in their forts. The cannonade continued; and, as evening fell, fireworks gave the battle a festive flavour. Later, as Captain Shadwell with fifty Royal Marines and a hundred French seamen under Commander Tricault tried to seize a foothold on the mud flats, the primitive Chinese gingalls – large-bore firelock muskets, seven and a half feet long and needing two men to fire – demonstrated their killing power at short range. Shadwell was wounded; Tricault and fifty men were killed. The armada eventually withdrew with difficulty. It had been an ignominious day for the West – but the last.

In his dispatch to London Bruce added a modern touch. Chinese success was due to the secret participation of Russian officers who were 'glimpsed on the walls in fur hats helping the Chinese and looking angry'. The explanation was disbelieved but a note of protest was delivered by Britain's ambassador in St Petersburg. News of the humiliation still brought down Lord Derby's government. He was replaced by Lord Palmerston, at seventy-five as hawkish as ever. 'We must, in some way make the Chinese repent the outrage,' he declared;[38] and *The Times of London* of 12 September 1859 thundered:

> We shall teach such a lesson to those perfidious hordes that the name European will hereafter be a passport of fear, if it cannot be of love, throughout their land. We must occupy Peking.[39]

While staying as a guest of the Queen at Balmoral Lord Elgin was ordered to return to China forthwith. In the meantime Lord John Russell, the new foreign secretary, instructed Bruce to demand unspecified reparations from the Chinese. There was now a telegraphic link between London and Shanghai and for the first time in history an order from London could be carried out (more or less) within weeks. The Chinese reply too was unusually prompt and, as Hope Grant described it, 'cheeky in the extreme'. Whatever his view about opium and the China problem, Lord Elgin was stung in his family pride and agreed to return.

* * *

The second march on Peking started on 2 June 1860 under a cloud of apprehension. Even though special boom-cutters were now carried by the fleet, the debacle of the previous approach was on everybody's mind. Misgivings were multiplied by the Chinese holding two British hostages, Sir Harry Parkes and *The Times* correspondent Thomas Bowlby. They had been sent under a white flag to negotiate about prisoners but the ensign was unrecognised – or ignored – and both were taken captive. Had the court in Peking recovered their nerve – and their cheek? Newspapermen tend to be nosey and Bowlby may have overstepped the mark; but Parkes was an old Harrovian like both Lord Palmerston and Lord Aberdeen. He could not be allowed to be roasted on a spit, a favoured Chinese way of executing foreign devils.

The advance nevertheless proved a walkover. Baron Gros suggested that Prince Gong had put his faith in holding the two hostages. It was also rumoured that the prince was plotting to bring about the downfall of his effete brother and replace him. The truth was different. The prince's troops had discovered a cache of opium that had been left behind in a deserted Buddhist monastery by the last opium-carrying caravan on its way to Peking. Why it had been dumped and the delivery abandoned nobody knew then or knows now. Traffic in opium was forbidden but the prohibition had long been ignored. Half the imperial household was thought to be addicted. Most likely, the merchants heard rumours of an approaching Taiping force. Never a desirable encounter, to be found in possession of a haul of opium would be a signal for extermination. Nobody in Prince Gong's army cared. The find was a miracle. It also proved to be the prince's and his army's downfall. And indeed, though this was not immediately apparent, his country's. Long starved of the drug, the troops were soon gorging themselves. Within a few hours many were unconscious. Officers were as hooked as the men. Over the next few nights what remained of the force melted away. Only about a hundred bodies remained, either dead or in a dream from which they would never awaken. Their happy smile puzzled the French troops, the first of the Western forces to arrive. 'It was,' Baron Gros reported to the Princesse Marie d'Angoulême, 'the oddest battlefield I have ever seen.'[40] One of the dead was Prince Gong, butchered by his own men.

Success bred concord. The only dispute to arise between the allies was about the relative height of the poles from which their respective flags should be flown. News came that in the light of Prince Gong's treachery the emperor and his court had once again departed for a hunting trip in the interior. Only a token garrison was left to defend the capital. Yet Elgin was troubled. 'What about old Parkes?' But as the armies encircled Peking meeting virtually no resistance even his mood began to lighten.[41]

The encircling meant occupying the emperor's Summer Palace outside the city walls. It was known as one of the wonders of the world but reality exceeded expectations. 'Palace' was an understatement: the complex was a city of palaces, a few larger than most European royal residences, others small but exquisite.[42] In the grounds, 'beautiful beyond belief', every region of China was reproduced on a reduced scale. There were lakes, mock mountains, gorges, bridges, copses, glades, a botanical garden and a zoo. Fabulous fountains were still spouting water. The gardens were put in the shade by the interiors. Not only were they the repositories of three thousand years of Chinese arts and crafts, but they housed tributes to a long line of emperors. Lord Macartney's gifts were there but looked paltry compared to those from the Mughal court, the Khans of Mongolia, the Shoguns of Japan and the Pashas of South East Asia. They included ceramics, carpets, textiles, banners, furniture, wall paintings, ingots and coins of gold and silver, bronzes, precious stones, an incomparable library and several rooms filled with the emperor's collection of pornography. As the looting began the comte de Tassin de Moligny, accustomed as he was to French upper-class luxury, wrote to his father in a state bordering on the orgasmic:

> Je prends la plume, mon bon père, mais sais-je que je vais tu dire? Je suis ébahi, ahuri, abasourdié de ce que j'ai vu. Les mille et une nuit sont pour moi une chose parfaitment véridique maintenant. J'ai marché pendant presque deux jours sur plus de 100 ou 1000 millions de francs de bijoux, de porcelaines, bronzes, sculptures, de trésors enfin non imaginables. Je ne crois pas qu'on a vu chose pareille depuis le sac de Rome par les barbares. [I have my pen in hand, dear father but I know not what to say. I am amazed, rapturous, swooning from what I have seen. The Thousand and One Arabian Nights now seem perfectly real to me. For the past two days I have strolled among a hundred or a thousand million francs worth of silks, jewels, porcelain, bronzes, sculptures, an infinity of treasure. I don't believe that anyone has seen anything like this since the Sack of Rome by the Barbarians.][43]

The only live creatures left in the palace were 200 imperial eunuchs, high on opium. With antique weapons they tried to defend the emperor's marble and jade private apartment. A dozen were shot. the rest were allowed to run. The French were given leave to keep whatever they could carry, the British only part of it. It still made them rich. Some of the rest was auctioned on the spot.[44] More than a hundred ox carts were needed to move the treasures, a mere fraction of the total, to Shanghai. De Montauban was to present the Empress Eugénie with a diamond necklace Parisian jewellers declared to be beyond price. Hope Grant refused his share of the loot but for his queen accepted a jade and gold sceptre.[45] A few dogs from a closely guarded breed were saved by an animal-loving looter. Bred to resemble the Chinese heraldic lion, brave, intelligent and spoilt by a long line of empresses, some survived the journey to England. They were inevitably named Pekinese. A puppy presented to the queen, and aptly renamed Lootie, yapped away at the ankles of visiting bishops at Osborne House until her deeply mourned death in 1872.[46]

* * *

After the arrival of special wall-crushing equipment, the reputedly impregnable walls of Peking were breached with ease and the city was taken. By a near miracle, or perhaps because of the fear of the inevitable retribution, Parkes and Bowlby escaped execution. Then, on 16 October 1860, Lord Elgin gave orders for the Summer Palace and its grounds to be torched. His motives remain a mystery.[47] According to Baron Gros, his friend was by then in a state of 'acute neurosis'. He kept repeating that a symbolic act was necessary which would cost the Chinese face but not lives. A weak man at heart, he was haunted by the reputation of his father, vilified for saving great art. Or perhaps he was influenced by the fate of twenty-two British and Sepoy troops who had been captured by the Chinese a few months earlier. They had been subjected to appalling tortures witnessed by Parkes and Bowlby before all but one had mercifully expired. Their mutilated remains had just been discovered.

The Summer Palace, still packed with a superb collection of art and treasure – some had been difficult to move – burnt for five days. The grounds too were systematically destroyed. It was, until the twentieth century and since the burning of the Library in Alexandria, the most spectacular act of vandalism in history. Lieut. Col. Garnet Wolseley, the future field marshal, wrote:

> When we entered the gardens they reminded one of those magical grounds described in fairy tales. When we marched out they were a dreary waste of ruined nothings.[48]

Another onlooker, recorded that

> The sun shining through the masses of smoke gave a sickly hue to every plant and tree and the red flame gleaming on the faces of the troops made them appear like demons glorying in the destruction of what could never be replaced.[49]

But in London the news made Elgin a hero.[50] At the next annual banquet of the Royal Academy he was guest of honour. Seated between the president (who happened to be Frank Hope Grant, brother of General James) and Sir Edwin Landseer, over port and cigars he made a speech:

> I am not a barbarian. I am deeply moved by the wonderful works of art surrounding us here in this noble building . . . and let me say that no-one regretted more than me the unavoidable destruction of the pretty pleasure houses and kiosks dignified in China by the name of Summer Palace . . . The whole complex of edifices was corrupt, redolent of the sickly smell of drugs. It had to be destroyed.[51]

He was given an ovation. A few days later the government rewarded him with their most coveted prize, the viceroyalty of India.[52] The conflagration of the Summer Palace was, in the words of *The Times*, 'a fitting end to the long-drawn-out squabble over an evil stupefying drug'. The principle of free trade had been upheld. The honour of a Christian empire had been defended.

CHAPTER 18

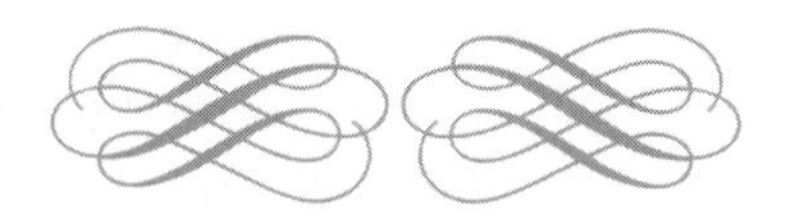

The Yellow Peril

The London Times was almost – but only almost – right. The fighting in China had ended and most newspaper readers felt that the squabble over opium, probably a triumph but just possibly a minor blemish on the task of building a great and benign empire, was now best forgotten. And yet, whatever the newspapers wrote, the opium wars were far from over: their long-term consequences would be as fraught as had been the armed conflicts and they would last longer.

This was not immediately apparent. So far as Britain was concerned, the objectives of the fighting had been achieved. Four years after the Peking Treaty seven-eighths of all of China's imports came from Britain or her colonies. In particular, between 1859 and 1882 opium imports from India to China rose from 51,000 to an unprecedented 112,000 chests.[1] Jardine and Matheson had diversified into other fields;[2] but their place was taken by the dazzling Sassoons. A prolific family of traders originally from Spain, they emerged on to the world stage in Baghdad, then part of the Ottoman Empire, in the 1820s. The founder of the family fortune was Shaykh Sassoon ben Sali, a shrewd banker whose policy of supporting destitute poppy farmers in years of hardship at low interest rates paid off. His son, David, later the family patriarch and father of twenty children, moved to Bombay, today's Mumbai, in 1832 and the firm began to spread. By the 1860s David S. Sassoon & Sons had branches run mostly by Sassoons at least by marriage in Bombay, Calcutta, present-day Istanbul, Singapore, Hong Kong and Shanghai. Their Judaism and tribal cohesion proved an invincible source of strength and the business soon dominated commerce between East and West not only in opium but also in silver, gold, silks, spices, cotton and 'whatever moves over land and sea'. But opium was the firm's mainstay, the family eventually being the virtual possessors and distributors of three-quarters of all opium grown in India. The average profit from their yearly shipment was estimated at £3 million. In 1874 the senior branch took up residence in London, Sir Albert Abdullah David Sassoon becoming a friend and obligingly amnesic creditor of the Prince of Wales, his son marrying a Rothschild heiress and serving as Conservative Member of Parliament for twenty years. It was a fairy tale, with the family's scholarly tradition – though not its involvement with opium – continuing.[3]

Against such a background pathetically worded requests from Peking to stop or at least to limit the opium trade were shuffled diligently between government departments until lost. When Peking tried to limit sales by increasing the sales tax, the legislation was squashed under veiled threats by the Liberal government in London. In 1870 Sir Wilfred Lawson introduced a motion in the House of Commons condemning the 'continued and deliberate ruination and poisoning of the ancient land of China' and he was listened to politely. But he was an amiable fossil. By then the Exchequer's income from opium exports to China had ballooned from a pre-war one-eighteenth to one-fifth of the revenue. Mr Gladstone, now in government, argued with his usual cogency that, in view of Britain's burdensome civilising mission around the globe, 'we could ill afford to lose such a resource'. The motion was defeated by 146 to 36.

As poppy cultivation spread in China, the trade for a time threatened to decline. Abandoning his former repugnance, the abolitionist Sir Rutherford Alcock, MP, with a heavy personal stake in David Sassoon & Sons, suggested flooding China with under-priced Indian opium until the natives were put out of business. The price could then be restored to its former robust level. He was anticipating the tactics of the Rockefellers, Carnegies and other twentieth-century American robber barons. In 1888 *The London Times* estimated that 70 per cent of adult males in China were habituated or addicted. Two years later, exhausted by futile protests, the fifteen-year-old emperor, under the thumb of his great-aunt, the Dowager Empress Cixi, former prostitute, imperial concubine, serial murderess and lifelong addict, revoked all laws against cultivation, trading and consumption. The laws had been so widely ignored that their revocation barely made a ripple.

* * *

For more than a century after the opium wars China ceased to be a country capable of providing her own people with the basic necessities of life. Landlords in the starving southern provinces were grasping, themselves often threatened with ruin. Taxes imposed by Peking were crippling but ruthlessly collected. Or collected from those still alive. Once the most highly organised state in the world, its roads, bridges and wayside shelters were crumbling away. Bandits roamed the countryside freely. As dams collapsed, floods devastated low-lying regions. Everything happened on a massive scale in China. Even well-intentioned ventures ended in disaster. Since the Peking Convention and mostly at French insistence Christian missionaries were once again allowed to spread the Gospel. They came from many countries and more than one denomination and they soon became aware of the ravages of opium. It was, as the German doctor-pastor Wilhelm Mott wrote, 'at the root of all the evils that beset my poor suffering flock'.[4] Unhappily, many missionary societies had picked up the notion fashionable in Europe after the discovery of morphine and later of heroin that these substances would counteract the habit-forming 'side effect' of the parent substance. There was a surface plausibility to this crass misrepresentation: since both drugs were active derivatives of opium they did abolish the craving for it. Today the naivety of the belief may seem criminal; but genuine hope of a cure (as well as hope of profits by the makers of morphine and heroin) blinded otherwise level-headed individuals. For several decades both morphine and heroin were judged to be a worthy cause for collections in churches; and, despite the incontrovertible refutation of 'morphine therapy' by George Calkins in his *Opium and Opium Appetite* published in 1871, crates of morphine paid for by the American Medical Missionary Association were still arriving in China in 1886. When at last the mistake unravelled the effect was devastating. In many parts of China addiction to 'Jesus opium' – that is, morphine or heroin – permanently undermined relations between the people and the missions. Faint echoes of the resentment are still occasionally heard.

Political turmoil added its toll. The Taiping were not finally crushed till 1864 when the 'Ever Victorious Army', a ragbag of European mercenaries and Chinese imperial troops under the charismatic command of Charles 'Chinese' Gordon, captured Changzhou Fu, the rebels' military base.[5] In defeat the Taiping spawned some of the most ruthless criminal organisations in the history of organised crime. A few still flourish.

* * *

Luckily or tragically, as conditions inside China deteriorated, demand for cheap labour in other parts of the world soared. So the coolie trade, soon nicknamed the 'pig trade', took off. It was to be less famous but no less grim than the transatlantic slave trade had been a century earlier. The first wave of migration was to South East Asia, mainly to the tin mines and rubber plantations of the Malay Peninsula. By 1910, 60,000 Chinese were living in Rangoon, 120,000 in Saigon and 200,000 in Bangkok. In Singapore they were the ethnic majority though without political representation. Desperately poor, their demand for opium overrode all other needs. By 1902 most of the adult population were addicted. On the bright side, licensed opium dens became a profitable tourist attraction and helped to lay the foundations of Singapore's prosperity.

But within a decade the coolies started to be shipped overseas. Hong Kong, Macao and Amoy were the main embarkation ports. The human cargoes were herded into corrals and imprinted with their letters of destination. (P, C and S were common, indicating Peru, California or the Sandwich Islands). Travelling conditions were similar to those endured in the past by African slaves. Or not endured. The British-owned vessel *John Calvin* lost 58 per cent of its cargo on its way to the West Indies. On American ships a mortality of 35–40 per cent was the norm. Woman were usually (and illegally even by already recognised international law) part of the trade. The British vessel *Inglewood* was about to sail from Amoy with a cargo of about a hundred girls aged under twelve. The mostly Chinese crew mutinied and their leader reported the shipment to the British Consul. Sir Harry Toogood, a devout High Church Anglican, took unauthorised action and ruined his career by arranging for the children to be set free and returned to China. But where would they go and who would escort them? A contingent of French nuns volunteered under the leadership of their superior, Mère Parduce. The group was rounded up by 'outlaws' a few miles outside the city. Mère Parduce was beheaded, the rest of the nuns were beaten and sent packing. Their charges was dispatched from another port.

Once abroad, the mass of first-generation Chinese lived short and depressed lives. Opium provided their only escape from physical and mental suffering. The drug was imported by Chinese traders but with a cut to the employers. However inhuman the slave trade had been, in most of the United States the Africans had enjoyed a measure of legal protection. The coolies working in the gold mines of California enjoyed none. When the most dangerous seams had been exhausted, they were cut adrift. Lucky survivors and those not incapacitated by crippling lung disease drifted into small-time opium dealing in the towns.

* * *

It was through the coolie trade that opium diffused around parts of the world previously untouched. It took some years for governments to recognise the mischief. In Australia, virtually opium-free before the mid-nineteenth century, 800 kilos of the drug were imported in 1890. By then Sydney's Lower George Street and Melbourne's Bourke Street were crowded with opium parlours. As usually happens, the development made the Chinese not only exploited but also hated. In 1888 the arrival in Port Melbourne of the SS *Afghan* with 250 Chinese migrants provoked a riot and the landing was blocked. What happened to the cargo? No records exist. Similar scenes were enacted in Lima, Cape Town, San Francisco, Amsterdam, Liverpool and presumably elsewhere.

What alarmed the host governments was the spread of the drug to the white population. In Victoria State by 1890, 700 'European' users were 'known to the authorities'. The development gave birth to a new expression. Nobody knows where the term was coined but it was soon understood everywhere. The Yellow Peril was the new monster instantly recognisable in a thousand cartoons. These showed a cadaverous evil apparition extending a grasping claw dripping with blood, sometimes plucking up little white figures of both sexes and all ages. Inevitably the fear and loathing were most virulent in countries previously uncontaminated. Australian Edward Dyson's story, 'Mr and Mrs Sin Fat' – typical names – described how a curious traveller accidentally discovers Mr Sin Fat's abode in an Australian city:

> Through the clouds of choking evil smelling opium fumes, I beheld debilitated Chinamen with faces like animals floating to hell in the midst of visions of heaven . . . and, worst of all, European girls of fifteen and sixteen, perhaps younger, decoyed in at the front door by the sheen of silk and the jingle of gold, and then left to percolate through that terrible place, irreversibly besmirched, to be finally cast out among the slime and rottenness of some back lane . . .[6]

Perth was edging ahead to become the biggest Chinatown, the future 'Dragon City', of the Antipodes.

* * *

In his last, unfinished novel, *The Mystery of Edwin Drood*, Dickens gave vent to his dislike of opium. (Or love–hate relationship. He had been dosing himself to calm his nerves during his last lecture tour in the United States.) The main character, John Jasper, one of his most rivetingly repulsive creations, is a respected choirmaster of Cloisterham Cathedral but leads a double life. An opium addict in his second and secret incarnation, he courts the betrothed of his disappeared nephew whom he may have murdered. All his and other people's money goes to feed his addiction. Dickens claimed to have researched opium dens, visiting East End slums with the police; but not many such dens in London actually existed at the time. No matter: they existed elsewhere and in Dickens's imagination. As Drood wakes from his opium-inspired dream and tries to collect his 'scattered consciousness' he finds himself in

> the meanest and closest of small rooms. Through the ragged window curtain the light of early day steals in from a miserable court. He lies, dressed, across a large unseemly bedstead that has indeed given way under the weight upon it. Lying, also dressed and also across the bed, are a Chinaman, a Lascar and a haggard young woman. The first two are in a sleep or stupor, the last is blowing at a kind of pipe to kindle it, shading it with her lean hands . . . 'Another', says an old crone next to the bed in a rattling whisper, 'Have another my dear!' He looks about him with his hand to his forehead. 'Ye smoked as many as five since ye come in at midnight', the woman goes on. 'Good boy. But poor me, poor me, my head is so bad. The business is slack, is slack! Few Chinamen about the docks and no ship coming in these days. Here's another ready for you, deary. But remember like a good soul, won't ye, that the market price is drefy high just now'.[7]

In Sir Arthur Conan Doyle's 'The Man with the Twisted Lip' and in Oscar Wilde's *The Picture of Dorian Gray* the incarnations of the Yellow Peril are Malays; and to Rimbaud they were often just 'petits jaunes'; but the message was the same. These were quality literature and a fraction of the total. Most outpourings were penny-dreadfuls; or rather, they cost a penny and were dreadful but they effectively mixed opium, crime, sex and orientals; and they were devoured. On a foray to Limehouse in east London the writer Arthur Henry Ward befriended a Chinese sailor known only as Wu; and 'my Scheherazade' implanted a series of gripping tales in his mind. One day Wu did not turn up for their assignation and Ward never saw him again; but the stories bore unexpected fruit. Publishing under the name of Sax Rohmer, Ward introduced the arch-villain, Dr Fu Manchu. The 'slit-eyed doctor' embodied all the evils the public saw in opium and orientals. Set on extinguishing the white race and dominating the world he uses weapons ranging from missiles to animal magnetism but his main strength lies in opium. He himself, 'like most Orientals', consumes the drug with impunity but he uses it with unequalled guile to enervate his enemies. Not, as an eminent physician was heard to comment, an ornament to our profession. The settings of his exploits switch from embassy balls, ocean liners and royal hunting lodges to repulsive dumps of the underworld. (The use of the sewers as a secret network of criminals as immortalised by Graham Greene in *The Third Man* was probably pioneered by Victor Hugo but first fully exploited by Sax Rohmer.) The doctor is always foiled by the brilliant Sir Denis Nayland Smith and

Dr Petrie (aided by a moronic troupe of Scotland Yard detectives) but he always survives to plot another day.[8] The stories became money-spinners: many were successfully filmed. They provided Peter Sellers with his last sparkling cinematic role.[9] Improbably, a few of the tales were based on real life, the fictional Kung being modelled on the notorious drug baron Brilliant Chang.

The Yellow Peril sold newspapers too. The Chinese population of London was only about 1,300 in 1910, and just 3,000 in 1921, minuscule compared to that of San Francisco or even Paris (where students at the Sorbonne included Mao Zedong, Chou En-lai and Ho Chi Minh) but that was enough to inspire a stream of first-hand reports. The accounts 'from life' of enslaved white girls in the grip of the yellow fiends and their drug would make a modern tabloid newshound blush.

The phobias and horror stories may have been reinforced by a lingering feeling of guilt. For about ten years after 1870 the Anglo-Oriental Society for the Suppression of the Opium Trade published a speciously titled journal, *Friend of China*, which combined pseudo-Christian breast-beating with horrific stories illustrating 'the vengeance' of the Chinaman descending on and corrupting white society. One of the Society's stalwarts, the Reverend George Piercy warned against any sense of 'false security':

> Those who have been claiming justice for China for the opium traffic at the hands of our government forget about the consequences of the supposedly retributive action . . . What could all this grow to but the plague spreading and attacking our vitals? . . . It begins with the Chinese in our midst but does not end with them.[10]

* * *

In the United States the Hearst newspaper syndicate fanned the flames of hatred. One circumstance helped. In 1874 America toppled over into an economic recession. The cause was, as usual, the greed of entrepreneurs in New York, Boston and Chicago. They themselves quickly recovered; but to ordinary Americans the temporary end of prosperity – they did not know that it would be temporary – was a seismic experience. During the despairing 1870s, as unemployment rose, it became accepted that the Yellow Peril not only corrupted white women but also robbed white men of their jobs. The charge contained a grain of truth. Building the Transcontinental Railways only coolies would risk spanning yawning ravines with heavy steel girders, blow tunnels in the rocks with cheap and untested explosives, trudge over scorching salt flats, face Indians trying to defend their land and be stung and eaten by despoilt wildlife. Commodore Cornelius Vanderbilt could not have found white workers to do these jobs, let alone paid them the coolie wages which made his obscene fortune. Opium was the key to the bargain. The Chinese addicts would do anything so long as the drug was provided. Even so, the cost in lives ran into tens of thousands. Of the 800 Chinese shipped in to work on the trans-isthmus Panama railway 468 committed suicide within three days of arrival. But to the Yellow-Peril press truth and facts mattered little:

> Most of us know vaguely [intoned a typical rant] about the colony of 'Celestials' who cluster about the lower end of our wonderful Bowery, but there are not many who know of the hundreds of American girls who are drawn into it each year from tenement houses and cigar factories to become associates and then slaves of the Mongolian. It is opium which captures the women and keeps them in slavery.[11]

And if the fiend could go after white women, what could keep him from seducing the white man's children?

> In our great city of San Francisco, young boys, yes, and little girls, with the look of cunning old men and women, sneak out of the vile alleys of the Chinese quarter into our beautiful sunshine and refreshing sea-breeze, with expressions of vice, duplicity and greed only to corrupt our own children. They carry the curse of China, opium, as their weapon. They and their poison must be rooted out before they will decimate our own youth and emasculate the coming generation of Americans.[12]

In times of economic stress any garbage is believed; and the gutter press was not the only organ to disseminate it. The labour leader, Samuel Gompers, a European immigrant still revered for his commitment to the American working man and woman, claimed to have found Chinese laundries in California 'pollulating' with white orphans and kidnap victims,

> tiny lost souls . . . forced to yield up their virgin bodies to their maniacal yellow captors . . . What other crimes are committed in these dark fetid places when these little innocent victims of the Chinaman's wiles were under the influence of the drug opium is too horrible to imagine. There are hundreds, aye thousands, of our American girls and boys who have acquired this deathly habit and are doomed, hopelessly doomed, beyond redemption.[13]

And a well-respected San Francisco doctor, Winslow Anderson, wrote:

> It is a sickening sight . . . young white girls . . . lying undressed or half-undressed on the floor or on couches, smoking with their Oriental 'lovers', men and women, in these Chinese smoking houses.[14]

* * *

In China the people's plight hardly changed after the proclamation of the Republic by Dr Sun Yat-sen 1911. His high-minded but softly-softly approach to the eradication of opium achieved nothing. By the 1930s his successor, Chiang Kai-shek, used opium taxes to bankroll his corrupt regime. Chiang himself and his wife loathed the drug and perhaps he had little choice; but the country's state of health was no secret from its neighbours. European observers, by contrast, watched with disbelief the collapse of the once-mighty empire under the Japanese onslaught of 1937. The Second World War saved China from total defeat;[15] but lavish American aid could not save the Kuomintang from being routed by the Communist armies of Mao Zedong. Was it faith in Marx, Lenin and Mao which made Mao's armies invincible? Or their strict outlawing of opium?

In 1950 the new Communist government forbade the cultivation, use and sale of all narcotics; and this time the ban was deadly. Traders and dealers in opium were executed – tens of thousands by any account – or sent to starve in China's Gulags. Users expressing a wish to reform were treated more humanely – detoxification clinics anticipated those of the United States – but hopeless or unwilling cases and relapses (which were the majority) were exterminated.

The fate through these convulsions of Wan Jung, wife of the last emperor, Puyi, was symbolic. She had begun to smoke opium when she was sixteen and eventually needed two ounces a day, enough to kill a beginner. After her husband collaborated with the Japanese by becoming puppet emperor of Manchuria (renamed Manchukuo) the Japanese encouraged her addiction. Her story was widely publicised in Europe and the United States, illustrating the moral decay of China. But her next captors, the Communists, had different ideas. In 1946 she and her husband were separated and her last days were spent in a cage. Descending into the hell of sudden withdrawal, she became an educational exhibit. Classes of schoolchildren were dragooned past her. A film crew recorded her sufferings. These were spectacular. She sobbed and screamed until other prisoners pleaded for her execution. Then she lapsed into violent delusions, imagining that she was back in her palace in Peking, ordering servants and courtiers about. Guards eventually refused to enter the stench and the filth. In 1987 Bernardo Bertolucci, Marxist-inclined director of the film, the *The Last Emperor*, was persuaded to cut the last scenes. No faint-hearted European audience, he was told, could bear to watch them.

* * *

In 1960 the Communist regime declared that the former Celestial Kingdom was free from opium for the first time in 150 years. For some decades at least this was probably true. The declaration marked the real end of the opium wars.[16] They had caused unimaginable suffering. They had been motivated by pure greed. In the West their memory has faded. In China it has not. Nor has their legacy – much of organised crime in the West and stubborn xenophobia in the East – disappeared.

CHAPTER 19

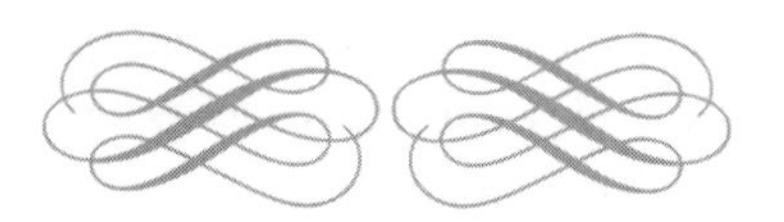

Doctors rule

ROUND DATES rarely coincide with significant historical events, but 1850 marks a convenient divide. The defeat of the revolutions of 1848–49 demonstrated the futility of the 'divine discontents' of the Romantics.[1] Romantic notions lingered on in literature, music and the arts – *Les Misérables* and the *Ring* were still to come – but whatever relevance they had ever had to everyday life had evaporated. What was left to guide ordinary men and women through the next half-century was science, displaying unbounded confidence, replacing woolly conceits like human brotherhood, liberty, the soul and even the victory of the proletariat.[2] At least science worked. The change affected all walks of life, none more so than health. Within a decade or two general anaesthesia and antisepsis transformed surgery from butchery into a healing art. Microbiology revealed the causes of some of the most feared medical diseases. In earlier centuries advances in the basic sciences had taken generations to yield practical results. By the end of the nineteenth century the clinical benefits were evident almost at once. Illnesses like rabies, typhoid and diphtheria became preventable, sometimes even curable. An open fracture no longer called for immediate amputation. (Tuberculosis and syphilis were the only two major infections which held out for half a century after the discovery of their causative organisms.) The replacement of a pleasant-tasting tincture of uncertain composition by a pure substance and the syringe and needle fitted this pattern of scientific precision.

* * *

Because morphine is only a 'kind of opium', the impact of these changes is still sometimes overlooked. Most significantly, the injection of morphine usually (though not of course necessarily) meant an increase in dosage. One ounce of good-quality laudanum contained approximately one grain of morphine. That meant that even opium eaters taking 2 ounces of the liquid a day were dosing themselves on considerably less than most injecting addicts. Though the latter generally kept themselves to a daily intake of 6 grains, some took as much as 20 or even 40. The last was the rough equivalent of 38 ounces of laudanum.[3] By comparison, the regular dose of the heavily addicted Coleridge was 38 ounces (2 pints) a *week*, as much as a long-term addict could afford to take. Injected morphine also eliminated the uncertainty of intestinal absorption. Even with a reliable tincture this rarely exceeded 80 per cent and could be less than 20 per cent. (In states associated with diarrhoea when the effect might actually be beneficial it could fall to zero.) And though morphine in tablet form was no better absorbed than the linctus, it was handier and, by the 1860s and '70s, significantly cheaper.

Two wars of the mid-century provided showcases for the new form of opium. Only professional armies fought in the Crimea, the numbers were comparatively small, and, except for Russia, the devastation took place thousands of miles from home. The suffering of the troops was often hideous; but less than one in fifty families in Britain and France had a son, husband or father among them. The use of opiates was never officially admitted. The War Office disapproved of 'stupefying drugs' but issued no outright prohibition. It was one of the few topics on which even Miss Nightingale never expressed an opinion, at least not in public. Privately she wrote to her friend, Harriet Martineau, that often nothing relieved her weariness but 'a curious newfangled little operation of putting opium under the skin with a needle'. She seems to have dosed herself throughout her stay in Scutari and perhaps for much of the rest of her long life as a professional invalid.[4] She also repeatedly asked for more painkillers in letters to her friend, the secretary of state for war, for her wounded charges. If there was more than patriotic morale boosting in the famous engraving in the *Illustrated London News* of 24 February 1855 which first showed her as the legendary lady with the lamp doing her round at night between two rows of quietly sleeping – or comatose – patients, then opium and morphine must have been liberally doled out to the sick as well. In various stages of hospital fever these patients would otherwise have been noisy and delirious: the scene would have been pandemonium.[5]

In his chilling dispatches William Russell of *The Times* mentions morphine in connection with one outbreak of cholera. He thought that the drug saved lives but there was never enough of it. Field surgeons in the French though probably not in the British or Russian army used injections of morphine. Drugs were also hawked around by a ragged army of camp-followers who materialised after each battle: Lord Cardigan estimated that occasionally they were three times as numerous as the men in uniform. They sold everything but mainly liquid happiness: the ancient poppy fields of Anatolia and the more recent ones of Bulgaria were just around the corner. In one of his orders of the day Maréchale Pélissier commanded that surviving but mortally wounded French other ranks should be given 'une grande dose de morpheine'. The Russian practice was to bayonet them.

The American Civil War was a catastrophe on a different scale. More than 620,000 were killed in the fighting, ten times more than in the Crimea; and 50,000 survivors returned home as amputees. Nor was the theatre of war some remote peninsula. It was the American homeland, and families suffered as much as the fighting men. It was also the first war in which morphine became a fixture on the battlefield. Garden patches of white poppies sprang up in both Confederate and Union territories, Virginia, Tennessee, South Carolina and Georgia becoming the main early providers. Without expertise the yield of opium was low but few tried to make a profit on it. And soon tablets and injectable morphine were available, at first imported from England but increasingly manufactured in small workshops on both sides of the political divide.[6] So were cheap metal syringes with fixed needles, difficult to clean and often blunt but bulletproof. By the end of the first year of the fighting morphine was administered on a massive scale in both armies, the supreme calmer of shattered nerves and broken bodies. Surgeon Major Nathan Mayer did not bother to dismount to dispense the liquid: he poured out the 'required dose' into his palm and let the wounded slurp it up. Or he dished out the tablets when available by the fistful. Exhausted, he kept the last ones for himself. Over 10 million pills of morphine and 2 million ounces of opiates as tinctures and powders were issued to the Union Army; and the amounts consumed on the Confederate side were probably not much smaller. As would happen after later conflicts, the effect of the practice would outlast the war. Tens of thousands of veterans returned home with lingering gastrointestinal disorders for which the only remedy was morphine. In 1868 Horace Day wrote in *The Opium Habit*:

> Maimed and shattered survivors from a hundred battlefields, diseased and disabled soldiers released from hostile prisons, anguished and hopeless wives and mothers made so by the slaughter of those who were dearest to them, found, many of them, [that] temporarily relief, the only relief, from their suffering is morphine.[7]

* * *

In other minor wars too opium was sometimes the only winner. The terminal whimpering of the wounded after the Battle of Solferino in 1859 horrified a visiting Genevese banker. What distressed Henri Dunant most was the lack of opium or morphine to ease the suffering even if little could be done to save lives. His protestations were met with incomprehension. Who would pay for the dope and who would deliver it to the wounded? The outcome was the foundation of the International Red Cross, its original function the provision of painkillers, especially morphine, on the battlefield.

* * *

In Britain as in most West European countries the pressure for some kind of regulation had been gathering strength. The Registrar General's earliest reports – his office, the first of its kind, was established in 1837 – presented infant mortality due to drug overdose as an everyday event. They elicited impassioned pamphlets from pressure groups like the Ladies' Sanitary Association. *The Massacre of the Innocents* and *Murder in your Backyards*, were not titles easily ignored. A reassuring whiff of class-consciousness clung to these publications. They suggested – almost certainly wrongly – that it was the working-class families, especially the young, who needed to be saved: the educated middle and upper classes could by and large look after themselves. After lengthy debates and two failed attempts the Poisons and Pharmacy Act was passed by Parliament in 1868. It was pioneering legislation but at the time pleased nobody. To make it acceptable the claims of too many legislators, doctors, drug manufacturers and other sectional interests had to be satisfied. The result was a legal construct of great verbal ingenuity but full of practical obscurities. It specifically excluded proprietary medicines containing opium, including most baby-calmers: they would not be brought within the law for another thirty years. Perhaps as a result of publicity infant mortality from overdose according to the Registrar General's figures nevertheless fell from 20.5 per million of the population in the mid-1860s to 6.5 per million in the mid-1880s. The prescription and sale of opium and its derivatives (among fourteen other substances designated as poisons) were meticulously regulated; but the penalties imposed for contravening regulations were laughable. Nobody would be prosecuted under the Act for twenty-eight years. For another twenty years fines were mostly ignored and never followed up.

This reflected not incompetence but confusion. Even dedicated anti-opium campaigners faced a dilemma. They regarded opium as an unmitigated evil. On the other hand, the idea that the law should prohibit the use of a drug by responsible adult citizens went against the liberal grain. Even to the most uncompromising pamphleteers the notion that the mere possession of the drug – or of any drug – should be a criminal offence would have seemed lunatic. It was clear that somebody would have to make difficult decisions; but who? In the event the Act achieved three changes. It introduced the skull and crossbones as a memorable logo for poisons. It re-endowed opium and morphine with the enchantment of forbidden fruit. And, most important, it made members of the medical profession legal arbiters.

The last was a new concept whose implications were almost certainly missed by the legislators. After 1868 a person caught in possession of a 'controlled' drug like morphine became either a criminal or a patient. Which it would be depended on the medical evidence. The innovation was part of a wider trend, the reason perhaps why its significance was not grasped. The criminal insanity plea to decide between a guilty or guilty-but-insane verdict was another new judicial wand to be wielded by doctors. It created a new status for the profession, giving them powers in areas in which the distinction between truth and opinion had always been uncertain. It was not a deliberate conspiracy.[8] Recently formed medical bodies like the British Medical Association, a trade union in fact, and the General Medical Council, a self-regulating authority, gave the profession political weight and a certain moral credibility; but they still faced competition. The similarly up-and-coming Pharmaceutical Society was keen to impress on both public and government that qualified chemists were a learned profession uniquely placed to control drugs. This, they agreed with their medical adversaries, should be taken out of the hands of grocers, tobacconists, wine merchants and pet-food suppliers but it should be handed to them, not to doctors. Doctors, they argued, were too busy, too distracted and, above all, too beholden to their patients. After all, they depended on their patients for their livelihood. As one leading pharmacist put it, this 'delicate relationship might unconsciously sway their much-preoccupied minds'. The argument was not without substance, but it failed. The alternatives were never debated in Parliament and not much outside it. The pharmacists scored initial victories but lost because their social prestige was lower than that of doctors. The powers which the Act bestowed on doctors were unspecified but enormous. They included – or would soon and inevitably include – a monopoly of the opium and morphine supply to patients, laws to punish any circumvention of that monopoly and an unwritten but impossible to resist power to coerce individuals to undergo medical treatment.

The outcome made a few thoughtful practitioners like James (the future Sir James) Paget uneasy. 'We are craftsmen, not priests, philosophers or policemen . . . We are not here to provide reach-me-down moral judgements in matters non-medical.' But the James Pagets were few. What made the development inevitable was timing. Medicine was riding on the crest of success. From being effective in dealing with previously fatal diseases like diphtheria or rabies it seemed only a small step to claim that addiction was a medical problem which nobody less than a fully qualified doctor was competent to deal with. Fifty years earlier such nonsense would have been ridiculed by the profession itself. Doctors were healers of the sick, not legal panjandrums or keepers of the nation's conscience. Now their new role was accepted as a matter of course. And it had a coincidental but far-reaching consequence. In Britain at least addiction to opium and morphine was henceforth to be categorised not as a habit, as an aberration, as a sin or even as a crime but as a disease.

This too was new. Alcoholism, except in some rare forms, had never been an indication for medical intervention. Family, friends, priests, solicitors and commanding officers often took a hand. Doctors rarely. Nor would cigarette smoking ever achieve the status of an illness. Of course the profession had much to say on both topics – some of it sensible – but dealing with them remained the responsibility of politicians, preachers, teachers and public-spirited citizens. Morphine addiction, by contrast, became – and more surprisingly remained – medicalised. More by accident than by design, several events reinforced this development.

In 1857 a French psychiatrist, Bénédict Augustin Morel, nominated 'la morphéine' as the overriding danger facing the human race. In his long and now unreadable best-seller, *Traité sur la Degénération*, translated into English within a year of publication, he asserted that far from progressing onwards and upwards towards an eventually perfect fitness, the species, or a large chunk of it, was being reduced to creeping dysfunction. Even ordinary citizens, apparently far removed from the world of drugs, could and would become victims. And that inevitably, he claimed in the very first paragraphs, would, within a few generations, lead to their extinction.[9] Today the grimly prophetic preamble might introduce a discourse on pollution, genetic engineering, terrorism, climate change or some other topical doom. In the post-Romantic decades the menace was 'paralysis of the moral sense'. Or rather, this was the *immediate* menace. The paralysis would not remain moral for long. It would be followed by initially slow but eventually precipitous physical decline. And the chief cause was plain. It was not just the stupefying action of opium and its even more deadly derivative, morphine. Critically, in Morel's view, it was their *trans-generational* effect. For the drug was destructive not only of the addict: more existentially, its imprint would be passed on to their offspring.

The doctor was not subject to debilitating modesty. To what he claimed to be his uniquely observant eyes – though not yet perhaps to the ordinary practitioner and certainly not to laymen – this insidious plague had been providentially revealed. True, hard evidence was still lacking; and the reason for that was clear. Nobody had been looking for it. But the catastrophe would soon be obvious. It might then be too late. The book elicited a frisson of recognition. Many could think of unrecognised examples of the fateful inheritance in their own circle of acquaintances. ('I always thought that Aunt Sybilla – or Uncle Philibert – was more than a little odd.') More dramatically, as the century progressed, Morel's doctrine seemed to mesh with the revolutionary new ideas of Darwinism. Eminent scientists suggested that what the Frenchman had uncovered was the flip side of evolution, the *decline of the unfittest*. It was up to doctors to stop the rot before it was too late.

The warning carried the more weight since it originated in France, hotbed of all unnatural vices (as well as home of the syringe). No legislation there outlawed drugs or drink till 1889; and, though what was finally enacted was on paper more stringent than regulations in Britain, compliance with the law across the Channel – as with strict prohibition on public spitting, another French innovation – was shockingly slack. Except in matters literary. Though illegal, opium was brazenly consumed in every cafe and bar; a collection of incomprehensible verse by M. Charles Baudelaire entitled *Les Fleurs du mal* (Flowers of Evil) (1856) was promptly banned.

Yet, despite formal and mostly ineffectual bans, the *fin de siècle* blossomed in France into one of the literary golden ages of the drug. A palmful of the avalanche of novels, short stories, biographies and verse remain literary gems. Guy de Maupassant experimented with morphine (among many other drugs) in his struggle with syphilis and his descent into madness; and his hauntingly beautiful *Sur l'eau* is among them.[10] A few, like Alphonse Daudet's *L'Evangeliste* and Maurice Taine's *La Comtesse Morphine* have literary merit.[11] The satire by Meg Villars and Willy, *Les imprudences de Peggy*, sheds malicious light on Colette and her morphine-addicted lover, Missy.[12] Jean Lorrain was a tragic and not entirely unworthy disciple of Baudelaire who styled himself 'l'ambassadeur de Sodom à tout Paris'. It was the sad conceit of a dying consumptive. 'I have the forgiving soul,' he wrote in a less grandiloquent mood, 'of a man who must die soon. These desperate injections of morphine must end.'[13] He was thirty-nine when an intestinal perforation ended his sufferings. Edouard Dubus, co-founder of the *Mercure de France*, lived a debauched life and died from an overdose in a *chalet de nécessité* in the place Maubert. He was thirty-one.[14] His friend, Victorien du Saussay's *La Morpheine*, is stark even for the age of Zola, mixing addiction, incest, adultery and sadism in about equal proportions.[15] As usual, the vast bulk of the avalanche was titillating trash.

Fin-de-siècle theatreland too on both sides of the Channel was soaked in drugs.[16] Sarah Bernhardt, the divine Sarah, was renowned for dosing herself for a variety of *douleurs* but in fact for first-night jitters. Sometimes opium made her forget her lines. In London she famously omitted a chunk of 250 lines from Dumas *fils*'s sensational *L'Etrangère*, shortening the play by twenty minutes and eliminating a vital clue. The author gallantly assured her that his creation was much improved by the mauling. He was probably right. But years later Sarah wrote in her autobiography:

> The opium I had taken made me arrive on the stage in a semiconscious state, delighted with the applause I received but not quite sure what the applause was about . . . My feet glided along the carpet without any effort, and my voice sounded to me as if it came from a great distance . . . I was in that delicious stupor that one experiences after morphine but afterwards I felt bad despite the applause and the curtain calls.[17]

Her acclaimed partner on the stage, Lucien Guitry, 'first and probably forever the greatest Cyrano' (according to the critic Jean Gauthier), was a user before performances though probably not an addict.

Oddly perhaps, the geniuses who still make the beautiful epoch loved outside Francophonia – Manet, Monet, Pissarro, Renoir, Sisley, Berthe Morisot, Rodin, Toulouse-Lautrec, Cézanne and other painters and sculptors – remained untouched by the narcotic tide: even Gauguin, social rebel living in faraway Tahiti, disdained drugs. And Degas grumbled about *morphinomanie* 'ruining the best cafes'. Was it the smell of turpentine which kept the *fée* at bay? Or the common sense which visual artists have always claimed to possess in greater measure than scribblers, thespians or musicians? Whichever, the tradition would survive into the twentieth century.

* * *

In Germany and Central Europe regulations were generally enacted piecemeal and often in the face of considerable resistance. In Berlin the battle was particularly embittered. Professor Rudolf Virchow, 'the pope' of medicine, was also a leader of the parliamentary opposition and on matters of health an oracle. Aware of the depredations of stupefying drugs he argued passionately for their medical control. To most deputies' surprise, the Chancellor, Prince Otto von Bismarck, would have none of it. What people cared to swallow or inject into themselves was their own business. The mantle of defender of individual freedom sat strangely on the great man's shoulders: not many realised that the colossus never ventured into the Chamber without fortifying himself with an injection of morphine.

* * *

Dosage was another circumstance which worked in favour of medicalisation. Unlike the window for most drugs (including drugs of addiction), the window for an effective but relatively safe dose of morphine was dangerously narrow. Cannabis and later cocaine users could and often did take many times their usual dose without coming to any harm. But only two or three times the usual amount of morphine could kill, usually from respiratory arrest. It made the drug the most favoured means of suicide in women and the second most favoured (after the revolver) in men. It was also the friend of murderers. Unlike killers with rat poison, strychnine or meat cleavers, they may often have remained undetected. Often but not always. In France Dr Edme-Samuel Castaing went on trial for poisoning with morphine his cousin Claude-Auguste Ballet to inherit the family fortune. In the course of the trial it transpired that his alleged victim's mother, father, uncles and brother had also died in similarly suspicious circumstances.[18] But absolute certainty was never possible: no blood or urine tests existed to detect opium, morphine and heroin till the 1930s.[19]

* * *

Though during a honeymoon period with the subcutaneous syringe doctors tended to teach their mostly middle- and upper-class patients how to inject themselves – not a great deal of tuition was required – by the 1870s the dangers of self-injection were being recognised. Not only was it easier with injections to administer or self-administer an overdose but the fabulous discoveries of the new science, microbiology, were filtering through, and the transmission of invisible germs became a subsidiary spectre. The latter was perceived as a threat especially to the labouring classes whose ideas of hygiene were considered by their social superiors to be deplorably lax. To the author and campaigner in good causes, Mrs Humphry Ward, née Arnold, 'my syringe and needle, like my toothbrush and prayer book, are sacrosanct. I would not dream of allowing even my best friend to use them'.[20] In fact, morphine never caught on among miners and factory workers as laudanum had done fifty years earlier. It was not fear of microbes. Drinking a sparkling liquid or swallowing a lump of jelly came naturally. Scratching the skin as in vaccination seemed no more than eccentric. Needles and syringes were something else. They were of course right.

Even among the middle and upper classes, doctors soon started to think, injections should revert to being a medical skill or at least be performed under medical supervision. One branch of medicine in particular claimed to speak with authority. On the continent by the end of the century the new specialty of *Nervenarzt* had become one of the most lucrative (and therefore competitive) divisions of private practice;[21] and it was beginning to flourish in Britain too. (A distinct name for it in English was never invented, the literal translation, 'neurologist', generally referred to specialists in organic nervous rather than in mental diseases.[22]) In 1877 a Berlin doctor and leading light of the new expertise, Edward Levinstein, who had been operating a clinic for addicts for twenty years, published a slim volume with the title *Die Morphiensucht*. It opened a floodgate of publications in a dozen languages. The title itself, rendered into English as *Morbid Craving for Morphia*, spawned a dozen new descriptive categories. 'Morphinists' were people who had acquired the habit through medical treatment and were unable to rid themselves of it. 'Morphinomaniacs' were those taking the drug for pleasure and disinclined to give it up. 'Narcomaniacs' were organically mad individuals whose mental disorder happened to focus on drugs. Other more or less appropriate designations cropped up from time to time. The trend would reach its terse culmination with the American 'dope fiend'.[23]

Even more impressive if also confusing to the layman was the claim of some *Nervenärzte* that they could recognise an innate constitutional predisposition to addiction in individuals not *yet* addicted. Those most likely to be tainted by latent insanity included most 'brain workers', homosexuals, alcoholics, Dr Lombroso's 'criminal types' and the hereditary feeble-minded. The catch-all initial diagnosis, hinting at an ethical as well as a medical dimension, was 'moral insanity'. But the insanity would not remain just moral for long. Pawning the family silver to satisfy the terrible craving would soon follow.

* * *

The half-century brought another change. Even among those who deplored all drugs laudanum was never without a hint of glamour. None of that ever clung to the injection of morphine. Perhaps it was the chill of the technique, the syringe and the needle so lacking in romance. Perhaps it was a shift in social attitudes. The Prince Regent and his cronies no longer set the fashion. Gravity and decorum were the natural habitat of the Victorian gentleman. Modesty was one of the highly prized virtues of their ladies. With the change came the sudden separation of private and public lives. What one did in private was one thing. What one did – let alone talked about – in public was another. The after-dinner cigar was the consecrated indulgence of gentlemen.

The sip of liqueur was the acceptable naughtiness of the ladies. Injecting morphine was, by contrast, one of the unmentionables. Even among the young and feckless being a slave to a drug was no longer *chic*.[24] This did not mean that addiction was on the wane. The reverse was true. Morphine by injection was spreading; but it was spreading with a blush.

Wilkie Collins, writer addict of the second half of the century, was nine when he overheard his mother talking to Coleridge. The poet was tearfully complaining about his struggle against opium. Mrs Collins, a no-nonsense woman who had spent much of her married life among no-hope bohemians, comforted him: 'Mr Coleridge, do not cry. If the opium does you good and you must have it, why don't you go and get it?' The exchange made a deep impression on the boy. A few years later he observed his father, William Collins, RA, finding relief in Battley's Drops, a form of laudanum, in his last painful illness. Unsurprisingly perhaps, Wilkie himself turned to the drug when he developed aches and pains in his limbs and eyes, probably a form of rheumatic fever. Seeking temporary relief turned into a lifelong enslavement first to opium and then to morphine. Incapable of holding a steady pen, he dictated *The Moonstone* to a secretary. Most of her predecessors had bolted, unable to face the writer now groaning over his craving, now crying over his attempt to resist it. That the thought of the drug never left his mind was understandable: the plot, complex but exciting and well crafted, turns upon the hero purloining a magnificent diamond, the Moonstone, in a drug-induced trance. But unlike Coleridge, De Quincey and other literary figures of an earlier generation, Collins derived no heroic aura from his addiction. His books were successful, Dickens befriended him, and he was never destitute; but short of stature, pot-bellied and bushily bearded, he never cut a Byronic figure. And he had no illusions about it. He boasted to his fellow novelist, Hall Caine, that morphine was a splendid stimulant of the imagination; but when Hall Caine asked him whether he would recommend it to a friend Collins protested: 'No! Never to a friend! Never! Never! It is a terrible menace! Don't ever believe anyone who tells you otherwise.'[25] Though medicalisation made addiction a disease, its status was unique. Doctors would advise but it was up to patients to cure themselves. What they needed was backbone, a body part admired by the Victorian professional classes above all others.

* * *

By the 1870 some doctors' leaders began to express concern. Getting hooked on injected morphine was more fraught than what now seemed like 'good old-fashioned laudi'; and withdrawal from it was more problematical. Too many youths from good families were becoming addicts. Too many physicians were beginning to regard the syringe and a box of phials as professional adjuncts, like the stethoscope or thermometer. In Germany Hans von Niemeyer noted in 1877 that it was common for practitioners to depart on their daily rounds with a syringe and a full bottle of morphine in their frock-coat pocket and come home with the bottle empty.[26] This he deplored as un-Hippocratic, as indeed it was. In Britain such cautionary comments were still rare. On moving to London the young up-and-coming Yorkshire physician T. Clifford Allbutt was both alarmed and incensed by what he saw as the 'London doctors' epidemic'. He was advised to leave well alone, but leaving well alone was not in his nature:

> Gradually the conviction forced itself upon me that injections of morphia, though free from some of the ordinary ills of opium eating, might nevertheless create an even worse artificial want and gain credit for assuaging a restlessness and depression of which it was itself the cause . . . If this is so, we are incurring a grave risk in bidding people to inject themselves whenever they feel they need it, and telling them that injected morphia can have no ill effects so long as it is done with their doctor's blessing and brings with it temporary tranquillity and well-being.[27]

Eminent colleagues chastised Allbutt for 'impairing the confidence of patients in their doctors', an infallible indicator of something seriously amiss. But in support of Allbutt factual reports began to describe cases of dramatic collapse of addicts misled by their doctors and the sufferings caused by attempts to cut the habit.

Addiction could be ruinous morally too. For the first time morphinism leading to a life of crime began to feature even in respectable newspapers. Few doubted at the time that Jack the Ripper, the East End serial murderer of prostitutes, was a high-born perhaps even royal addict. And little publicised, indeed deliberately suppressed, one particular trend was beginning to cause concern among the upper echelons of the medical royal colleges.

It seemed that doctors were both carriers of the plague and its most frequent victims. In 1885 the American physician J.B. Mattison made the shocking statement that most morphine addicts in the United States were doctors and suggested that about a third of all licensed practitioners in New York were addicted.[28] In Britain reputable surveys, notably one by the opium expert and apostle of the 'painless renunciation by faith', O. Jennings, concluded that doctors, nurses and their families formed the bulk of the seriously habituated.[29] Their life was often stressful, their access to drugs easy and in times of crisis the temptation to seek 'chemical relief' hard to resist. Jennings knew: he had been an addict himself. The problem suddenly appeared more widespread than anyone had suspected. One survey conducted by Dr Louis Lewin, the Berlin psychopharmacologist, estimated that doctors in Germany accounted for 40 per cent and their wives for another 20 per cent of addicts.[30] Lewin was an honest practitioner, not a publicity seeker, and his revelations hit the headlines.

But headlines did not solve problems. No simple way of stopping the medical epidemic was in sight. Denial was an option but not, in the long term, an effective one. From time to time a scandal would lead to demands for 'drastic action'. Insofar as such demands meant (or mean) anything, they meant stricter controls. The suicide of Professor Eberhard Sacher of Vienna, authority on women's diseases, became a *cause célèbre*. Sacher was not himself an addict but blamed himself for turning his teenage daughter into one (she suffered from attacks of abdominal pain after a botched abortion). In 1891 he killed her by accident and himself by intent; or perhaps both deaths were accidents; or perhaps both were suicides. Adolphus von Stolz, Bishop of Graz, suggested that any prescription for morphine by a doctor should be countersigned by an ordained priest: a special congregation might even be founded for that purpose. Fortunately, nothing was done in a hurry in the Habsburg Empire and, though half-baked ideas had a better than average chance of being speedily acted on, the proposal fizzled out.

* * *

The espousal of a trade by the criminal underworld is always a sign of its profitability; and by the end of the century budding anti-drug legislation was making the illicit opium business profitable for the first time. New ports of entry were being developed. Sunny Nice had long been the last-hope station of the tuberculous. Now, with Marseilles conveniently near, it also became a morphine distribution centre. 'This is not the Côte d'Azur any more,' the young Aubrey Beardsley, addict and dying of tuberculosis, quipped, 'but the Côte d'opium.' The habit was no longer regional or confined to any class. Shrouded in deafening secrecy, the high-born were among those widely suspected. The Emperor Francis Joseph forbade his physicians to prescribe anything more powerful than *Bouillon mit Knödel*; but his composure after the suicide of his only son was widely attributed to 'tranquillising injections' by a quack. And did not the double tragedy of Mayerling itself reek of opium? The Crown Prince Rudolf had long been suspected by his unhappy wife, the Archduchess Stephanie, of being a user. In St Petersburg courtiers whispered that the success of the monk Rasputin in calming the Tsarina was due to his use of opium disguised as holy water. Most of this was tittle-tattle; but the splutterings of royal doctors against charlatans supported the rumours.

Doctors' protests usually fell on deaf ears, part of the decline in the moral authority of the profession. Financially and socially it never enjoyed a higher status; but the higher the status, the more old-fashioned virtues – like compassion – seemed to shrivel. Or so many ordinary people felt. The waspish Mr George Bernard Shaw was described by doctors as a clown; but clowns often speak the truth; and his dismissal of the Harley Street elite as 'licensed murderers' struck home.[31] Of course it was only the old parable of the camel and the eye of the needle restated; but the hypocrisy of widespread medical addiction coupled with the preaching against the evil of drugs had something to do with it. But many of these concerns would surface with greatest force not in Britain but in the United States of America.

CHAPTER 20

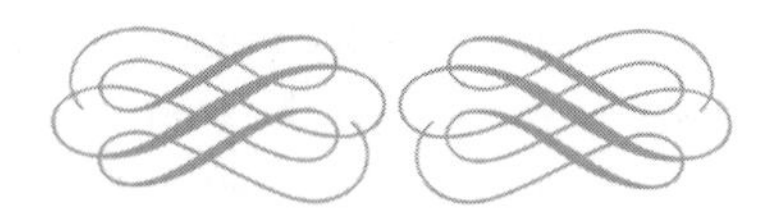

American voices

AFTER THE STARRY-EYED first decades of the nineteenth century the United States developed into the fastest growing market for opium and opiates. Even during the years before the Civil War New Englanders imported enough of the drug – about 16,000 kilos in 1840 – for the business to come to the attention of the United States Customs. Customs slapped a duty on the merchandise, raising the price from less than $1 to $2.80 per kilo, a trend that was to continue. Despite such unpopular impositions, imports rose between 1840 and 1850 to 44,000 kilos a year. In addition to old-fashioned laudanum, still preferred by traditionalists, morphine tablets, ointments, tinctures, clysters and a variety of specialities of the house began to make their commercial appearance. By the time the first shots were fired at Fort Sumter in 1861 imports had risen to a yearly 60,000 kilos and by 1870, as a legacy perhaps of the war, they had reached 250,000 kilos. These were official figures. An unknown additional amount was smuggled into the country by enterprising masters of West Coast schooners and during and immediately after hostilities the poppy was cultivated by American farmers. Native American Indians never took to it; but war veterans returning home spread the habit to all parts of the country.

There was another splendid novelty: as in Europe, subcutaneous syringes and needles were getting cheaper and better. Professor H. H. Kane of Harvard waxed lyrical:

> The glass cylinder and metal cased instrument is excellent: it is strong, the fluid in the cylinder can be seen, bubbles of air can be detected and most dirt readily removed . . . Messrs Codman & Shurtleff of Boston are making the most excellent instruments, so tempered and joined as to be absolutely accurate.[1]

Although labelled and intended to be 'subcutaneous', veins began to be hit. Among the unwitting pioneers Dr George Jones of Cincinnati described the initially alarming but eventually gratifying experience of treating a patient for a frozen shoulder:

> The patient threw up her arms, complained of suffocation, giddiness, faintness, a severe tingling sensation flowing the course of the circulation. Her countenance at first turned livid, then flushed; the eyes became unusually brilliant, muscular twitchings, with profuse sweating and cold extremities ensued. After a few minutes she lapsed into a deep sleep which lasted a few hours. When she awakened she felt 'ever so much better'.[2]

The accident was clearly worth considering as a deliberate therapy: it required a degree of skill but with perseverance it could be mastered.

* * *

But the dangers and ill-effects too were beginning to surface. Only twenty years after the optimistic pronouncements of the Reverend Colton,[3] William Rosser Cobbe, morphinist and newspaper editor, wrote: 'Because it is insinuating, because it makes no noise in its progress, men have been silent about its infernal ravages. Alas! Its very silence is ominous of its strength. The silent forces of nature are those which are most potent.'[4]

Cobbe was one of many to sound a note of alarm. Past the turbulent mid-century warnings, excoriations and messages of doom were beginning to be uttered by both victims and campaigners. They included doctors, preachers, politicians and men of letters ranging from the saintly to the psychopathic. Fitz Hugh Ludlow, described even in his lifetime as 'the American de Quincey' and more recently as 'the pioneer of inner space', occupies a niche near the saintly end.

* * *

Born in 1836 in New York, Fitz Hugh was the son of an abolitionist minister at a time when abolitionist ministers were far from popular even in the North. Only a month before Fitz Hugh's birth the family were driven from their house by a furious anti-abolitionist mob; and by the time they returned their home had been plundered. The moral heritage Fitz Hugh would always cherish and try to live up to was not always attainable. It also inspired his first published work, a poem entitled *Truth on His Travels*. Truth in the composition wanders the earth searching in vain for a band of people who will respect him. It is a sad and not particularly original idea but suffused with noble sentiments. In his first printed volume he would introduce himself as a bookish, sickly young man, 'musing and reading on the sofa when I should have been playing cricket, hunting or riding . . . My thirst for adventure was easily quenched when I could ascend Chimboraco with von Humbold and chase hartebeests with Cumming quietly eating muffins.'[5]

At university, first at the College of New Jersey (now Princeton), and then at Union College in Schenectady, studying philosophy, he did well, writing some of the rousing songs which are still sung at commencement and other solemn occasions. He also developed an interest in pharmacology, acting as an anaesthetist at what must have been one of the earliest operations performed under general anaesthesia. But the same interest nudged him to ransack the pharmacy of a friend for 'inebriating and inspiring substances which would enliven and stimulate my imagination'. The result of his experimentation was *The Hasheesh Eater*, the book which he published at the age of twenty-three and on which his fame largely rests. Apparently written at the behest of his physician to help him through a period of withdrawal, it is a curious confession, a hotch-potch of observations and ruminations, fascinating but rarely totally credible.

> I had caught a glimpse through the chinks of my earthly prison of the immeasurable sky which should one day overarch me with unconceived sublimity of view and resound in my ears with unutterable music . . .[6]

The fantastic experiences recounted differ strangely from those of ordinary cannabis users, even of addicts consuming massive amounts. On the other hand, they vividly recall De Quincey's opium dreams, which had clearly impressed Ludlow. Were his own exultations more literary than physical? Or was hasheesh, as for Baudelaire, a code for opium? In contrast to the dulcet 'opium', hasheesh sounded exotic and vaguely menacing. Whatever the truth, the book was a *succès de scandale* and opened to the author the doors of New York's bohemia. It was a boisterous world more akin to Murger's Latin Quarter than to more sedate literary London. Its leading denizens – Walt Whitman, Fitz-James O'Brien, Bayard Taylor, Thomas Bailey Aldrich, Edmund Steadman, Artemis Ward and their boys and girls – met in Pfaff's beer cellar on Broadway and on Saturday night at Richard Henry Stoddard's and their outpourings launched a new generation of magazines. Swimming in this pool Ludlow felt comfortable. He passed the examinations necessary to practise as a lawyer but decided to devote himself to professional and full-time writing instead. He delighted in the fact that New York was tolerant of the notoriety he carefully cultivated: 'No amount of eccentricity surprises a New Yorker or makes him uncourteous. It is difficult to attract even a crowd of boys on Broadway by an odd figure, face, manner or costume.'[7]

He wrote prolifically for the new magazines, earned respectable royalties; then he travelled west and threw himself into the literary life of San Francisco. He was bowled over by the young Mark Twain: 'he imitates nobody . . . he is a school by himself'. On a slightly more restrained note Twain reciprocated. Ludlow also encountered and described the ravages of opium addiction among Chinese immigrants:

> I shall never forget till my dying day that awful Chinese face which actually made me rein my horse at the door of the opium hong where it appeared, after a night's debauch at six o'clock in the morning . . . It spoke of such a nameless horror in its owner's soul that I made the sign for a pipe and proposed in 'pigeon English' to furnish the necessary coin. The Chinaman sank down on the steps of the hong, like a man hearing medicine proposed to him when he was gangrened from head to foot, and made a gesture palms downward toward the ground, as one who said 'It has done its last for me . . . I am paying the matured bills of penalty.'[8]

Perhaps it was experiences like this which for the rest of his life made Ludlow into one of a heroic breed, an addicted anti-addiction fighter, heroic and tragic. As he churned out pamphlets and articles to the end, opium became the focus of his life and writings: 'my life's ruling passions is – a very agony of seeking to find – any means of bringing the habituated opium-eater out of his horrible bondage, without, or comparatively without, pain'. In 1868 he contributed an essay, 'What shall they do to be saved?', to *The Opium Habit* by Horace Day, the first book in the United States to try to deal in a detached way with the drug predicament.[9] It was also one of the first publications to recognise that the problem was growing into a national menace. Ludlow expanded his contribution in a paper, both utopian and despairing, 'Outline of the opium care', envisaging free and accessible drug addiction treatment clinics:

> The opium addict is a proper subject not for reproof but for medical treatment. The problem of his case need embarrass nobody. It is as purely physical as one of small-pox . . . He is suffering under a disease of the very machinery of volition; and no more to be judged harshly for his acts than a wound for suppurating or the bowels for continuing their peristaltic motion.[10]

He was not a felicitous phrase-maker like De Quincey; but there was – and still is – an air of genuine concern about his writings that led thousands of addicts to write to him for advice. He seems never to have left a letter unanswered: some of his responses are still coming to light. He also devoted his last years to treating addicts as an unqualified but dedicated doctor. A friend wrote:

> I have known him to go for three weeks at a time without taking off his clothing for sleep, in attendance upon the sick. His face became familiar in many hospital wards . . . He supported out of his scanty means a family of which one of the members had become a victim of opium. This family had no claim on him except for the sympathy which such misfortunes always excited in him. The medicines and money for this family could not have amounted to less than a hundred dollars and this case was only one in many.[11]

But Ludlow himself could not break the habit. He also, some time in his early thirties, developed pulmonary tuberculosis, opium's coughing twin. He married an older woman with a family from an earlier marriage, and she and his sister Helen looked after him with the devotion consumptive addicts often received from their family. This was as well since compassion was rarely a notable attribute of their medical attendant.

> Dr Smith said the other day that there was no use in his wasting his strength and time treating Mr Ludlow, for he took a teaspoonful of morphine in a glass of whiskey at least once a day, and while he was doing that it was only time and strength thrown away treating him.[12]

Another friend wrote:

> While he was doing so much to warn and restore others from the effects of this fearful habit, he himself was still under its bondage. Again and again he seems to have broken it: only those with him knew how he suffered at such periods . . . I recall a night he passed with me . . . and [after going to bed] he was a long time making his preparations. At length I suspected he was indulging his old craving and for the first and only time in my life I spoke to him harshly and . . . characterised his treatment of his friends as shameful . . . He replied depreciatingly and, turning down the gaslight, came to bed . . . Feeling reproved in my harshness, I said 'Think, Fitz, of your warning on the subject, and of your efforts, on behalf of others'. In a tone and pathos I can never forget he answered with the quotation: 'He saved others, himself he could never save.'[13]

His consumption deteriorated and he left for Europe with his sister and wife. After resting and visiting museums in England they travelled to Switzerland.

There sanatoria were mushrooming in remote and unfriendly mountain regions; and he died in one of these doom-laden encampments in 1870, on his thirty-fourth birthday. In a passage of 'What shall they do to be saved?' he had written: 'Over the opium-eater's coffin write thank God! A wife and sister can stop weeping and say: "He is free." '[14] It could have been his own epitaph.

* * *

Near the other end of the saint-to-psychopath spectrum Anthony Comstock represented a tradition that stretched back to seventeenth-century witch-hunters and foreshadowed Senator Joe McCarthy of the twentieth century. A Connecticut farmer's son, his education was cut short by the Civil War; and at twenty-six, in 1870, he was a veteran. But an unusual one. He was remembered by his comrades chiefly for the memoranda addressed to their commanding officer complaining about the 'profanities uttered by my brother officers in the course of and after battle'. He was by then a self-declared 'Roundsman of the Lord', soon to be the creator of the New York Society for the Suppression of Vice. He had considerable gifts of persuasion. In 1874 he successfully lobbied Congress to pass 'Comstock's Law' making the delivery or transportation of 'obscene, lewd or lascivious material . . . any method or information pertaining to birth control' and 'all kinds of inebriating substances' illegal. He was soon loathed by civil liberties organisations but passionately supported by religious fundamentalists of the more rabid kind. A master of the political manoeuvre, with the help of church groups he was made a Special Agent of the United States Postal Service with police powers including the right to carry and use arms. He now travelled wherever he wished and could apprehend smut-mongers and arrest chemists and doctors suspected of dealing in drugs. Although he is remembered today mainly as a pursuer of obscene publications, including anatomical atlases and any art book featuring nudes, and as the self-styled 'weeder in God's garden', opium and other drugs in fact came highest on his list of targets:

> Fathers and mothers, look into your child's face and when you see the vigour of youth failing, the cheek growing pale, the eyes lustreless and sunken, the step listless and faltering, the body enervated and the desire to be alone coming over him much; when close application to study becomes irksome and the buoyance of youth gives way to peevishness and irritability, look for a cause and remember the Devil's favourite word: DRUGS![15]

His campaign as God's champion drove at least fifteen people to suicide and he professed to be 'mighty jolly' whenever that happened. An old lady in

Greenwich Village whom he had twice checked into the Tombs Prison for publishing her opium-inspired trysts with the Archangel Michael heard that he was going for a third warrant and put her head into the gas oven. This was another 'good day' to him but the incident provoked critical headlines and for a time donations to his anti-vice campaign declined. But these were temporary setbacks. He rarely bothered to prosecute pharmacists suspected of stocking opium, morphine, contraceptives or abortifacients, preferring to go in with an axe and destroy the premises himself. He became the self-styled 'Terror of Opium Dens', demolishing buildings which he suspected because he had seen 'shifty-looking Chinks' entering. He claimed that through his campaigns he was responsible for the destruction of 15 metric tones of books, 284,000 plates of obscene material, four million pictures and uncounted bottles of opium and morphine. He also took credit for 4,000 arrests and some 500 convictions.

Yet, despite his many triumphs and the adulation of millions over a career lasting more than thirty years, his was, by his own account, not a happy life. Though he never expressed despair in public, the evils he fought, especially the spread of drugs, could not be contained. On the contrary. The number of users, addicts, pushers and dealers continued to rise. The habit was also transgressing social barriers. Christians fell victim to it as well as atheists and Jews. Almost beyond belief was the fact that addicted women were beginning to outnumber addicted men. The opium and morphine lobby was fighting back, not shrinking from physical violence. In his later years his vigour declined, the result perhaps of a blow to his head by an anonymous assailant. For this he was obliged to take frequent and unspecified painkillers. But he was overreaching himself. In 1902 he campaigned successfully to have Bernard Shaw's *Mrs Warren's Profession* banned from all New York's public libraries and Shaw's response managed to damage his reputation as well as to wound him:

> Nobody outside of America is likely to be in the least surprised. Comstockery is the world's standing joke at the expense of the United States . . . It confirms the deep-seated conviction of the old world that America is a provincial place, a second-rate country town civilisation after all.[16]

'Comstockery' passed into the language. He continued to lecture and to write to the end but increasingly he was the butt of jokes. Yet Comstock was never without a few up-and-coming acolytes and his legacy was not lost. The gifted young lawyer Edgar G. Hoover was one of his last disciples. He died in 1915 aged seventy-five.

* * *

Comstock may have been a monster; but others, more level-headed, also saw the looming danger of drugs. In 1867 Dr Joseph Parrish, director of one of the more successful clinics for treating addiction in Media near Philadelphia wrote without much exaggeration:

> The Opium-appetite, which even in the United States has attained stupendous dimensions, lying ambushed beneath the foundations of health and happiness, has hitherto failed to command from medical writers more than a passing notice; yet the doomed victims grope their way by thousands through their life of semi-oblivion, without a solitary ray to illuminate their . . . path towards resistance and self-conquest.[17]

Only in one respect was he overstating his case. Voices *were* being raised and books on addiction *were* being produced. In 1870 Alonzo Calkins, a New York practitioner, published *Opium and the Opium-Appetite*, a classic in a field in which trash predominates. Enlivened with quotations from world literature and based on laconically but vividly recounted case histories, it is an all-round survey of the rising American drug habit. Comparing opium and morphine with alcohol, cannabis, cocaine, tobacco and other substances of addiction Calkins was in no doubt that opium and its derivatives represented by far the greatest threats to his country's future:

> Opium and its product morphine are the most powerful exhilarants, inebriants, anodynes and toxins. Under habitual use they let loose the imagination to the detriment of judgement: the more one soars the more the other is depressed . . . And not only judgement but the will too is chained down . . . The mind becomes engrossed with hallucinations and relapses into a world of dreams . . . The peculiar morbid influence, subtle and slow in the beginning, cumulates with progress. The reaction is primarily on the brain and stomach but ultimately results in a general vitiation of the organism and premature and progressive decrepitude. Hyperaesthesia, alvine torpor, perturbation of sleep and venous congestion grow into prominent symptoms . . . With vacillation of purpose comes a sense of moral degradation . . . a paralysis of conscience, a torpor of affections, and a stolid indifference alternating perhaps with the recklessness of despair.[18]

He was scathing of those who boasted that they were constitutionally immune. 'Against the temptation to habitual use neither individual constitution nor social status and conditions provide immunity . . . self-imposed reductions upon established measures are among the rarest exceptions.'[19] The

1 Ripe pod of the opium poppy, scored and exuding opium.

2 Nefertiti, queen of Egypt's reforming pharaoh Akhnaten in the fourteenth century BC, presenting her husband with a poppy. From a carving in a pyramid.

3 Goddess with a tiara of poppy pods, from an excavation in Knossos, Crete, twelfth century BC.

4 Page showing the poppy, from an Arabic translation of Dioscorides's *Codex Vindobonensis Medicus Graecus,* the prescriber's bible for centuries both in the Islamic world and later in the Christian West.

5 Paracelsus, genius, charlatan or both, who introduced the name 'laudanum' into European medicine.

6 *The Lady's Death*, concluding scene of William Hogarth's series of engravings, *Marriage A-la-Mode*. Every detail tells a tale but most eloquent is the upturned empty bottle of laudanum on the floor, the lady's final route of escape.

7 One of the cathedral-like opium drying sheds in Patna, India, towards the end of the eighteenth century. Much of the produce was beginning to be directed toward the Celestial Kingdom of China.

8 Samuel Taylor Coleridge, poet, sage and charismatic addict who nevertheless survived until the age of seventy-four.

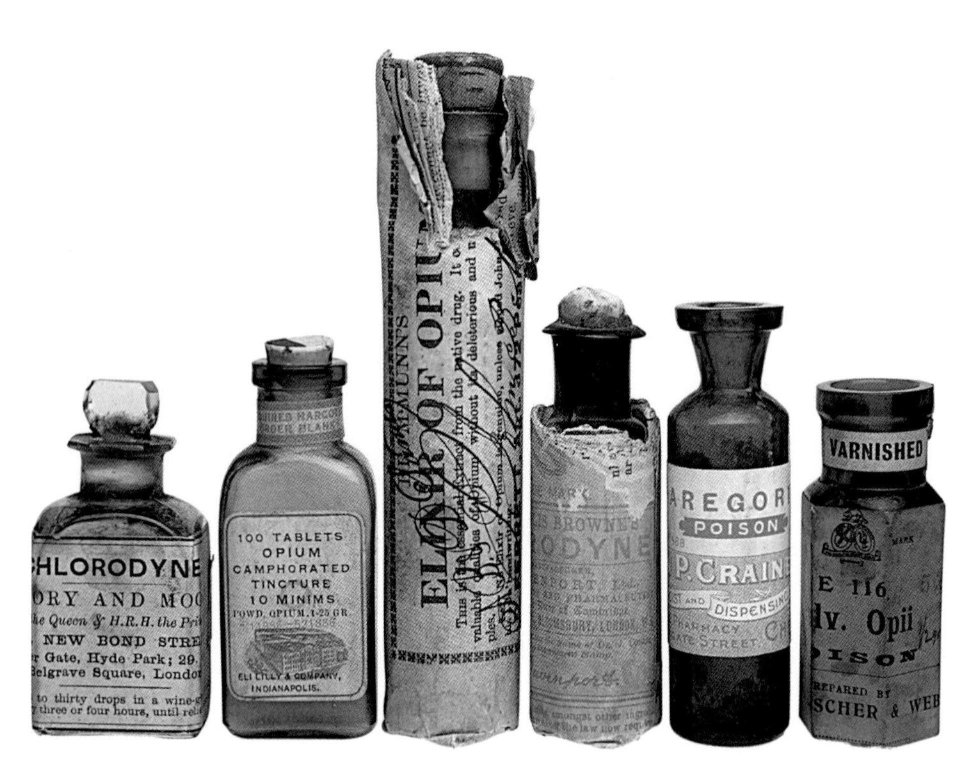

9 A selection of a vast range of proprietary opium-based medicines. Originally a treatment against cholera, chlorodyne – desperate housewives's 'old chlory' – was for a long time the market leader.

10 The Opium War as widely perceived in Europe, the United States and by high-minded politicians, clerics, schoolmasters and writers in Britain.

11 The teeming bay of Canton at the mouth of the Pearl River, until the end of the Opium Wars the only portal of entry of Western goods allowed by the Imperial Government. Representatives of the Western trading nations and their families were strictly confined to the 'Factories' on the side of the river opposite the walled-in city.

12 'The Lady with the Lamp', the famous illustration in the *Illustrated London News* showing Florence Nightingale on a night round in the military hospital in Scutari during the Crimean War. The patients, mostly wounded from the battlefield, must have been heavily dosed with opium to present such a peaceful picture; and so, by her own account, was the lady herself.

13 Fildes's illustration to Dickens's last unfinished novel, *Edwin Drood*. The hero-villain, John Jasper, is depicted in an opium den.

14 *A Drug Addict Injecting Herself* by Eugene Grasset. The artist was a brilliant but now little-remembered chronicler of the Belle Époque.

15 Is she on a high? The enchanting barmaid at the Folies-Bergère as depicted by Edouard Manet looks unfocusedly into the middle distance, seemingly oblivious of the crush around her and of one particularly pressing customer. The original is in the Courtauld Gallery in London.

16 An advertisement for Mrs Winslow's Soothing Syrup, one of hundreds of opium-based baby-calmers aimed at every social class on both sides of the Atlantic and on both sides of the Channel. English preparations enjoyed a wide international reputation and were de rigueur in aristocratic households on the Continent.

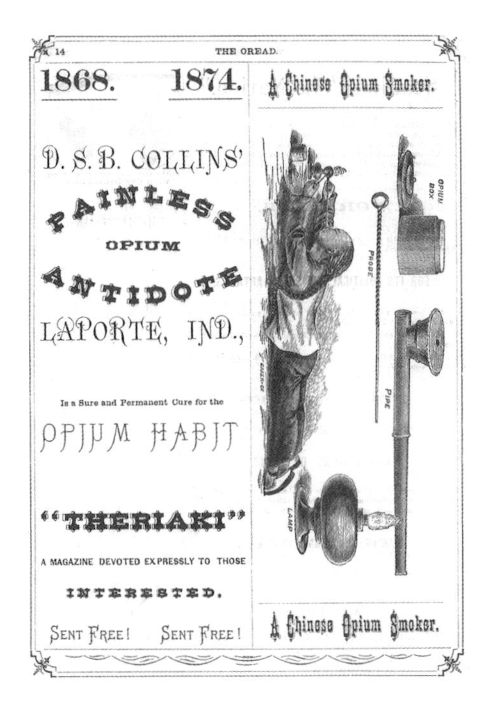

17 Advertisement for a 'Sure and Permanent Cure for the Opium Habit', one of many. This particular advertiser was 'Professor' Samuel Collins of La Porte, Indiana, a bricklayer by training, whose bottled cures sent by mail guaranteed to work even in the most hardened cases. His house magazine, *Theriaki*, was full of thrilling 'before and after' stories, many penned by grateful offspring of former addicts. Collins made millions and retired in glory. The contents of his always 'personalised' remedies were never revealed.

18 A neat subcutaneous morphine injection set similar to the one advertised by Harrods and by Savoury and Moore in October 1914 as 'an ideal Christmas gift to friends at the Front'.

19 One of countless depictions of the 'Yellow Peril' monster printed as late as the 1930s.

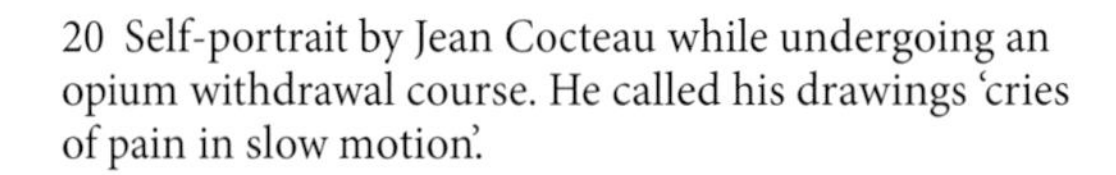

20 Self-portrait by Jean Cocteau while undergoing an opium withdrawal course. He called his drawings 'cries of pain in slow motion'.

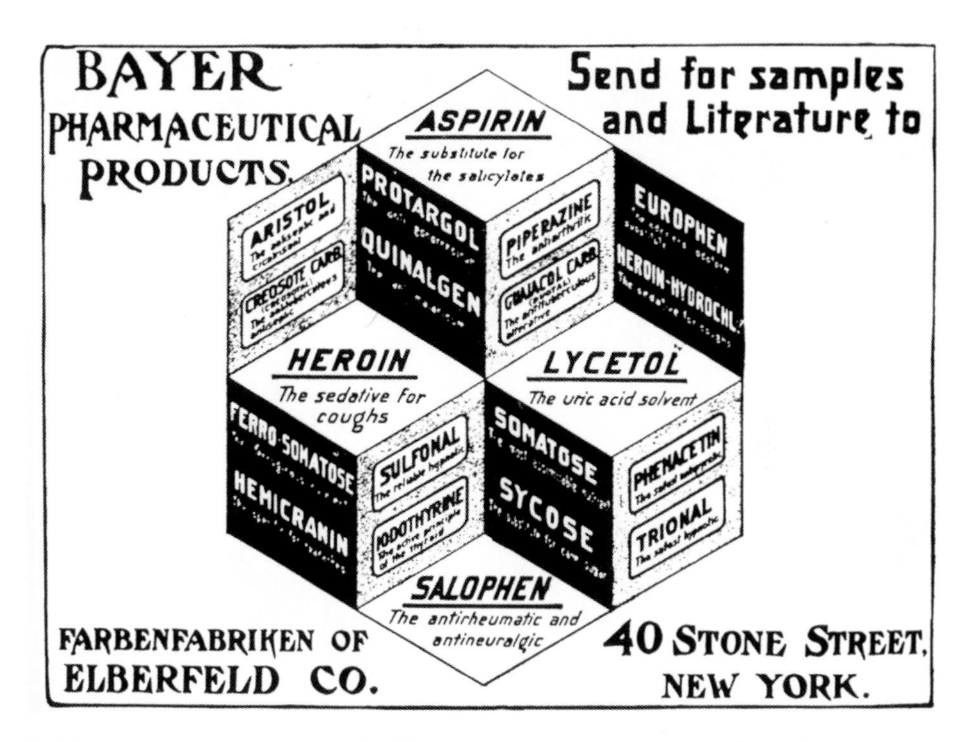

21 No more coughs and colds with the 100 per cent safe new Bayer product, 'heroin'.

22 Harry Anslinger with 'intercepted contraband', in characteristic pose.

23 One of the penny-dreadfuls based on the opium traffic and the evils of addiction. William Lee was the pseudonym of William Burroughs, and *Junkie* was his first published work (1953). It was based on his initiation as a user and pusher in Greenwich Village (partly under the influence of Allen Ginsberg and Jack Kerouac). *Narcotic Agent* by Maurice Hellbrant was a reprint: novelties and reprints were often combined to make an irresistible bargain. Both books were heavily 'expurgated' to eliminate not so much references to drug use as passages of explicit homosexuality. *Junkie* did not become available uncensored and over the counter until published by Penguin in 1977; and its companion piece, *Queer*, did not see daylight until 1985. But Burroughs ended his life as a member of the American Academy and decorated by the French State.

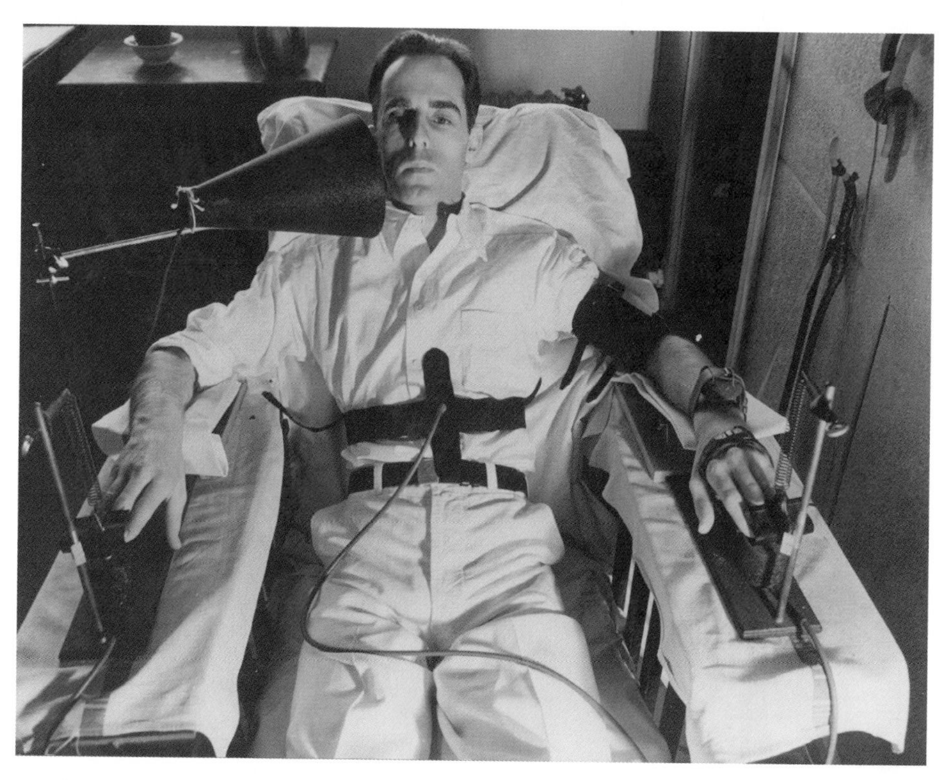

24 The Darrow photopolygraph was used at the Lexington Farm to measure a patient's mental and physical reaction to slang references to drugs. In this test a researcher in an adjoining room shows the addict words such as 'dope' and 'informer'. The patient's facial reactions, pulse, blood pressure, breathing and galvanic skin response (a change in the skin's electrical conductivity due to stress) are monitored. Doctors listen to the patients's verbal responses via the microphone. Such experiments were part of early attempts to understand the psychological factors in drug addiction and relapse. They were often conducted on inmates under the influence of narcotics.

25 Cicely Saunders, pioneer of modern palliative medicine.

26 U.S. soldiers spraying a poppy field in Afghanistan in 2008. Eradicating the poppy was later recognised as incompatible with the winning of hearts and minds.

only hope was expert supervision in some of the specialised clinics like the one established by Dr Parrish.

Yet Calkins, like the majority of informed Americans in the mid-century, had little faith in sledgehammer legislation. This was partly based on a reading of history. He recalled how the mighty Shah Abbas in 1582, despite his great power, was forced, under protest from users, to 'recede from his decree against opium'. The numerous and thunderous edicts issued by a succession of Chinese emperors would have failed even without Western intervention. More recently

> Army Surgeon Smith observed in Pulo Penang that penal enactments had failed to stem the tide, and that thefts, robberies and murders too were being perpetrated again and again and increasingly in numbers, so impetuous and uncontrollable has the perverted appetite for opium fumes become . . . The wise legislator recognises both the facts of history and the deductions of science; the unballasted enthusiast urged forward by zeal would often oppose to intemperance of one species intemperance under some other guise but no less damaging . . . Sadly, often the one-idea philosopher is no wise alarmed at the threatening hurricane unleashed by himself. Prohibition once declared, monopoly must follow, with all its enticements to collusion, corruption and fraud. Is there any advantage in making the few rather than the many custodians of the public conscience; and where shall the ultimate supervision be reposed?[20]

Like many before him and since, he was crying in the wilderness.

* * *

By 1871 when Calkins published his book, newspapers and magazines as well as the medical journals were overflowing with cautionary and alarming tales.[21] The dread 'diathesis' of the European medical literature was crossing the Atlantic. Contradicting Calkins and the American perception, it suggested that some unfortunate people inherited a mysterious predisposition to nervous weakness which in turn made them fall prey to drugs. But why the sudden increase in numbers? As William Rosser Cobbe observed, 'scientific men tend to charge on heredity the physical tendencies which they cannot explain and which do not reflect credit on the race'.[22] Americans in particular liked to think of themselves as fit, dynamic, innovative and unafraid of new ideas: indeed, they sometimes seemed to court danger for the sake of the untried. Obsession with 'hasheesh' was part of the search for novelty. As the staid *Knickerbocker* magazine remarked only half in jest: 'We *Yan-ne-kees* always want to try things, from a new mechanical device to a new patent medicine to a new drug.'[23] Injections of morphine combined all three, an irresistible combination.

As happens after calamities, the years after the Civil War saw a frantic chase after new thrills. The younger generation wanted to leave painful memories behind. 'We're quick-thinking and restless people prying into everything: opium cannot pass us by untempered with,' one young addict declared.[24] Among their elders such pronouncements raised fears not dissimilar to those that had troubled Thomas Trotter in Scotland a century earlier. Were the emotional demands made by technical progress too heavy? Progress should have created happiness. What it did create was the temptation to 'increase energy'. When the striving failed, it led to disappointment. Disappointment led to drugs. Racist undertones began filtering through.

> We think and we exhaust ourselves, we scheme, imagine, study, worry and enjoy and proportionately we waste away. This is far less true of rude and primitive people. Unlike us, the barbarians are not nervous. That could give them a huge advantage over us. That must stop.[25]

The last decades of the nineteenth century in America were both triumphalist and riven with anxiety. To question the benefits of technical progress was heresy; but thoughtful men and women were troubled. Whatever the value of technical advances as a whole, the effects on society's less sturdy members could be mixed. The hopes voiced by the likes of the Reverend Colton only thirty years earlier now seemed fatuous. As the use of morphine and the subcutaneous syringe spread, in the eyes of many drugs were outstripping liquor as an existential menace:

> Life grows more complex with each development of science. New faculties are brought into activity, new forces are called into being when knowledge of their relation to life and health is still wanting. Hence inebriety, insanity and a host of new diseases and above all drugs! Every advance or refinement brings conflict which has to be paid for in blood and nerve and life.[26]

These stresses were causing concern in many countries; but the virtues of American society magnified the dangers. Apart from feverish spasms like the Napoleonic Wars, when a field marshal's baton and a ducal coronet were carried in every private's knapsack, class rigidities in Europe were still undented. True, social barriers could be pierced by wealth, however amassed, and the passage of a generation or two; but almost never by upstarts in their own lifetime. It was, by contrast, the pride of every American that no circumstance of birth prohibited rising to the top. This was as it should be. But worriers felt that unlimited opportunities could give rise to unlimited expectations; and when unlimited expectations remained unfulfilled – as for some reason they almost always did – this could lead to despair. What in a moment of inspiration Fitz Hugh Ludlow foresaw was coming true: 'We must achieve . . . and if we cannot achieve we must pretend to achieve. That often means drugs.'[27]

During the Civil War a prescient doctor wrote:

> After the war more freedom would prevail but also more rootlessness and discontent . . . The American dream of always searching for what is new and what is better is a noble aspiration, it is what makes our nation the hope of the world; but oh how often it is bound to disappoint. And for consolation in our disappointment what will be easier than to reach for the blessed drug of oblivion?[28]

CHAPTER 21

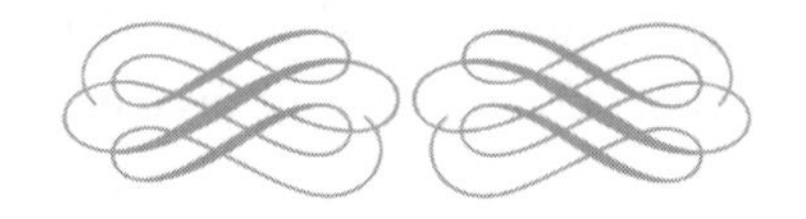

Nervous waste

As the nineteenth century was drawing to a close, it seemed that it was America's youth which was falling victim to drugs. Anxiety was stoked by ignorance. The term 'American disease' was cropping up in the newspapers. What was it? An epidemic? Why had it suddenly taken a grip? Dr George M. Beard called it *neurasthenia* – literally, nervous waste – and claimed to know most of the answers.

The son of a Connecticut minister, Beard graduated from Yale in 1862 at the age of twenty-three and by then had firmly decided on his future career. He was to devote himself to an illness which was common but about which doctors knew nothing. He was certain of that since it had afflicted him as a medical student and his medical attendants had been useless. It did not even have a name. To fill the gap he invented 'neurasthenia'.

Today neurasthenia takes up no more than a few patronising lines in medical dictionaries and textbooks. One standard work defines it as 'complaints of chronic weakness, easy fatiguability and sometimes a vague feeling of exhaustion . . . the patient's complaints are improbable and lengthy but distressing to him and there is no evidence of seeking secondary gain other than attention . . . There is no treatment.' This is a long way from the crushing list of physical infirmities itemised by Beard as part of the syndrome:

> tenderness of the scalp, intermittent dilatation of the pupils . . . sick headaches, pain, pressure and heaviness in the back of the head, a debilitated look in the eyes . . . floaters in the visual field, occasional double vision and frightening intimations of blindness, noises in the ears, an 'atonic' voice, meaning 'softness, faintness, want of courage and clarity of tone' . . . a deficient mental control over speech and action, inability to concentrate, heterophemy

> [saying one word when meaning another], irritability, excessive hopelessness, morbid fears for no discernible reason, a foretaste of death, palmar hyperidrosis [sweating], facial flushing . . . insomnia, twitching of the eyelids a variable tenderness of the entire body, irritation (rather than plain itching) of the skin, lancinating pains or dull ache in the spine or head . . .[1]

The list is formidable but by no means complete. The complaints were not only itemised but also discussed. To Beard they represented the real and important manifestations of a disease which needed to be recognised. And, he was sure and loudly proclaimed, it *could* be treated! In this last assertion, based mainly on a variety of electrical stimulations, he was less convincing; and in the long term this damaged his reputation. But at the descriptive level he was a pioneer of psychosomatic medicine.[2] He was also an acute analyser of the causes of that 'most terrible man-made pestilence and the commonest cause of nervous waste, drug addiction'.

> The necessary evils of specialisation in the new industrial societies both narrow and intensify everyone's life and cause constant worry about performance . . . One cause of the increased nervous diseases or neurasthenia and the flight to drugs is that conventions of society require the emotions to be repressed while our civilisation gives unprecedented freedom and opportunity for the expression of the intellect.[3]

Half a century before Freud, Beard castigated the repression of 'biological drives' whose denial only led to anxiety, discontent and a recourse to drugs. He was on first hearing or reading extraordinarily persuasive and he quickly acquired followers. One of his admirers, J.B. Mattison wrote with concern about the merits of self-restraint, but also about its long-term pitfalls:

> Man is enabled by will-power to contain himself within the bounds of decorum and decency. It is by the will that man is an abstemious animal and when given to overindulgence in drugs he is simply evidencing atavistic principles which so often crop up during the evolutionary process. Civilisation teaches man to 'wish' and to 'wait' for many things in his life upheld by the hope of their fulfilment in some future existence if not this, and also that by restraining his passions and appetites he is serving his own long-term if not immediate interests. But sometimes such restraint can have dire ill-effects: indeed it can be more dangerous than the underlying noxious drives and actually destroy the individual. A middle course must therefore be found.[4]

Determination is often described as one of the defining American virtues but a steady stream of memoirs relating to the slavery of addiction underlined the difficulties. For addiction was nothing less than slavery. 'It is not the man who eats opium but opium who eats the man,' Nathan Allen wrote;[5] and William Rosser Cobbe in his sensationally successful book, *Dr Judas*, suggested: 'It is not *pleasure* that drives forward the confirmed opium eater but *necessity*, no more resistible than Fate. Not even the mythological gods could strain against that.'[6] In their self-revelations addicts insisted on their desperate efforts to be cured and become normal members of society. Some who could not break the habit sought freedom in death. 'It held no terror for them equal to life without opium.'[7] But most eventually gave up trying even if it meant becoming social outcasts. The recognition of failure was always a deeply distressing moment. 'There is no situation more tragic than that of an opium eater who had been determined to quit only to be forced back into the habit and realise that life to him must forever be that of a walking shadow . . . that he must languish out his natural existence locked a prisoner in the arms of a grisly demon.'[8] Or, as another opium user told his doctor: 'Truly the name "fiend" is aptly used. Opium will make a villain of the best of us. There have been times when I believe that if I could not have obtained the money for opium otherwise, I would have taken the shoes off my wife's or child's feet and sold or pawned them.'[9]

The claim of past addicts, revelations of new and exotic worlds, was becoming suspect. Periods like people have their distinctive priorities. Americans of the late nineteenth century tended to value the concrete over the abstract, facts over fancy and reality over dreams. Knowledge and social skills acquired through working with other people were more highly prized than poetic visions and metaphysical speculations. This view was shared by most addicts themselves. 'All the pleasures I can derive from morphine would not equal the enjoyment a well man receives from normal animal spirits alone; and all the poetic force I obtain from opium stimulation can never approach that which would have been mine from my own free and natural condition.'[10] To new immigrants becoming part of the community mattered. If to old-established Americans achievement was important, then they wanted to achieve. Addicts too strove to accomplish something their fellow citizens could admire. The American literature of addiction is full of stories of users resolved to write great books, discover momentous scientific truths, make advances in their profession or become business tycoons. In the end the dream always had to do. As the educator David Starr Jordan wrote, voicing the verdict of the average man:

> To cultivate sensation for sensation's sake with no purpose beyond it, whether it be in art, music, love or religion, is to live a purely sensuous life, a life of weakness, decadence and, above all, social uselessness.[11]

But uselessness was not all. Neurasthenia also made addicts develop qualities ordinary Americans feared and despised. After a time they became deceitful, self-absorbed, uninterested in public affairs, indifferent to the future, preoccupied with their own one overriding need. They developed a kind of low-grade craftiness. Consulting their doctor or pharmacist they would ask for and buy a dozen items to conceal their desire for only one; and while being served they would purloin a flask or a box from the shelves. They would rarely be caught; but after a time they became known as thieves. It was often the beginning of a vicious circle. Awareness of public distrust would further turn them inward, and this would make them even more untrustworthy. They were seen – as indeed they were – as indifferent to the needs and aspirations of others: tomorrow to them did not exist in a land that had eyes fixed on the future.

Addiction threatened marriage and family, two anchors in the turbulence of frenetic technical progress. 'The trouble is', Dr Charles W. Earle wrote severely, 'that addicts are so oblivious of their responsibilities in life and so prone to get wrapped up in the exhilarating effects of their drug, avoiding all the troubles, pains and demands of normal life, that they cannot see that it is their duty both to themselves and to their friends and family to stop the habit.'[12] But even if they did – as many were determined to do in moments of drug-induced euphoria – how were they to sustain it?

* * *

As in Europe, doctors (who were not themselves addicts) knew little about addiction. The subject was not touched upon in the medical curriculum; and by and large teachers shared the general distaste for drug users. What they did know was that treatment almost always ended in failure; and no profession relishes a succession of disasters. But professional indifference left the field open to charlatans. The entrepreneurs benefited from the changes that were transforming post-Civil War America. Newspapers now accepted cheap advertisements and were carried to every hamlet. Falling postal rates made ordering medicines through the mail affordable. The plain brown wrapper was meant to and did – more or less – hide the addict's condition from family and neighbours. To naive recipients printed government patents and copyrights – anything with an official flavour – on the labels seemed to mean official approval of what was in fact a wholly unregulated and unsupervised trade. The new generation of quacks skilfully used the trappings of science, medicine, and technology to promote their ware but they also appealed to their would-be patients' suspicion of reputable medicine. In the Midwest and the South this was strengthened by endemic resentment of the East Coast, citadel of the medical establishment as well as of oppressive financial and political power.

Samuel Collins of La Porte, Indiana, was a bricklayer by training who developed a thriving business selling through the mail bottled cures guaranteed to work even in the most hardened of cases. As his clientele grew he built an impressive business centre, a ranch-style residence for himself and accommodation for supplicant clients. (The establishment did not qualify as a sanitarium because it had no medically qualified staff.) He also published a magazine, *Theriaki*, bursting with thrilling before-and-after stories and testimonials from grateful families. Letters from children were particularly moving: Collins's creation, eight-year-old 'Mary Stubbs' with her pigtails and dimples, an early incarnation of Dorothy of the Rainbow, became an icon. All letters to her (care of Professor Sam Collins) were answered in her charmingly unformed handwriting (also by Professor Sam Collins). The procedure was honed to a fine art. Each client – never 'patient': they were not ill – filled in a detailed questionnaire about his or her history with meticulous attention given to past drug habits, dosages and ill-effects, if any. A 'wholly unique and personalised' regime would then be designed. The standard liquid which was eventually dispatched always contained small amounts of opium in addition to bromides, chloral, alcohol, sometimes cocaine and other sedatives that allayed the distress of withdrawal and gave the impression for a time that the client was slowly but surely reducing his or her dependence. Daydreams of wish-fulfilment can be infinitely more powerful than reality. Clients would report progress, and many reports were exultant. And, even in cold print a century later, pitiful. But they did not greatly arouse the professor's pity. Indeed, in his own estimation he was a fine specimen of American business enterprise. Others, more hidebound, demurred. Professor Albert Prescott of the University of Michigan suggested that vendors of anti-addiction cures were like spiders hunting and devouring flies and that service in chain and ball would not be too severe a retribution for them. But who listened to grumpy old academics? The trade did not come under official restraint till 1906; and, when it did, legal obstacles only stimulated the ingenuity of those whose flair and vocation was to circumvent them.

* * *

At the other end of the scale from out-and-out charlatans was the sanatorium – sanitarium, in American parlance – movement identified with the latest advances in science, technology, psychiatry and pharmacology.[13] More than a hundred such institutions catering for all kinds of clientele had sprung up in the United States by the end of the century. The inspiration of some was Christian, others were modelled on a Tibetan lamasery; some offered to pamper the pampered, others advertised their 'austere but charitable intent'. In general terms their aim was to return addicts to normal society by teaching them to manage their tensions and anxieties. Even establishments in the middle of bustling cities tried to conjure up a rural ambience, testimony to the belief that it was the urban-industrial lifestyle which fuelled the desire for drugs. The Battle Creek Sanitarium in Michigan used self-grown organic food (a novel idea at the time) to create a flawlessly natural milieu. The Bornmouth Sanitarium pretended to be a farm where horseback riding was the backbone of therapy. (In all ages the benefits of the proximity of horses seemed to mesmerise a section of the medical profession.) Shultheiss in upper New York State masqueraded as a frontier encampment, complete with covered wagons and nursing attendants dressed as Indian braves and their squaws.

But successful though many of these institutions were for a time, the movement reached its apogee with the clinics founded by Leslie F. Keeley, the self-professed most controversial doctor in the United States. Mystery surrounded his origins but he claimed to possess a medical qualification from Rush College, Chicago, and that he had served as an army surgeon in the Civil War. His noble left profile with a reassuringly bushy eyebrow was widely reproduced but he rarely appeared in person even to his classes. He never attended conferences or gave interviews. His crusade was primarily directed against alcoholics but in a higher price bracket his institutes also catered for compulsive smokers and drug addicts. The first complex he founded at Dwight, Illinois became a mecca and the model for over a hundred franchised Keeley Institutes in every large city in the United States and in many capitals of Europe. Their vaguely modernistic architecture was as distinctive as McDonald's eateries are today. Apart from intense one-to-one soul-searchings, no-holds-barred group discussions and courses of physical exercise, the nub of the regime was a solution of 'bichloride of gold' devised by Keeley after lengthy experimentation and ceremonially distributed three times a day to resident students.[14] It tasted reassuringly unpleasant; but what exactly it contained apart from the gold was never revealed. Franchised clinics were provided with their supply from the laboratory in Dwight. Although the 'freshly prepared medication', like freshly pressed fruit juice, was described as 'undoubtedly superior', outside the centres bottles could also be bought for home consumption. Sold always in pairs, slightly different brews were available for different forms of addiction. Tobacco users paid $5, neurasthenics

$8, alcoholics $9 and opium addicts $10. The shape of the bottles, flat on one side and bulbous on the other, the label designs and other accoutrements were protected by copyright but Keeley never sought protection for the formula. That would have necessitated specifying its content. Despite – or perhaps because of – the secrecy and the expense, his success was prodigious.[15] During the year 1892–93 more than 14,000 people took the Keeley cure; and 'no thanks, I've been to Dwight' became an acceptably polite decline of a drink in New York's clubland. Keeley died at fifty-five in 1900, still a reclusive millionaire, but some of his institutes survived till the 1920s.

Rival establishments were based on other theories and concoctions. Dr George E. Pettey's clinic in Memphis was founded on the idea of a latent toxaemia which needed to be exorcised by prolonged and copious sweatings. Dr Ernest S. Bishop in New York was strong on antibodies and heroic and prolonged purgation. Dr William E. Waugh of Chicago waxed grim about gastric poisons generated by drugs. Gastric suction, long before it became a surgical commonplace, was the answer. All had advocates and all claimed cures. Many were profitable. None achieved reproducible results.

* * *

As the drug 'epidemic' – an increasingly used though inapplicable term – showed no sign of abatement, the pressure for 'action' grew. The phrase that in public life almost always foretells disastrous decisions, that *something must be done*, was increasingly heard. Its urgency was intensified by a new menace that would change the history of drugs.

CHAPTER 22

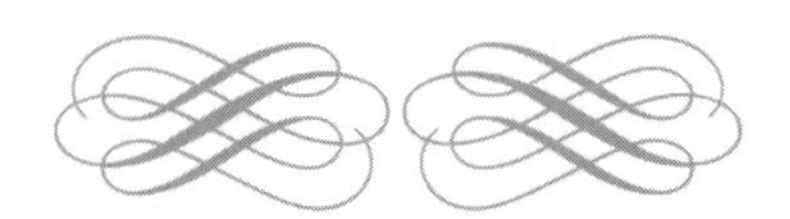

A heroic substance

THE STORY BEGINS in London at St Mary's Hospital, Paddington, where in 1874 Dr C.R. Alder Wright, a chemist and future Fellow of the Royal Society, was trying to find a derivative of morphine which might be effective but not habit-forming. A fairly obvious possibility was to try a chemical reaction known as acetylation, the transfer of an acetyl group, -CH2-COOH, or several acetyl groups, from acetic anhydride, a reactive form of acetic acid or vinegar, to the compound to be modified. The experiment resulted in a greyish powder which Alder Wright rightly assumed to be diacetylmorphine. He fed it to his dog which in the past had shown itself partial to morphine and might appreciate a slight change. But the animal became violently sick, then hyperactive and nearly perished. To an English dog-lover this was, in today's idiot political jargon, 'unacceptable'. He threw the powder in the bin but, at the urging of an assistant, reported his findings in a short paper in the *Chemical and Pharmaceutical Journal*.[1]

There the matter rested for more than twenty years and might have rested indefinitely if a young professor of chemistry from Göttingen University in the part-time employment of the small and struggling pharmaceutical firm of Bayer in the Rhineland had not been a compulsive browser through old journals. In the course of his browsing Heinrich Dreser came across Alder Wright's communication. The concept of acetylation struck a chord: this was the chemical reaction which his small team was trying on salicylic acid, the standard drug used in rheumatic fever, to make it more palatable.[2] Excitedly Dreser instructed one of his postdoctoral students, Felix Hoffman, to repeat Alder Wright's experiment. Hoffman had his hands full but professorial suggestions in Wilhelmine Germany (and for a long time afterwards) were not topics for a full and frank exchange of views. In the event Hoffman had no difficulty in recreating the Englishman's grey powder; but, instead of trying it on dogs, some time in September 1898 Dreser fed it to rats and rabbits. They liked it. Dreser then persuaded four young workmen in the workshop – Bayer was not yet a factory – to give it a trial. No ethical committees existed at the time and nothing by Dreser was done by halves. His human guinea pigs were given huge but luckily non-fatal doses of the new substance. The absorption of oral diacetylmorphine is even less predictable than that of morphine, but much of it must have got through. Had the experiment been a fiasco, the powder would have been cast back into obscurity, perhaps for ever; but even the most sceptical among the four youths stated that its effect was nothing short of miraculous. It made the meekest among them resolute and confident, a second Horatius Cocles.[3] The after-effect was a slight depression, headache in two subjects and a tinge of blueness of the lips in one; but all four were ready, indeed eager, to repeat the experiment. One of them described the action of the powder as 'heroisch' inspiring Dreser with a name. He had been toying with something implying the 'miraculous' ('wunderlich'); but this might offend Catholic members of the Bayer Board. A better alternative was *heroin* from *heroisch*.

* * *

Though controlled clinical trials were still in the future, both Dreser and the Bayer Board prided themselves on their concern for the safety of their customers. The firm subsidised two general practitioners in Berlin who would distribute the drug at random among their patients and see what it did to them. (The randomness was considered the last word in scientific rigour.) The doctors were allowed four weeks to come to a conclusion – a generous allowance at the time to reach a verdict – but a few days were enough. Their reaction was enthusiastic. Most striking was the effect on 'chestiness', sore throats and coughs, distressingly common in Berlin in mid-December. Symptoms disappeared. Two chronic patients with pulmonary tuberculosis stopped coughing up blood. Most unexpectedly, even the most dejected found that the new drug gave blessed relief from mental perturbation. With its cessation, splitting headaches subsided. Hope, that most precious of commodities, was kindled. The festive season was enjoyed by all. In one case a long-established craving for morphine seemed to evaporate. Even a terminal cancer patient seemed to face the future with new serenity. Seeing these wonderful effects more than one relative and friend begged to be given the drug. Strict adherence to the protocol meant their requests were refused, an exception being made for the Countess von Roon, a kinswoman of the Chancellor. It later transpired that she fed the drug to her depressed dog which, contradicting Alder Wright's experience, instantly regained its sunny disposition. Complications and serious after-effects were none.

It would have been unethical to withhold such an elixir for longer than absolutely necessary. Dreser reported the discovery to the Congress of German Naturalists and Physicians where it earned him an ovation. He gave credit to Alder Wright's preliminary experiments; but what mattered was the future. This was to be bright for millions of sufferers from coughs and colds and incidentally for Bayer's. The new product was claimed to be ten times more effective than codeine and eight times less toxic. Compared to morphine it was said to be five times more potent and totally non-habit-forming. (Nobody questioned how the last invaluable attribute was established after a trial lasting four weeks.) Even the head of the firm, the dynamic but sceptical Carl Duisberg, was caught up in the excitement. Together with the new acetylated salicylic acid, recently launched under the name of *aspirin*, heroin would make the firm's future.

It was sold in boxes of vials for injection and in attractive little sachets stamped with a lion and globe, the labels on both stating that during chemical processing the new drug had been cleared of the addictive properties of morphine. By subcutaneous injections the effect would be felt 'at the end of the needle'; but the most popular form was the pastille – 'heroin cough lozenges' – soon to be sold by the million. An elixir in a glycerine solution was also on sale and was said to be absorbed from the stomach within minutes. In whatever form, the new drug could be safely given to children of all ages, even to infants, and to pregnant women.

Distribution in Germany was followed by an aggressive export drive and reactions from abroad, especially from England, were even more enthusiastic. Indeed they revealed a new indication and potential benefit. Some English doctors found heroin a uniquely effective cure for morphine addiction. One practitioner from Brighton recounted the case of an advanced morphine addict, 'a former officer in Her Majesty's Indian Army now incapable of leading a normal life or even reading *The Times* newspaper' whose craving for morphine was 'miraculously cut short' after a course of four-times-daily heroin injections lasting two weeks. (The long-term effects do not appear to have been reported.) Propelled by such triumphs the new remedy quickly crossed the Atlantic and the first and until then the only representative of Bayer's in the United States, a certain Friedrich Muller of New York, reported that 'no remedy was ever heralded and received in this country with such enthusiasm'. Nor were expectations disappointed. The drug 'had everything to offer at no cost in health and at a magnificently low price'.[4] Demand exceeded all hopes.

* * *

As the most potent and widely used opium product heroin has now survived for over a hundred years and there is no immediate prospect of it being supplanted. It may therefore be appropriate to digress briefly and consider its production and, more important, its clinical action.

Contrary to what is sometimes featured on both the large and the small screen, no sophisticated laboratory equipment is required to convert crude opium to heroin. Morphine is first extracted with lime dissolved in hot water. The former precipitates organic waste, leaving the morphine and a few related compounds in the supernatant suspension as a faint white band. The band is drawn off and reheated. After the addition of concentrated ammonia the morphine solidifies and sinks to the bottom. It is collected by filtration, now weighing about a tenth of the original material. Equal quantities of morphine and acetic anhydride are then heated for six hours at 85° C and chloroform is added to precipitate remaining impurities. The solution is drained. Added sodium carbonate makes the heroin sink. It is further purified with activated charcoal and alcohol. The alcohol is gently evaporated and the residual pure heroin is converted to a greyish amorphous water-soluble hydrochloride salt. The whole procedure requires no more equipment than can be found in any small hospital laboratory and depends for success on conscientious application rather than on specialised skill. No chemical procedure is foolproof; but the high-powered heroin laboratories featured in some blockbusters are fantasy.

In the bloodstream heroin is rapidly converted to morphine but, being significantly more fat-soluble in its native state, some of it penetrates the central nervous system at once. Its immediate effect is therefore stronger than that of morphine but it is of shorter duration. The skewed time-curve makes a direct comparison difficult but weight for weight heroin is reckoned to be five to eight times more powerful as a happiness-inducing agent and as an indirect analgesic. The analgesic effect is 'indirect' because it is largely due to the psychological response to pain rather than to the objective suppression of experimentally induced painful nerve impulses. Even after a hundred years much of this effect is incompletely understood.

* * *

A good deal of the early heroin literature makes uncomfortable reading today. In cases of cough or sore throat, the first 'absolute indications' for the drug, four or five daily doses for four or five days were recommended to suppress the symptoms permanently. This would have been enough to make discontinuation of the injections or pastilles a wrench, the first observation that should have sounded alarm bells. It did not. A few unfortunate accidents were recorded. The wife of the local pastor in Bielefeld, not far from the Bayer establishment, who had been denied continued injections against her attacks of migraine, committed suicide. Such mishaps were dismissed as exceptional and to be expected with any drug in highly strung individuals. But by 1902 the evidence of risks was getting stronger. At least a dozen cases of addiction and some infant deaths had been published from various parts of Germany. By 1905 the evidence was overwhelming. One elderly and anxious member of the Bayer board suggested that instead of selling 'pure heroin' as a Bayer product, the drug should be sold to other pharmaceutical companies to be incorporated into their patent cough medicines as an 'adjuvant'. There was no need to mention heroin. The suggestion was voted down as being arguably against the law. Much simpler and perfectly legal, reports of toxicity, addiction and death would be suppressed. As a gesture, the heroin content of the best-selling cough lozenge was reduced from 82 to 65 per cent. It continued as one of the best-selling drugs over the counter. Ignorance was later claimed to have been the reason why Bayer's did not withdraw heroin from the market or stopped advertising it as entirely safe until 1913. When it was finally discontinued, no explanation or advice was given to established users. As late as 1912 J.D. Trawick, a Kentucky physician, was quoted as saying that casting aspersions on this 'wonderful drug' was like 'questioning the fidelity of a good friend'.[5]

* * *

Dangerous, perhaps; but as a recreational drug heroin had much to commend it. It was more immediately effective than morphine; and many maintained (as many still do) that as an inspirer of instant confidence it was superior. Paradoxically, considering that it was a morphine derivative, its street price was for long and in many places lower than that of the parent drug. There were several reasons for this. As an off-white amorphous powder it was easier to 'cut'. Even cut the immediacy of its effect seemed to make it acceptable. It was slightly more easily concealed than morphine: trafficking in it was simpler. For some reason still not entirely clear it was less constipating, or so its devotees claimed. It was – and is – less prone to cause nausea and gastric irritation.[6] Relaxation after the high was said to be less likely to generate sleeplessness and nightmares. As with most addictive drugs, the whole truth has always been difficult to establish. Both drugs readily cross the placental barrier but heroin probably more so. Babies of mothers who continue to take the drug through pregnancy can be born severely addicted. It is a condition distressing to treat. The newborn is no more ready or inclined to be weaned from the drug than are grown-ups. Traces of heroin are excreted in the mother's milk.

A large body of experimental work has explored the effect of both heroin and morphine on animals. There are considerable species differences but even within one species the two drugs have significantly different effects. Most (but not all) dogs, like Alder Wright's terrier, tend to dislike heroin but not morphine: in some breeds heroin is safe but in others it is fatal even in small doses. In cats heroin causes convulsions, morphine does not. Hamsters and rats like both. Horses and greyhounds are significantly but not blatantly slowed down and tend to lose important races. Testing for either drug in the bloodstream, tissues and urine is not particularly difficult; but for opium nothing like a breathalyser test exists.

One property of heroin has never been in doubt. Of all opiates, indeed probably of all drugs, it is the most addictive. This is not an easy characteristic to measure. There are individuals, even if not many, who can maintain themselves on the same dose over a lifetime. Others need to vary the dose slightly and generally upwards but suffer little or no ill-effects. But by the most obvious measure of addictive power, the suffering caused by withdrawal, heroin is ahead of all other mind-benders. The description of its horrors in Robert S. de Ropp's *Drugs and the Mind* cannot be bettered but is not for the squeamish:

> Eight to twelve hours after the last dose the addict begins to grow uneasy. A sense of weakness overcomes him, he yawns, shivers and sweats all at the same time while a watery discharge begins to trickle from the eyes and inside the nose which he compares to 'hot water running up into the mouth'. For a few hours he may fall into an abnormal tossing, restless sleep known among adults as the 'yen sleep'. On awakening, fifteen to twenty-four hours after his last dose, the addict begins to enter the lower depth. The yawning may become so violent that it can dislocate the jaw, watery mucus pours from the nose, tears from the eyes. The pupils are widely dilated, the hair stands up and the skin itself is cold and shows that typical goose flesh which, in the parlance of addicts is the original 'cold turkey'.
>
> Now, to add further to the addict's miseries, his bowels begin to act with fantastic violence: great waves of contractions pass over the stomach, causing explosive vomiting, the vomit being frequently blood-stained. So extreme are the intestinal contractions that the surface of the abdomen appears corrugated and knotted, as if a tangle of snakes were fighting beneath the skin. The abdominal pain is increasingly severe . . . As many as sixty large watery stools may be passed in twenty-four hours . . .
>
> Twenty-four to thirty-six hours after his last dose the addict presents a truly dreadful spectacle. In a desperate effort to gain comfort from the chills that rack his body he covers himself with every blanket he can find. His

> whole body is shaken by twitchings and his feet kick involuntarily [the origin of the term 'kicking the habit'] . . .
>
> Throughout this period the addict obtains neither sleep nor rest. His painful muscular cramps keep him tossing on his bed or on the floor. Now he rises and walks about. Now he lies down. Unless he is an exceptionally stoical individual [and stoics do not become addicts], he fills the air with cries of misery. The quantity of secretions from eye and nose is enormous [as is] the quantity of fluid expelled from the intestines. The profuse sweating keeps bedding and mattress soaked. Filthy, unshaven, dishevelled, befouled with his own vomit and faeces, the addict at this stage presents an almost subhuman appearance. As he neither eats nor drinks he rapidly becomes emaciated and may lose as much as ten pounds in twenty-four hours. His weakness may become so extreme that he cannot raise his head. At this stage many physicians will start fearing for the lives of their patient and give them an injection of the drug. This almost at once removes all the dreadful symptoms but only for a time.[7]

Death from withdrawal does occur but is rare. The patient may beg to be taken out of his or her misery but they are usually too weak to take their own lives. If no additional drug is given the symptoms begin to subside on the fourth to sixth day but the patient is left deeply exhausted, nervous and restless. He or she may continue to suffer from stubborn colitis. Today the clinical course can be significantly modified with tranquillisers, opiate analogues and expert hypnotherapy and psychotherapy but it remains one of the most painful experiences – and one of the most effective means of torture – in the technically advanced world.

* * *

In general the honeymoon with the new 'safe' opium derivative was slightly shorter than after the introduction of morphine to replace laudanum. In 1906 the Council of Pharmacy and Chemistry of the American Medical Association added a warning to its listing of heroin among 'new and non-official cough and pain-killing remedies' that 'the habit is readily formed and can lead to deplorable results'. Such perceptions take time to percolate through the medical profession and took even longer a century ago. For a few critical years the warnings bypassed the lobbyists and politicians who now clamoured in Washington for all drugs of abuse to be stamped out once and for all. Campaigners with their sights fixed on such luminous objectives sometimes miss seemingly unimportant details. For some years heroin was such a detail.

CHAPTER 23

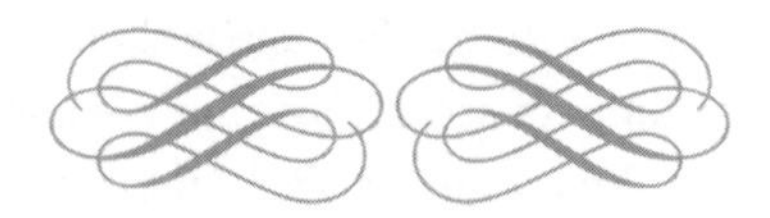

The birth of a crusade

THE CAMPAIGN – BUT crusade may be the more appropriate term – to suppress all drugs of addiction everywhere was the first in which the United States chose to lead the world. Once again international politics and opium interlocked. After triumphantly concluding the Spanish–American War – the entire Spanish Pacific fleet was sunk in Manila Bay by Commodore George Dewey on 1 May 1898 – the United States found herself to the amazement of many an imperial power. To some this was profoundly unwelcome. Imperial conquests were not envisaged by the Founding Fathers. But the consequences of victory are no more escapable than those of defeat; and those who were uneasy unexpectedly found themselves in a minority. The *Washington Post* of 1902 summed up the national mood:

> The policy of isolation is dead. A new consciousness seems to have come over us of strength . . . and with it a new appetite, a yearning to show our strength . . . The taste of empire is in the mouth of the people, even as the taste of blood in the jungle.[1]

The most tangible result of America's new strength was the acquisition of territories which would never become part of the Union but were clearly destined – in their own best interest – to be guided by the United States. This was a somewhat ambiguous circumlocution but better to circumlocute than to face a slightly unpalatable fact. As booties of war the United States now possessed what other nations might well describe as colonies. And from the start these possessions proved to be as troublesome as other imperial powers had found them. In the Philippines in particular, a collection of some 7,000 islands, an independence movement against Spain had been smouldering for years; and the insurgents had established a republic immediately in the wake of Spain's defeat. This, President McKinley concluded, was not an option. What concerned him was the probability, even the certainty, that abandoned to themselves they would be grabbed by a colonial power. An independent Philippines ruled by local savages was unthinkable. 'There was nothing left for us to do,' the President reported to an assembly of fellow Methodists, 'but to educate the Filipinos and uplift, civilise and Christianise them ourselves.'[2]

Of this, there was to be little sign over the next decades. The guerrilla forces who had been gathering to fight against Spain now turned against the Americans. The war was fought with extraordinary brutality on both sides. General Jacob H. Smith, affectionately known as Hell Roaring Jake, famously instructed Major Littleton Waller: 'I wish you to kill and burn, the more you will kill and burn the more you will please me. I want all persons killed who are capable of bearing arms against the United States.'[3] More representative perhaps was the opinion of an army doctor. 'What a thieving, treacherous, worthless bunch of scoundrels these Filipinos are . . . You can't treat them the way you do civilised folks. They're really animals.'[4] His antipathy, as the Reverend Stephen Stegner, a Methodist missionary, reported to his superiors, was reciprocated by the Filipinos, or rather by the Filipino elite. This was a small but influential Spanish-speaking urban class who held non-Spanish-speakers, including the large Chinese minority, in contempt. The antagonism did not augur well for the future, least of all for the solution to what was the besetting problem of the islands – opium.

More precisely, opium became the besetting problem with the American occupation. For half a century the Spaniards had operated a lucrative opium monopoly to cope with the demand of the 700,000 Chinese. Contracts to sell the drug were auctioned to the highest bidder with the proviso that they would not be sold to Filipinos. Both the bidders and the operators of the public 'dens' were Chinese and this was a condition with which they readily complied. Rich Chinese, too fastidious to smoke in the company of the common herd, were permitted by the Spaniards to establish more exclusive clubs. The system had raised the equivalent of $600,000 a year for the colonial government; and the price had by and large deterred native Filipinos from contracting the habit. The arrangement was promptly repudiated by the American authorities as 'utterly repugnant to American practice and theory of government'. There were to be no more auctions, opium shops were to be shut and opium smoking was prohibited. Consumption at once began to rise. In theory the abolition of the monopoly was an honest effort to suppress the trade. In practice and with the unstoppable smuggled inflow of the drug its price fell sufficiently for poorer Filipinos to give it a try. Soon many developed a taste for it almost as avid as that of the Chinese. When a particularly fierce outbreak of cholera struck Manila soon after the new prohibitions came into force, the governor, William Taft, proposed a temporary reversal of the imposed ban. The news, slightly garbled, leaked out in the United States and triggered an outcry. Was undue pressure being brought on the governor by wily Chinese? Taft had presidential ambitions and reversed his proposal. There was to be no going back. The Philippines were to become the testing ground of a 'great force for good' that had been gathering strength and was now ready to strike.

* * *

The original focus of the movement aiming at a worldwide prohibition of narcotic drugs was the International Reform Bureau, a missionary organisation based in Philadelphia but with numerous outposts in the eastern United States. Its director, the saintly Reverend Wilbur Crafts, had been lobbying for many years for a worldwide ban on both liquor and opium, 'especially among the child races that civilised nations are essaying to Christianise'. He had always included the Chinese among the last but it was the acquisition of the Philippines that changed official reaction. This had been respectful, as it would be for an impeccably moral movement which had no hope of succeeding. But, unlike the Chinese and other unspecified children, the Filipinos were now the responsibility of the American people. Their real or potential addiction and its consequences were dangers that could not be ignored.

Other developments contributed to a change in attitude. Theodore Roosevelt's White House was alive with rampaging children, boisterous pets, despairing domestic staff, rushed-off-their-feet secretaries and the occasional bellow from the master himself, a notably different place from the quiet, sometimes slightly dowdy establishment of his predecessors. The new president had no doubt that it was time for the United States to assert its might. Elihu Root, his secretary of state, agreed. He immediately and enthusiastically embraced the programme of the International Reform Bureau: 'As to opium in China and among the savage races, they are a disgrace to civilisation.' Others in positions of influence concurred. In 1902 the President's congressional ally, Henry Cabot Lodge of Boston, pushed a resolution through the Senate forbidding the sale of opium and alcohol to all 'aboriginal tribes and uncivilised races'. The provisions of the Act were soon expanded to banish the sale of narcotic and inebriating substances in America itself to 'uncivilised elements' such as Indians, Alaskans and railroad workers (meaning the Chinese) as well as to the native inhabitants of Hawaii and, above all, the

Filipinos. A presidential commission was to be dispatched to the new colonies to study the opium problem and to make recommendations. The choice for the post of the Commission's chairman, to be appointed Episcopalian Bishop of Manila, was inspired.

Charles Henry Brent was already known for his pastoral work in the black ghettos of Boston and Philadelphia. The son of a canon of Toronto Cathedral, he was also recognised as a man of deep Christian faith. Retrospective glory is often heaped on reformers who succeed. On balance and many years later Brent looked back on his ministry in the Philippines as a failure. So in a sense it was. He made friends and even a few converts; and he himself learnt most of what there was to know about the opium trade. But within twenty years of the abolition of the Spanish monopoly the opium habit among Filipinos quadrupled. Yet Brent was among the few anti-opium campaigners who were admired by his opponents. He was a persuader rather than a preacher but of his commitment there was never any doubt. It was rooted in a personal tragedy. Ten years earlier his brother, Willoughby, a country doctor on the coast of Nova Scotia, had become an opium addict. William cared for him until the doctor killed himself. No wonder then that the bishop's hatred of the drug occasionally roused him to uncharacteristic outbursts:

> Your rich prodigal beauty no field crop under the sun can match. The flowering poppy is vivid, dramatic and passionate, like some superb adventuress alluring troops of lovers, and, vampire-like, sucking out their souls with her kisses.

But he had an intuitive understanding of what drove suffering and backward people to the drug and he did not think that coercion could ever bring about a change. He had no personal ambition. He had not been keen to come to the Philippines but, once there, he declined offers of some of the most prestigious episcopal sees in the United States. Yet it was soon clear that, with opium flooding into the islands from every direction, a purely local solution would never be found. In this, though in little else, he was in agreement with the most vocal member of his commission, the man who would later be dubbed Father of the American Antinarcotics Laws.

* * *

Hamilton (known to all as Ham) Wright was a medical graduate of McGill University, handsome, clever and a liar of the dangerous 'transparently honest' variety. Even when what he said was patently contradicted by the facts, it was hard not to believe him. His early researches on the bacterial origin of beriberi – a disease caused by vitamin B_1 deficiency – were widely acclaimed but quickly disproved. He found a new mission as the standard-bearer of the worldwide war on opium. In this role he became brash and bullying, two exemplary attributes in the Theodore Roosevelt entourage. The President himself became convinced that here was a man who knew the answers to the drug problem; and since most foreign problems had a drug dimension, this was a useful expertise. Foreigners by contrast found the doctor arrogant, patronising and untrustworthy, a dislike which the doctor reciprocated. With the exception of the Chinese whom he perceived as America's backward but natural allies, he regarded all non-white Americans as belonging to a lower evolutionary order.

His choice of China as a potential ally was shrewd. The empire was the pre-eminent example of what opium could do to an entire nation. It produced sloth instead of energy, dreams instead of achievement and illusions instead of realism. 'The whole world,' Wright wrote, 'has regarded with a shudder China's prostration underneath the curse of the drug; and American shudders have been the most vigorous.'[5] The empire could never become great again until it cast off the curse; and it was the mission of the United States to help it do so. What America could offer was not aid, goods or even religious faith but technical know-how. This the Chinese would soon recognise as God's blessing. It would make them overflow with gratitude. Technological advances would help them overcome their abject craving for the poppy. Wright was not entirely phoney. He was one of the first generation of his countrymen to believe that under American leadership worldwide prosperity for all deserving to prosper was attainable. He also had no doubt that opium and its derivatives were the most immediate obstacles in the way. But in his means he was totally unscrupulous, brushing aside reservations that might have troubled those less single-minded. He also underestimated the enemy, neither the first nor the last to do so.

He was a gifted speech-maker though his message to civic, political, religious and trade audiences varied little. First came the drug problem itself. The *New York Times* journalist E. Marshall reported one such speech. Did Wright's audience realise that 'of all the nations in the world the United States consumed the most habit-forming drugs per capita . . . Did they realise that opium, the most pernicious drug known to humanity, is surrounded in this country with fewer safeguards than any other nation in Europe fences it in with'?[6] Both statements were untrue but few were inclined to check. Then came the race card, steadily rising in intensity: 'Did his audience know that a fast-increasing number of cases are being reported of Negroes in the South becoming addicted both to opium and to a new form of vice . . . coke sniffing . . . which irresistibly compels them to rape white women . . . What makes the menace even more horrible is the fact that the drugs actually improved their marksmanship.'[7] Finally came Wright's trademark theme: the gratitude of the Chinese for the support of the United States. This was boundless, especially since 1894 when Parliament in Great Britain had shamefully repudiated the recommendations of their own Royal Commission and failed to ban the Indian opium trade. What a contrast to the resolution shown by the United States Congress to outlaw the smoking of opium by the Filipinos and to offer systematic detoxification for all Chinese addicts on the islands. 'This,' according to Wright, 'has raised the spirits of the best of the Chinese statesmen mightily.'[8] Indeed, so powerful was the impact of United States policy on the Chinese that before long a movement was afoot in Peking 'to suppress the drug in the empire altogether'.[9]

Circumstances helped to ensure that among businessmen at least Wright's advocacy fell on receptive ears. American traders had never been able to acquire more than a miserable 8 per cent of China's opium trade, an irritating anomaly attributed to underhand practices by their European competitors. It was time they were taught a lesson. American expertise in railroad technology, an immensely lucrative gambit, would be enthusiastically received by the opium-sodden mandarins. It would make them see the light and recognise their indebtedness to the United States. But there were wheels within wheels.

So long as anti-opium legislation was directed against faraway islanders United States Congress was a willing partner. It was different with addiction at home. A significant number of congressmen and senators felt that federal laws on moral issues – and drug addiction could be seen as one – were unconstitutional and thoroughly undesirable. Their enforcement would, they argued, require the action of the police; and law enforcement by the police was traditionally a matter for the individual states. It was a gut reaction but a powerful one. Legislators, or at least a significant number of them, would baulk at giving the federal administration new powers. But Wright glimpsed a way around this. If a comprehensive ban on drugs were part of an international treaty ratified by Congress and signed into law by the President, then individual states would have to comply. The idea of an international conference to explore such a possibility had been mooted some years earlier by religious anti-opium campaigners; but the project was then brushed aside as pious nonsense. In 1908 Wright espoused it with ardour. Formerly of no official status, the President now appointed him Opium Commissioner of the United States. He should prepare for a grand concourse of nations.

* * *

The First Opium Conference, which met in Shanghai in February 1909, was a milestone in both international relations and the fight against opium. It was also something of a farce. The opening was delayed by a week ostensibly as a mark of respect for the recently deceased dowager empress, a lifelong opium addict. It also gave delegates a chance to sample the delights of the city. These were exotic, numerous and laced with opium. One Italian diplomat involved in a brawl in a notorious den had to be sent home in disgrace. American Marines had to disperse lobbyists and demonstrators trying to buttonhole the delegates. (The United States and European powers enjoyed extensive extraterritorial rights including the right to station troops to protect diplomatic enclaves.) Bishop Brent was eventually elected chairman. By then delegations had arrived from China, France, Germany, Great Britain, Holland, Italy, Japan, Austria-Hungary, Persia, Portugal, Russia and Siam. The United States delegation tabled the key motion, that 'there is no non-medical use of Opium or its derivatives that is not fraught with grave dangers'. But agreement was half-hearted. The British would not say anything of substance in the presence of an unaccredited representative from India and the Indian delegate remained wrapped in tight-lipped silence in the presence of a British lord. France protested that French, the 'tongue of diplomacy', was not an official language. Persia, along with India the major supplier of foreign opium to much of Eastern Europe, was represented by a merchant heavily involved in the opium trade: he smiled blandly, seemed to listen with polite interest and said nothing. Most delegates with nothing to say, by contrast, spoke at length, recommending stringent regulation at some time though not yet. This was more or less agreed by a show of hands counted by Bishop Brent. The Dutch delegate put forward the idea of another meeting. This was enthusiastically supported. It was an anticlimactic climax but Wright got what he wanted. The 'strong consensus of opinion' which he claimed had been achieved made it 'incumbent on the United States' to take a step that would convert the 'Declaration of Shanghai' into international law. No less important, the proposal of a follow-up conference to be held in The Hague in the Netherlands would keep up the momentum.

Intense politicking followed. Huntingdon Wilson, head of the Far Eastern Division of the State Department, regarded Wright as a windbag; but Wright found a supporter in the new secretary of state, Philander C. Knox. They jointly drafted a bill which was mild, almost symbolic at one level but draconian at another. On the surface it merely called for the registration and payment of a modest, almost token tax on all opiates and other drugs of addiction and the registration of all drug sellers; but violators would face a far from token five years in jail and a fine of up to $5,000. Unlike imperial fulminations in China and elsewhere and in contrast to the largely toothless legislation in several European countries, this was to be the first truly effective anti-drug law in the world. If, of course, it was passed.

To introduce the Bill, Wright first tried to recruit James R. Mann who some years earlier had carried through the Mann or 'White-Slave' Act, the first federal law governing morals. When Mann declined Wright approached Representative David Foster of Vermont who agreed half-heartedly. During the deliberations of the 'Foster Bill', and carefully timed, the report of the Shanghai Conference, written entirely by Wright, was released. The 'agreement' presented opium as a worldwide evil which was now not only poisoning the child races but spreading to the adult world as well. This global curse 'America had heroically taken on to defeat'. In addition to opium the report reserved especial venom for 'that most appalling drug', cocaine, which had not in fact been discussed in Shanghai at all. Wright hoped that the racist horror stories – 'opium and cocaine are increasingly the direct incentives to rape by the Negroes in the South' – would overcome the resistance of Southern congressmen whose state-rights sentiments represented the greatest obstacle to the Bill. Even so and despite President Taft's appeal to Congress, the Foster Bill stalled and was eventually killed. What killed it were two special interest groups in alliance: the pharmaceutical lobby, because the Bill provided no exemption for patent medicines containing opiates; and the brewers because they perceived it as a prelude to anti-liquor legislation. But it was a temporary setback. A new sponsor was found in the person of Francis Burton Harrison, one of the notorious drunks on Capitol Hill. The drunkenness was a valuable attribute. Congressional acts, as brewing millionaire Zacharia Knobloch explained, could be treacherous when interpreted by tight-lipped judges. Harrison's sponsorship was a guarantee that no surreptitious clause curtailing liquor would be slipped into the text. While banning morphine and cocaine the new Bill also exempted patent medicines containing less than half a grain of heroin. In effect, it would be possible to purchase heroin over the counter for another ten years.

But before the final act in Washington the first Hague Conference convened in 1912. There was a short delay. Britain objected to a procedural point and Wright vented his irritation on the mild-mannered Randall Davidson, Archbishop of Canterbury, a lifelong abolitionist. Brent insisted on Wright apologising but the atmosphere remained sulphurous. Representatives of Portugal, Japan, Russia, Italy, Germany, Persia, the Netherlands, the United States and China finally expressed their agreement to global legislation in principle but with reservations. Portugal was in favour of a ban with the exception of trade based on the Portuguese enclave of Macao. This in practice would have given Macao a global monopoly on opium trading. But a similar demand was made by the Netherlands on behalf of the Dutch East Indies. Germany insisted on protecting her fast-growing pharmaceutical industry by exempting heroin from any provision against 'opium derivatives'. Japan denied exporting syringes and needles to China or anywhere else: the range of samples produced by the Chinese delegation were forgeries. Russia and Persia insisted on continuing to grow poppies for 'legitimate purposes', whatever they were. The French delegate protested against any infringement of the 'age-old rights of Indochina to trade in anything'. The conference eventually patched up a communiqué in which delegates committed themselves to 'endeavouring' to enact domestic legislation and to control all phases of opium, morphine and heroin production with the exception of patent medicines containing stated (but in practice unenforceable) amounts of morphine, cocaine and heroin. Since, however, only 12 of the 48 states in attendance at the Shanghai Conference were represented in The Hague none of the agreements could be binding until more signatures were collected.

None of this could prevent Ham Wright from proclaiming an international agreement. His career was reaching its peak. In a long and wide-ranging interview with the *New York Times* he was hailed as

> the person who knows more about opium than any other living man ... earnest, energetic, nervous, magnetic ... often referred to as Dr Opium in the White House and proud of it ... He is the leader of the world wide crusade, from world's end to world's end his fame is known. Everywhere he is regarded as the man who single-handed has accomplished most.[10]

After this breathless introduction Wright could expatiate on the iniquities of all those, including the pharmaceutical industry, who opposed legislation:

> The industry has already lost much of its dignity by fattening on drugs of addiction. One druggist out of every five lives from the profits of evil addictive drugs ... Lack of foresight has already made this nation the greatest drug fiends in the world, not excluding the Chinese ... No other country in the world today is either faced with such an opium problem or finds its efforts to meet the problem so viciously resisted.[11]

The Philippines had provided a great example of what could be achieved. There

> the vice had gained such foothold that it would have utterly destroyed the entire population but for the intervention of Mr Taft, Mr Root, and the Opium Commission. They deserve great credit ... The real problem now lies at home ... Every year we consume 500,000 pounds of the drug in

> contrast to the combined consumption of six European nations of 40,000. Only prompt and drastic action can save us from the same dismal fate that has overtaken the Celestial Kingdom of China. Fortunately, this is at last being realised by the President, the Secretary of State and a few others who can see beyond their own abject self-interest.[12]

Neither the Hague protocols nor any other manifesto of uplift could prevent another outbreak of racist hysteria. This time the official organ of the American Medical Association joined in, editorialising that 'Negroes in the South are reported as being addicted to a new form of vice – that of "cocaine sniffing" . . . It appears to induce blacks to rape white women.'[13] On 8 February 1914 the *New York Times* published an article by Edward Huntington Williams, entitled 'Negro Drug Fiends Are the New Southern Menace'.

> Murder and insanity due to dope is fast increasing among lower class blacks. Southern sheriffs had increased the calibre of their new weapons from 0.32 to 0.38 calibre to be able to bring down crazed Negroes rampaging under the effect of drugs.[14]

Two more conferences in The Hague followed, the second in 1913 and the third opening in May 1914. The last brought Wright's personal ascendancy to an end. Roosevelt having split the Republican Party which refused to nominate him for another term, the 1912 elections brought a Democratic administration to Washington and with it a number of Wright's personal enemies. In particular he was loathed by William Jennings Bryan, the new secretary of state, a mixture of evangelical fervour and dubious politics. His accusation, almost certainly false, that Wright was an alcoholic was a signal for a concerted attack on the doctor. When he refused 'to take the pledge' he was fired.[15] But even Wright's downfall could no longer impede the progress of Harrison's Bill. It sailed through Congress and was signed into law by President Woodrow Wilson on St Valentine's Day, 1914.

* * *

Like some other momentous Acts, once passed the Harrison Act acquired a life of its own. It had originally been argued that its purpose was clearly indicated in its preamble: it was essentially a means for regulating the taxation of 'narcotic drugs'. Its subsidiary and even more innocuous objective was to 'gather reliable information'. But it contained the mild-sounding clause that a doctor was allowed to prescribe and use the drug 'in the course of his professional practice only'. After 1917 this was interpreted as meaning that doctors were not allowed to prescribe opiates to addicts as part of the treatment of the addiction since addiction was not a disease. As a warning shot half a dozen practitioners known for their opposition to the Act were arrested and one was imprisoned. The debate might have dragged on if, with exemplary speed, the Supreme Court had not ruled that the Harrison lobby's interpretation of the Act was entirely correct. Shortly thereafter the court confirmed in *Webb* that medical practitioners could not prescribe narcotics purely to help addicts to cope with their 'self-inflicted moral infirmity which was not a disease'.

* * *

A critical uncertainty remained. To police the Act's provisions the only federal force available was a small contingent of less than a hundred officers who had been assigned to the Treasury Department to investigate income tax evasion. The Harrison Act had in fact been presented as a tax measure largely because this squad of 'T-men' was the only federal force which might be able to enforce it. The question now arose whether to interpret the Act in a narrow sense which could be policed at state level or to expand this 'T force'. The debate aroused high passions as federal-versus-state powers always did. Ironically, the decision was eventually made not in Washington or even in the United States but in Paris. There, in 1919, the peacemakers in a moment of pious intentions but inadequate thought incorporated the protocol of the last Hague Conference on opium into the treaty documents. The protocol, concocted in a hurry in the summer of 1914, was a mishmash of vague hopes and imprecise recommendations, not a blueprint for legislation; but to the treaty the United States was a signatory. This gave those who had promoted federal enforcement an automatic lever. International law must be obeyed and only a federal force could ensure that. No other country had in fact interpreted the Hague protocol in the same literal way; but few legislators had read the document in full. What amounted to a confidence trick worked.

The Narcotics Bureau of the Treasury was duly given the task of implementing the Act. Despite post-war retrenchment in most departments of state, within five years the Bureau's complement of agents trebled. During the creeping unemployment of the early 1920s recruitment presented no difficulties. Apart from youth and 'dynamism', the only absolute requirement was the applicant's 'visceral' hatred of druggies. It was the appointment panel's pride to be able to spot the genuine article. Few liked the organisation, which seemed to be accountable to nobody, but nobody could accuse it of wasting public money. Between 1920 and 1925 more than 2,000 people, including

124 doctors, were arrested for 'illegally disseminating or dispensing opium derivatives'. Only eight of the doctors were found guilty and only four were sent to prison but many were disgraced. Dr Wilbur Muffet, a well-respected Boston practitioner aged eighty-four, committed suicide before the verdict. The crusade was on the march.

CHAPTER 24

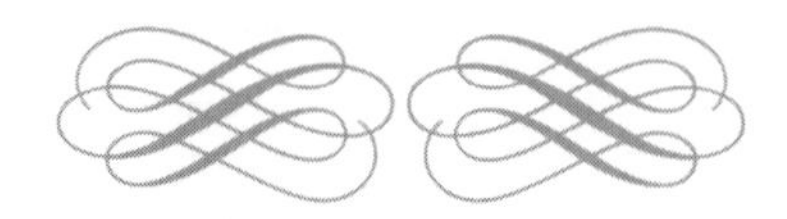

War and Peace (of sorts)

BUT THIS IS LEAVING behind historic events elsewhere. Two fatal shots by a consumptive Serbian student at an Austrian archduke and his wife on 28 July 1914 led to a war that, with uneasy intervals between periods of wholesale killing, would occupy thirty-one years of the twentieth century.[1] On the outbreak of hostilities on 3 September 1914 negotiators in The Hague dispersed amid lamentations but promised to reassemble next spring when this regrettable *bêtise* would be over. In the meantime existing provisions relating to opium would have to be shelved. Neither the need for opiates nor the practical implications of those needs were high on anybody's list of priorities. But, though no precise figures exist, anecdotal evidence suggests that consumption in the belligerent countries started to increase almost at once and continued to increase throughout the next four years.

Despite the reciprocal blockades and the U-boat campaign which in some countries dangerously reduced the stocks of essential drugs like quinine, supply of opiates rarely ran out. Turkey, Bulgaria and to a lesser extent Hungary met the needs of the Central Powers. The Western Allies received most of their supplies from India and Egypt. Turkish opium was also smuggled to Italy. The hundreds of thousands of red poppies of Flanders Fields, nurtured by young blood, yielded no narcotics.

When morphine and heroin did begin to attract attention, it was their effect on troop morale and discipline which jolted ministerial minds. By the end of 1915 it began to dawn on some of the war leaders that the sheer volume of machine-gun fodder might ultimately become the decisive factor in the war; and numbers at the fronts were beginning to be affected by the tide of desertions and continued incapacity after periods in hospital. Field Marshal Franz Conrad von Hötzendorf, chief of the Austro-Hungarian General Staff was the first to issue explicit orders to prohibit the selling of narcotics to soldiers. The penalty for breaking the prohibition increased fast from a few days' detention and loss of leave to death in February 1916. Ten offenders, five soldiers and five civilians, including an orthodox priest, were executed in Galicia on 15 March 1916. Pictures of the bodies on the front pages of Budapest newspapers appeared the following week. But in the face of death the death penalty tends to lose its deterrent clout. Geza Csath, doctor, poet and wonderchild of Hungarian letters, had enlisted at the outbreak of war and was working in a military hospital in the Carpathians. 'War is pain, war is fear, war is despair, war is death'.[2] While doling out quantities of morphine to patients mutilated and shattered in mind and body, he kept himself going by consuming increasing amounts of the drug himself. Within six months he was addicted.

Other army commanders followed von Hötzendorf's example. Withholding supplies of morphine was the immediate cause of the first mutiny in March 1917 in the French Army. It was not brought to an end until the order was rescinded, Surgeon General Boissier de Brune warning the High Command that the entire medical corps would desert if morphine was not provided to field hospitals. In a chapter of the war still inadequately researched some of those judicially murdered for cowardice in the British Army, including Second Lieutenant John Wilson, MC, had probably committed their 'offence' under the influence of drugs.

Only marginally less worrying was the spread of narcotics at home, both the illicit sale to army personnel on leave or convalescing and among the civilian population. The Defence of the Realm Act (DORA) was enacted by Parliament on the outbreak of the war with hardly a debate, abolishing or curtailing freedoms that had been cherished for centuries. Sir Malcolm Delevingne, deputy undersecretary at the Home Office, now proposed that drug dealing should be brought under the umbrella of this catch-all legislation. The need seemed to become daily more pressing. In February 1916 two upmarket department stores in London, Harrods and Savory & Moore, were fined for ignoring existing legislation by advertising in *The Times* prettily gift-wrapped packets of morphine in the form of 'lamels', 'useful and attractive presents for friends at the front'. As Sir William Glynn-Jones, prosecuting counsel in the Harrods case observed, it was 'exceedingly dangerous for a narcotic drug to fall into the hands of men on active service': not only could it make them inattentive on guard duty and thus imperil the lives of their comrades but the habit could render them 'sullen, morose, bad-tempered and even insubordinate'. In the light of such threats and the grim lengthening of the casualty lists public agitation grew; and regulation 40B added to DORA made the selling of intoxicating drugs to members of the armed forces a criminal offence. On 10 February Horace Kingsley, an ex-soldier, and Rose Edwards, a London prostitute, were the first to be sentenced to six months' hard labour under the Act for selling drugs to Canadian troops billeted in Folkestone on their way to France. The case was exciting, but it demonstrated the limitations of legal constraints. An elaborate investigation involving a decoy and a chase by policemen on bicycles had been called for to catch two small fry out of probably thousands. At the same time Captain J.B. McMurray, in charge of the Folkestone military hospital, testified that forty soldiers were already under treatment to try to rid them of the habit. Some might prove to be permanently unfit for active service. Other arrests followed. Sir Edward Henry, Metropolitan Police Commissioner and Sir Edward Lloyd, General Officer Commanding London District, urged immediate action. But what action?

The medical response to such splutterings was usually guarded; but in the heat of the war it became positively enthusiastic. The *Lancet* editorialised about conditions prevailing in a great war once again securing overdue legislation which had been advocated by doctors but ignored for many years of torpid peace.[3]

But even under the new regulation the eradication of drug trafficking in English cities proved impossible. Spectacular operations like those employed against Kingsley and Edwards could be mounted occasionally but not on a daily basis. The police themselves were overstretched. When William Johnson, avuncular head porter of the Café Royal and widely known as the brain behind Soho drug dealing, was arrested the only charge which could be proved against him was possessing a single sample of a proscribed drug improperly labelled. The case was thrown out by the magistrate with a biting comment about the mental age of the prosecuting police sergeant. Worse was to come. Horrific examples with names and places of soldiers driven by drugs to violence began to circulate. Dr C.W. Saleeby in the *Daily Chronicle* described soldiers crawling into chemist shops in France begging for their favourite drug. 'Could the spread of morphine be a Hunnish plot to bring the Empire to its knees?' It might have been pointed out (but was not) that the same warning in reverse was headlined in the *Berliner Zeitung*. More rigorous regulations followed and once again met with general approval. The *Lancet* in particular expressed the view that 'In future as the new regulations begin to bite, the morphine addict will depend on supplies from clandestine sources; and, as special efforts are being made to stop the smuggling, the cost will soon be so high that, we may assume, addicts will become an extinct type.'[4] Rarely has a medical journal been so prescient and silly in the same sentence. But the agitation in the Northcliffe press for the instant hanging of dealers on the nearest lamp-post was resisted.

* * *

Armistice Day changed the drug scene. In the defeated countries – Germany and the fragments of Austria-Hungary – military units were disbanded and told to make their way home as best they could. Geza Csath was by then a patient in a detoxification unit in an army hospital in Transdanubia. One day food, candles and fuel ran out and everybody was discharged. In a dangerous state Csath walked home to Baja in southern Hungary, some eighty miles away. He fell into the arms of his wife Olga and then shot her. He was arrested, then died from an overdose of morphine.

His was not the only drug-related murder and suicide. In March 1919 the supply of morphine ran out in Munich and adulteration reached a new peak. Thousands of addicts were admitted to hospitals jaundiced, in acute liver failure. A bacterial epidemic was first suspected but the truth soon emerged. The adulterant was never identified or at least never revealed; but, while some of the patients died quickly, others made an apparent recovery only to relapse after a few weeks and die a more lingering death

* * *

On the winning side the end of the war shifted the drug scene to the world – or half-world – of the smart and fashionable. In London Miss Billie Carleton, singer and actress, was found dead in her flat from an overdose. At her inquest and his trial Reggie de Veulle, her friend and dress designer, explained that he had been arranging parties in the West End at which drugs were freely available and at which Miss Carleton was much in demand. Such jollifications tended to be on the relaxed side, with ladies and gentlemen, as *The Times* put it, 'frequently divesting themselves of their outer garments'. 'Unholy rites' were mentioned with a shudder. It was unfortunate that Miss Carleton, 'a frail beauty, too delicate for this world', had overindulged (or perhaps she was abnormally sensitive), but there it was: no pleasure in life was without risk.[5] At the inquest of suicide or murder by heroin of Miss Freda Kempton, a twenty-three-year-old dance instructress in West End nightclubs, the glamorous Chinese drug dealer, Brilliant Chiang, appeared in person. And wholly unlike the Chinese stereotype of newspaper cartoons, he was a person indeed, handsome and radiating power. He was mobbed by a crowd of amorous young ladies, 'one, more daring than the rest', as the *Empire News* reported, 'fondling and stroking the Chinaman's smooth black hair and even insinuating her fingers under his richly embroidered shirt'. 'Drug Sign',

'Cocaine Eyes', 'Morphine Twitch' and other drug jargon started to pepper newsprint and then everybody's conversation. The American 'drug fiend' made his bow.

Britain's most notable contribution to the menagerie was 'The Great Beast' in the person of the writer and poet Aleister Crowley. The descendant of Plymouth Brethren, he became, in his own estimation, 'The Wickedest Man in the World'. He was expelled for minor offences from two public schools but then inherited his father's fortune from the family brewery and did well at Trinity College, Cambridge. But he became interested in the occult and to promote mystical, sadomasochistic and erotic visions started to indulge in morphine and heroin. He was a prolific writer of well-plotted and torrid autobiographical novels culminating in *Diary of a Drug Fiend*, published in 1922 when he was forty-seven. The book became an international best-seller. Some critics with personal knowledge of narcotics always claimed that his 'first hand experiences under the influence' are fictitious; but that remains questionable. In a lawsuit which he brought for defamation against a newspaper the presiding judge stated:

> What I have learnt in this case is that we can always learn something new. I have never heard such dreadful, horrible, blasphemous and utterly abominable stuff as that which has been produced by the man who describes himself as the greatest poet of our age. It makes an ordinary person's flesh creep.[6]

This was great publicity; but what exactly Crowley did apart from tormenting goats, celebrating black masses, preaching sexual freedom and advocating the use of liberating drugs – 'oh glory to you divine heroin' – is not clear. Like most of his kind he went out of fashion when real bombs started to rain on London and other cities; and he died in a cheap Hastings boarding house in 1947 mourned only by his elderly tabby cat.

* * *

'It is strange to reflect,' Mr Garrett, the Bow Street magistrate commented in the Reggie de Veulle case in 1919, 'that until recently [before DORA] these horrible drugs could be bought by all and sundry, even by children, in ordinary shops like a grocery or petshop.' This was a slight exaggeration but sufficiently true for Sir Malcolm Delevingne to write in a memorandum to the Home Secretary: 'Such a deplorable state of affairs must never be allowed to recur. Even in peace-time no relaxations of the regulations . . . must be permitted.'[7] In Parliament Members demanded harsh punitive action against traffickers like Reggie and blasphemers like the creepy Crowley. Referring to de Veulle's vile 'Chinese associates', Colonel C.E. Burn, MP called for the immediate reintroduction of flogging: 'I have never taken any personal pleasure in such practices but those aliens are frightened by nothing else.' Sir Malcolm managed to frustrate the soft Department of Health and the sluggish Board of Trade and kept drug control in the stern hands of the Home Office.

* * *

Soon a wider stage opened on which to debate reform. From the ruins of the First World War there emerged what President Wilson, with a glint of his pince-nez, described as a 'New Hope of the World', an international organisation destined 'to prevent future wars by establishing relations on the basis of justice and honour and to promote co-operation, material and intellectual, between all countries'.[8] It was to be named the League of Nations; and one of its earliest and boldest actions was to inaugurate a permanent 'International Advisory Committee on the Traffic of Opium and other Drugs'. The body first met on 4 August 1924, exactly ten years after the beginning of the First World War. It was then described as a 'foster child' of the League rather than a legitimate offspring, the term allowing the all-important participation of the United States and perhaps of other non-League members. But as the Belgian chairman insisted in his opening address, this was still an historic moment. It heralded the end of the immoral and evil trafficking in stultifying drugs and the terrible habit which over the centuries had blighted the lives of millions. The Queen Mother of Spain in the audience was moved to tears.

The first act of the Committee was to issue a list of sixty-two countries (always described as 'powers' in League language) which, by signing the Versailles Treaty, had automatically signalled their adherence to the final Hague Convention. Only Argentina, Lithuania, Paraguay, Persia, Mexico, Monaco and Russia had not yet ratified for reasons which varied from oversight (Paraguay, Persia and Monaco) to preoccupation with the world revolution (Russia and Mexico). The next step was to ascertain the current state of the trade by calling for information from countries in East and South East Asia, traditional centres of poppy cultivation. The replies foreshadowed the magnitude of the task ahead.

Though India had officially ceased exporting opium to China, it was still the largest producer of the drug, about 75,000 kilos per year, of which about half was exported. Indian opium was low in morphine and therefore ill suited to extraction but 'ideal for smoking and oral consumption'. Beyond these facts information was vague. The India Office in London, apparently piqued by not being consulted by Sir Malcolm Delevingne, issued an 'unofficial' paper, *The Truth about Indian Opium*, which stated:

> The prohibition of growing and eating opium in India we regard as impossible, and any attempt at it as fraught with the most serious consequences to the people and the Government . . . Centuries of inherited experience have taught the people of India complete discretion in the use of the drug . . . Even if it were possible to legislate the suppression of opium in India, of which there is at present and in the foreseeable future no prospect whatever, geographical and political limitations would place it beyond our power to prevent illicit export, import and cultivation.[9]

This caused deep concern, but no immediate solution suggested itself. Baron von Mettau, the Latvian delegate, wondered if poppy growing might be prohibited under threat of instant execution of the culprit and his or her entire family. One of the baron's ancestors successfully used that approach. Sadly, perhaps, it was pointed out to him that twentieth-century India was not seventeenth-century Latvia. Who exactly would pay the compensation? There was no rush of volunteers.[10]

The position in other Asiatic countries appeared a little more tractable. Japan claimed that due to stringent precautions by the Imperial Council of Ministers the country was entirely free of 'the vicious habit of opium smoking'. In Korea the situation was similar. There were rounds of applause. But in the Straits Settlement where nearly half the population of about a million was Chinese the distribution of opium was a lucrative government monopoly; and both the manufacture and the consumption of the drug were rising. In Sarawak and Brunei too consumption was on the increase; and so it probably was in Ceylon, today's Sri Lanka. The trend arguably reflected a gratifying rise in living standards: opium was, after all, a luxury even if not a desirable one. Not all delegates agreed. A great deal of opium was consumed in the Kingdom of Siam, today's Thailand, where the trade was the foundation of the fortunes of some of the country's most powerful families. But a quarter of the national opium tax revenue of about $7 million was being put to educational use. There was, as Monsieur Nicolae Titulescu of Romania put it, much here that remained 'opaque' to European observers.[11] The Portuguese enclave of Macao derived its entire state revenue from gambling and opium – brothels paid sweeteners but no tax – and prohibition of opium was not to be contemplated. The Portuguese government was unaware that allegedly about a tenth of the world's opium trade passed through the colony: the figure was in any case dwarfed by the trade through Hong Kong whose prosperity was entirely based on narcotics. Sir Malcolm Delevingne protested half-heartedly; he did not think much of the stewardship of the Colonial Office either. The French delegate explained that French Indochina bordered on countries in which poppy growing was practically *inéradicable*. The frontiers ran over impassable mountains impossible to police. The native population of about 20 million consumed about 64 tonnes of opium a year, not a negligible amount, but no serious health problems had arisen as a result. Indeed, French Indochinese youth were exceptionally fit: some had won prestigious scholarships to the Sorbonne. The allegations that high French officials were heavily implicated in the opium trade were disgraceful. In the Dutch East Indies the comparatively small Chinese population were incorrigibly opiophile but otherwise law-abiding and industrious citizens. One could not readily envisage Parliament in The Hague agreeing to ban opium or any other drug.

By all accounts China was still the centre of the opium trade and its position was described as 'many-faceted'. The chief Chinese delegate, Dr Sao-Ke Alfred Sze, was a Harrow-educated Harvard graduate, one of the few of the insalubrious cosmopolitan crowd who were polluting the clean air of Lac Leman to be received by Genevese society in the rue des Granges. But despite his charming smile, occasional twinkle and classical French his utterances about the depredations of the opium wars and the post-war policies of the European powers, 'most of them represented in this hall', had somnolent Western delegates jumping from their seats in apoplectic rage.[12] 'At a guess,' Dr Sze said, his country produced about 15,000 tonnes of opium a year. The French delegate immediately suggested that the figure should be multiplied by a hundred. Dr Sze smiled enigmatically.

Not surprisingly perhaps, the eventual resolutions were misty, in places incomprehensible. (Multilingual translation was not yet the art it is today.) Second and third conferences were, however, planned and, happily, an American delegation would then attend. After much argument Russia too was invited. The arguing had been wasted. The Kremlin's reply was prompt:

> The People's Commissariate of Foreign Affairs wishes to point out in the first place that the Union of the Soviet Socialist Republics is far ahead of all the other European countries in the campaign against the diffusion of narcotic drugs and has taken a number of effective and strong measures which other governments at present still in power in other countries have not dared to adopt owing to their fear of private interests. By its own efforts and through its own initiative the Soviet Government has obtained results which the Conference [in Geneva] can only contemplate as an object to attain in the distant future.[13]

Not much *pravda* there, Monsieur Edouard Herriot of the French delegation was heard muttering;[14] but not much anybody could do about that either. Time still seemed to be on the side of the talkers. The real victims were the ordinary people in the defeated countries (not eligible for membership of the League till the mid-1920s) struggling to survive economic ruin, moral collapse and political violence.

CHAPTER 25

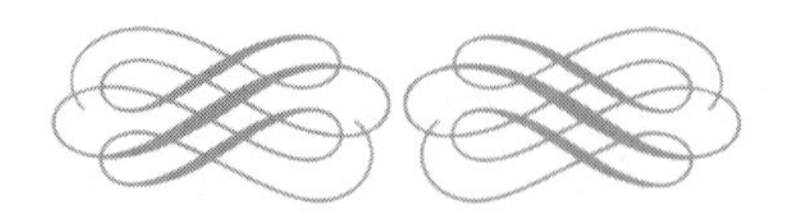

Victims and survivors

THE YEARS OF hyperinflation that hit the Weimar Republic in Germany after the war, the threat of impossible reparations and the occupation of the country's industrial heartland now tend to be remembered for their *Galgenhumor*. It was the golden age of the cabaret and of sexual emancipation, of surreal films and of anything goes. In the world of the super-rich champagne flowed and *der himmlische Pravaz* had become a fashion accessory. Morphine and heroin were not cheap but for dollars everything was affordable. To ordinary folk uninterested in cabarets and exotic sex it was less fun. Their world had collapsed, their gods had deserted them. The novel *Kleiner Man was nun?* (Little Man, What Now?) is not much remembered today but it documented like no other their existential plight.[1]

The book's author, Hans Fallada, came from a well-to-do, austere but not loveless middle-class family in Saxony.[2] After evening meals his father, a judge, read aloud from the treasures of world literature. Shakespeare and Schiller remained Hans's favourites, especially the history plays. The Bible was for Sundays. Theirs was the class which had been comfortably off before and even during the war and now found their salaries and pensions not worth a tram ticket. Even more incomprehensibly, their patriotic war bonds had become scraps of paper. They somehow survived – or most of them did – but this was the time Hans would later recall in his early work. Then a romantic schoolboy's duel gone horribly wrong – the duel culture had survived the war – led to the death of a friend. Hans could not be legally charged but the ensuing publicity forced him to change his name from Ditzen to Fallada, the character of a wise talking horse in a Grimms fairy tale, and to become a farm labourer in a distant part of the country. ('Hans' too was an invention based on Grimm: his given name was Rudolf.) He also started to inject himself with morphine and the habit became an addiction. References to the drug crop up in most of his novels; but most revealing are the unpublished fragments found after his death. A *Factual Account of Luck in the Life of a Morphinist* gives a glimpse of the Berlin of the 1920s and provides one of the believable personal accounts of an addict's predicament:

> It was during those bad Berlin times when I got hooked on morphine. For a few weeks all was great. I had a good temporary job with a newspaper, good contacts and had no difficult in obtaining 'Benzin', as we called the stuff between ourselves. But my upbringing and background made me slightly worried and I decided that after a last feast, that is a last hefty dose, I would dump the habit for good . . . But when I woke next morning and there was nothing on my bedside table to inject I realised that I simply had to have another dose, simply had to, at whatever cost. My whole body seemed to be consumed by a kind of painful agitation, my hands were trembling, and insatiable thirst filled not only my parched mouth but my whole body. I snatched down the telephone receiver and rang Wolf. I did not give him time to speak, I just croaked: 'Have you got Benzin? Come at once, I'm dying.' I hung up and lay back. I was overcome with a feeling of profound blissful anticipation. The mere thought of the injection to come within a few minutes made my body relax. The telephone shrilled. Wolf. 'Why did you hang up? I can't bring you Benzin. I haven't got any myself. I am just going out to chase some up.' 'But one injection, one injection only. Wolf, I can't survive without it.' 'But I'm telling you, Hans, I haven't got any.' 'But you have, you have, I know it.' 'I swear . . .' 'But I hear it in your voice: you've just been shooting up. You sound so satisfied.' 'This evening at nine o'clock . . .' 'But . . . but . . . I can't wait. I . . .' But it's no use. 'All right, but you won't deceive me! Swear!' 'Don't talk nonsense, Hans. Nine on the Alex [Alexandra Platz].'
>
> I get up slowly. Getting dressed is hard. My limbs feel weak and are trembling uncontrollably. My brain no longer believes that I'll be getting morphine. But I will. Kind Wolf comes and picks me up by car. We drive to the pharmacy on Alex. Wolf gets out, limping. He has some inflamed glands in his groin. I relax. In a few minutes I'll have my Benzin. In the nick of time too. My whole body has started to hurt again. I can't swallow. I try to imagine the needle slipping under the skin. O happiness! At last! At last! A few minutes only now. Here comes Wolf. I see at once that he hasn't got it. He is pale and sweating. He sits behind the wheel, mumbles another address. But they're not human! They're beasts! Animals! How can they let a man suffer like this? 'I thought it was a certainty,' I stutter. 'No, the old man wasn't here,

only some uncouth young lad. Sharp as a razor. No stuff.' The car stops at another pharmacy. Wolf disappears, then he's back. 'Nothing?' 'Nothing.'[3]

And so it goes on, page after page. There is a fit of uncontrollable sobbing. A doctor is summoned by Wolf. The terror of being sent to hospital is overwhelming. New special units had recently been set up by the Berlin Municipality where truculent patients were said to be chained to the walls. But the doctor is old, Jewish and untroubled by police regulations. A journey to his down-at-heel surgery. The injection at last. Indescribable bliss. A small charge. The piece has no real ending, just as Fallada's addiction had none.

Fallada was halfway through a detoxification course when Hitler came to power. Suddenly he found himself one of Dr Goebbels's targets for wooing. Regrettably, the doctor confided, many of Germany's most successful writers had been Jews. Not surprising, really, with all the publishing houses and the press in the hands of international Jewry. But Herr Fallada was still there and he was impeccably Aryan. His previous novels had described the degradations under the Weimar Republic. His follow-up novels should describe the country's rebirth under the Nazis. A few anti-Semitic episodes would earn him the favour of the Party and the Reichs-Propaganda-Minister; and one form the favour would take would be access to goods not normally available. Top people in the Reich were not constrained by ordinary laws. Herr Fallada writing novels in the Nazi interest would automatically find himself among them.

The trouble was that Herr Fallada loathed the Nazis. His best friends had been humiliated. Some had gone into exile. A few were in concentration camps, unreachable, perhaps dead. The rewriting of German history and literature by Dr Goebbels's panders revolted him. The beloved Heine was now 'ein Dichter unbekannt' (an unknown poet). When, in the late 1930s, pressure from Goebbels became outright bullying Fallada and his wife and children secured exit visas for a 'holiday' in England. They put their house on the market and their modest belongings into storage. The Hollywood version of *Little Man, What Now?* had been a box-office hit and Fallada was now a marketable commodity in the West. (He hated the idea of the book being made into a tear-jerker and never went to see the film.) Putnam's, his English publishers, arranged the family's visas and sent them train tickets. A few minutes before leaving the family home Hans went for a walk along one of the lakes around his smallholding of Carwitz. There, for the past five years, he had tried to immure himself in writing, the family living largely off the land but also selling surplus produce. To outsiders the place, fifty miles north of Berlin on the featureless Brandenburger Plain where the wind howled from the Urals, has little charm. But to Hans it was *zu Hause*. Returning from the walk, he told his wife to unpack. He could not live anywhere else. But nor could he write the books Dr Goebbels wanted him to write. Yet surely this Nazi madness could not last. In the meantime he would write fairy tales. The superb political thriller, *Alone in Berlin*, was strictly for the back of his desk drawer. But at least in Germany he was assured of his modest but regular daily supply of the stuff on which he was now dependent. Morphine had been many things to many people. To him it would be the walls of his internal exile.

Fallada survived the war, occupation by the Russian Army and the beginning of Communist East Germany. The cultural supremo of the new regime, J. B. Becher, was a Moscow-trained Communist but also an old admirer. Becher did his best to promote his friend; but sodden in alcohol and drugs Hans was beyond redemption. He finished his last book, *Every Man Dies Alone*, but it gave him no pleasure. 'I have no appetite for life left,' he wrote to his mother from the detoxification unit of a hospital. But he still lectured to a group of medical students about his addiction and the life of an addict; and on one of his audience at least, Werner Goetze, later professor of medicine in Berlin, the emaciated and yet incandescent figure made a lifelong impression. 'It was the most gripping medical talk I have ever heard,' the professor told the present writer in 1960.[4] Fallada died shortly afterwards, as he told his audience he would. He was fifty-six.

* * *

Millions of others were destroyed by morphine, some famous, the vast majority not. And yet such a fate was never inevitable. Morphine and heroin were more rapidly destructive than laudanum or the opium pipe and addiction to them made a productive career more difficult; but it was sometimes possible to fill responsible jobs and to live lives society would later judge to have been successful.

Sir Thomas Lauder Brunton, physician to St Bartholomew's Hospital and discoverer of the beneficial effect of amyl nitrite against anginal pain, described the case of a Member of Parliament who in Brunton's experience took the largest regular dose of morphine, a minimum of 24 grains and a maximum of around 39 grains subcutaneously, every day.

> In the House of Commons while sitting with his arms apparently crossed he stuck the needle through the sleeve of his morning coat into his biceps and injected morphine; or, his hand lying apparently quiet on his thigh, he made the injection into his quadriceps. He carried three or four syringes ready charged in his waistcoat pocket. This man was not to blame at all . . . because it was the very fineness of his character which led to his acquiring the habit.[5]

So at least Brunton believed. A barrister by profession, the MP had lost his wife in childbirth; and began injecting the drug during the serious and eventually fatal illness of his daughter. She was eighteen when she developed a brain tumour. 'He feared that his anxiety and insomnia were preventing him from arguing his cases successfully in Court and that he was letting his clients down.' His need for the drug steadily increased.

> He made a very brave attempt to give it up, and the suffering he went through was simply awful. He was told that he would have to go through hell to rid himself of the habit, and the poor man said to me one day: 'Yes, but there are grades of hell and I have got down to a very low one.'[6]

Yet the politician, now thought to have been Sir Frank Lockwood, became a respected Solicitor General in Lord Rosebery's government, admired for his clarity of mind: most colleagues saw him as the next lord chancellor but one. But, there was another side to him. His close friend, the actress Madge Kendal, noted that he was 'obsessed with death'. It was not a 'morbid obsession, more a kind of nostalgia'. It overcame him when he was spacing himself out, delaying the next injection. During such periods he was unapproachable, absent in all but body. Eventually he gave himself another shot and became his usual witty and sociable persona again. Except that he had no time for socialising. Under the influence he worked without a break; and next day he might dazzle the jury in court or make a thoughtful speech in the House of Commons. None of his staff suspected that his life depended on a drug though they noticed that he shunned engagements not essential and work-related.[7] He died in 1897 when he was forty-nine. Medical details were not mulled over in Victorian obituaries.

* * *

No less remarkable were the achievements of America's greatest surgeon. At thirty William Stewart Halsted was the all-American hero, well-born, former captain of football at Yale, sociable, outgoing, rich (but not too rich) and, according to his friend, William Osler, a first-class mind and a virtuoso pair of hands. As expected from up-and-coming young doctors, in 1891 he departed with a few friends for a study tour of Europe. There they visited the leading surgical centres of the day – with Theodor Billroth operating in Vienna and the first Lord Moynihan in Leeds the decade saw the birth of modern abdominal surgery – but what impressed the visiting Americans was the recent discovery of a young and unknown Viennese ophthalmologist called Carl Koller. In 1884 Koller reported to a meeting of the German

Ophthalmological Society in Heidelberg the analgesic property of cocaine on the usually highly sensitive cornea of the eye.[8] Halsted grasped the potentially wider implications of the discovery. He showed that, while cocaine had no effect on the intact skin, it acted as a powerful local analgesic when injected immediately below the outermost layer. He and his colleagues then embarked on an orgy of experimentation on themselves and each other. Their findings were brilliant. When cocaine was injected into a sensory nerve the whole area supplied by the nerve was anaesthetised. Not only ocular but dental surgery too could be made painless. Abdominal operations in poor-risk patients could be performed under local analgesia. But their findings were deadly too. None of the young men understood the danger of the drug even when injected locally; and, using each other as guinea pigs, all became addicted. Three never recovered, dying within a few years; and Halsted would have gone the same way had he not been rescued by a friend.

William Henry Welch, professor of bacteriology and dean of the newly founded Johns Hopkins University Medical School in Baltimore, had known Halsted from Yale and had tried to recruit him to his creation, 'my baby'. Halsted accepted before departing for Europe. The details of Welch's rescue operation were not revealed till 1989 when Osler's private diaries were made public.[9] Welch, a pussycat of a man, had hired two goons to kidnap his friend and take him to a private cruiser. The cruiser took Halsted on an enforced holiday. All comforts were provided, including, it was rumoured, a French chef; but the purpose of the tour was treatment. The doctor on board had sworn to Welch that he could cure anyone from cocaine addiction in a month; and so he did. Unfortunately, his method was to calm his patients'/victims' agitation with morphine. Halsted returned to occupy the new chair in Baltimore cured of his craving for cocaine – but a morphinist.

The new professor's formerly ebullient personality had changed beyond recognition He was cold, aloof and unpredictable though at times and to old friends unexpectedly and extravagantly generous. Small surgical errors, especially offences against the aseptic ritual and the 'no-touch technique' in the operating theatre might throw him into a towering rage but serious emergencies left him calm.[10] Generally taciturn, one never knew what he did or did not notice. Surprising his team who regarded him as oblivious of those milling around him he enquired one day why one of his accustomed theatre nurses was absent. He was told that she had had to move to another department because her skin was sensitive to the new antiseptic fluid the professor had introduced for sterilising instruments. A week later Halsted arrived with two pairs of specially made thin rubber gloves, such as had never been used in surgery before. Wearing those she could return to the operating theatre and two months later he married her. She was Caroline Hammond from an old Southern family who had lost their fortune in the Civil War but still lived in the half-ruined family mansion in Cashiers Valley in the Blue Ridge Mountains. Halsted purchased the property, restored it and on annual holidays became a competent kitchen gardener. It was an odd marriage, husband and wife inhabiting adjacent but separate establishments and meeting, it seems, affectionately but infrequently.

To Welch's relief, the change in personality in no way impaired Halsted's gifts as a surgeon. Almost, it sometimes seemed, the reverse. Innovations, some revolutionary, and most requiring intense preliminary study in the dissection room, emerged almost every year from his department. His scope embraced the whole of surgery. His fame spread. His reported post-operative recovery rates raised sceptical eyebrows until the sceptics saw him in action. 'I believe him now,' Lord Moynihan remarked after watching 'the prof' remove a thyroid gland: 'he is a master. Why, I think he is as good as Moynihan.' The 'invention' which spread his fame worldwide was the radical removal of the cancerous breast and its lymphatic drainage in one 'block'. The huge and mutilating operation, known as radical mastectomy, is hardly ever performed today but it was the first and for many years the only treatment which was sometimes effective in a deadly disease.[11]

Throughout his career Halsted continued to inject himself twice daily with a high dose of morphine, organising his professional tasks around these times and regularly disappearing into his private rooms when the effect of the last injection was wearing off. Caroline's role in this regime remains uncertain: some believe that she too became a low-grade addict: such husband and wife teams were not uncommon. Yet only a few close friends like Osler, Welch and the obstetrician Howard Kelly knew about 'Stu's burden' until the secret seeped out after his death. It seems that, unlike the great majority of addicts, he varied his daily dose slightly but resisted steadily increasing it. He could not have survived otherwise. But nor did he attempt to give up; or, if he did, nobody knew about such attempts. He remained intensely private, eating his midday meal on his own, attending committee meetings but shunning convivial gatherings. Visiting surgeons were welcomed in the operating theatre but few were asked to sample Caroline's culinary skills. Halsted himself hardly ever travelled. Yet his influence on the evolution of surgery was immense and, for his time, almost entirely benign. He was sixty-five when America entered the First World War in 1917 and two-thirds of the surgical staff and senior students at Johns Hopkins volunteered for active service. Even cutting back on the teaching his daily workload doubled, but he coped: indeed, during 1917–18 he published two important papers on thyroid surgery. He died from gall bladder disease on the eve of his seventieth birthday on 7 September 1922.

* * *

While individuals like Halsted waged private wars against addiction, a few successfully, most not, the League of Nations in Geneva went on deliberating how to make the plague a thing of the past. The Second Narcotics Conference in 1929 was dominated by the newly arrived American contingent, admired but not liked. Under the messianic leadership of Congressman Steven Porter the delegation included the saintly but now sick Bishop Brent (soon to die in Lausanne), Dr Rupert Blue, a self-important medical expert, Edwin Neville, a former United States Consul in China and authority on all things Asiatic and, most dedicated of them all, Mrs Hamilton Wright, widow of the pioneer. Neither she nor Porter were in any doubt that complete international control over opium production could be achieved in no time if only other countries showed the same commitment as the United States. In the United States, Porter claimed mendaciously (though perhaps he believed it himself), the problem of narcotics was about to be solved by determination and a firm hand. Why could not other countries follow America's lead? Such questions did not endear him to his bumbling fellow delegates.

Lord Cecil, leader of the British Empire delegation, on the other hand, was long-winded but popular:

> Perhaps I may be allowed to remind the Conference that in England it is a tradition and I believe a very wholesome tradition, to give full weight in the administration of our far-flung possessions to the man whom we call 'the Man on the Spot'. And what the Man on the Spot tells me is that the opium trade is a very complicated problem . . .[12]

Awakening from his slumber M. Aristide Briand, peace-monger and gourmet, congratulated the Englishman on his *bonne sense Anglaise*. In the fullness of time both would receive the Nobel Peace Prize, as did the League collectively on more than one occasion. Its exemplary anti-narcotics labours were regularly cited.

But words evaporate, diplomas gather dust. Statistical tables showed that worldwide over the years 1925–30 traffic in opium had steadily increased. By 1930 at least 90 tonnes of morphine in excess of legitimate medical requirements had escaped into illegal channels. This, a Scandinavian delegate reminded the conference, represented 1,161,156,000 grains of morphine, one grain being 'a very adequate daily dose for a beginner and 10 grains taking a lot of working up to'. It was indeed food for thought, though not for much action.

Then came a series of political shocks, some with an opium subtext. In the 1920s Japan had been the knight in shining armour whose rigorous anti-narcotics policy was held up as an example to all. But in 1931, after invading Manchuria without warning, the invaders started to use opium as a weapon. The effect was horrifying.

> Traffickers set up clinics at village fairs [in conquered territories] advertising their skill of curing tuberculosis and other diseases. The medicine sold was always the same: heroin and morphine. When patients returned they were registered and given more narcotics. It was a deliberate policy of making addicts[,] pursued with Japanese thoroughness.[13]

When relayed to Geneva the report caused outrage. The use of opium as a means of enfeebling the enemy was described by Lord Cecil as 'the most iniquitous act of any country in the history of war'. But, short of declaring war on the aggressor, the League could do nothing: indeed, its charter did not even provide for an act of expulsion. In the event expulsion was unnecessary. Contemptuously Japan walked out. Soon Hitler's Germany, rearming at speed, gave notice of secession; and Italy, in a huff over the invitation to the emperor of Abyssinia, staged an operatic exit. Several South American countries followed.

* * *

While most countries paid lip service to the League – League associations flourished in England, their garden parties worthy of notice by the *Tatler* – individual governments found it necessary to pursue their own drug policies. In Britain after much discussion an advisory commission was set up by the Ministry of Health under the chairmanship of the president of the Royal College of Physicians of London and an international authority on alcohol-related diseases, Sir Humphry Rolleston. Its composition was entirely medical and so were most of the witnesses. This could be faulted – and was. No addicts were called, only two middle-ranking policemen and of course no criminals. But discreet pressure from the Home Office and Sir Malcolm Delevingne for exemplary severity (as well as the blandishments of mentions in the Birthday Honours), were resisted.

When in 1926 the Commission published their report it was immediately credited with establishing something called the 'British System' as distinct from the 'American System'. Neither had ever been defined but 'Rolleston' resolutely eschewed the penal attitudes of the Harrison Act favoured by the Home Office. It rejected an all-out ban on heroin. It agreed to the establishment of 'Medical Tribunals' jointly staffed by civil servants and doctors; but they would not, as the Home Office and Sir Malcolm had suggested, police doctors. What they *would* do was left reassuringly vague; and the bodies soon withered away. All-important, as the *British Medical Journal* pointed out, was a restatement of the British view that addiction was a disease requiring treatment rather than a crime calling for punishment. Compulsory treatment and various methods of enforced withdrawal regimes 'when these become available and are shown to be effective' could be part of the medical management but every effort should be made not to drive addicts into the arms of quacks and criminals. Crucially, doctors would not be prosecuted for providing addicts with drugs, even with heroin, provided that it was part of a treatment programme.

The apparent 'softness' of the recommendations elicited outrage from hardliners. 'A prescription of masterly inactivity in the face of a non-existent problem,' the journalist David Downes mocked. But he was wrong. The problem was non-existent only in the sense that the Commission recognised that it would not go away in the face of draconian legislation or indeed of any legislation unsupported by other measures. The negative merits of the report were easy to overlook, as negative merits usually are, but they were impressive. Its recommendations were never debated in Parliament but were described as 'valuable' by the Department of Health though not by the Home Office. Over the next decade they were adopted by many Western European governments.[14]

* * *

One important consequence of 'Rolleston' was incidental. Though none of the Commission were psychiatrists, a 'psychological approach' and 'psychiatric support' were mentioned in the report; and addiction was again and again described as a 'neurosis' or 'mental illness'. From that time psychiatry began to shift from a purely institutional role, a legacy of the nineteenth century, and search for a new, individual clientele. For several decades the bulk of that clientele in Britain would be addicts. Expert witnesses in court too would be increasingly specialists in mental health. Many, including Sir Malcolm Delevingne, found this irritating. Why should specialists in mental diseases know anything about addiction, 'a physical ailment insofar as it was an ailment rather than a moral failing'? But despite its apparent blandness 'Rolleston' would shape British attitudes until the 1970s and its humane spirit still occasionally stirs.

* * *

While individual countries developed their own ways of coping (or not coping) with drugs, the League continued to deliberate grand international options. It admonished and praised and collected more data. Major events in the real world like the rape of Abyssinia, the *Anschluss*, the Munich Conference and the dismemberment of Czechoslovakia passed it by: none of the powers consulted it. Then, on 1 September 1939, Hitler's Panzers invaded Poland, quickly followed by those of his friend Stalin; and the Hope of the World went into hibernation.

CHAPTER 26

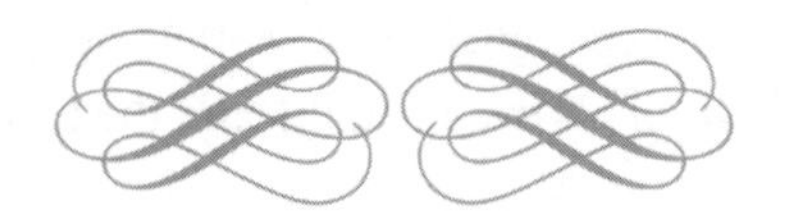

Unholy alliances

NOBODY WHO HAS experienced the Second World War, whether carrying arms or as a civilian (and civilians suffered much the heavier losses for the first time since the Thirty Years War), would ever doubt that the suffering would have been infinitely greater without morphine and heroin. In Hans Fallada's words, 'morphine was the only ministering angel who, without ideology or war aims, served mankind impartially in its madness'. The drugs were available in most armies (and in the few prisoner-of-war camps visited by the International Red Cross) and supplies ran out only under exceptional circumstances. When they did, the consequences were catastrophic. In besieged Stalingrad in January 1943 when the last morphine ampoule had been injected, the desperation triggered a wave of suicides among the German troops. It also prompted Field Marshal Paulus's surreal last order of the day making suicide a capital offence. But the Field Marshal himself had his personal cache almost intact when he was captured by the Russians.[1]

In occupied countries where Resistance movements began to organise in 1941 morphine continued to play a clandestine role. Emmanuel d'Astier de la Vigerie, black sheep of an aristocratic French family and founder of Resistance Liberation-Sud, maintained that in the early days one could become a resister only if one was already maladjusted. This in his case, and presumably that of others, included being an addict. But he underwent a rigorous course of detoxification after joining the Underground and became one of its heroes. A petty drug trafficker named Georges Deshayes was blackmailed by the Marseilles police to act as a double agent and was largely responsible for the collapse of the important *Mediterranée* resistance group. Among those betrayed was Jean Moulin, de Gaulle's agent in occupied France: he was arrested, tortured and executed. 'Avoid any contact with drugs and drug addicts' became a rule of the Resistance, but too late. In Holland and Norway, resistance and the black market in opium remained closely linked, some of the most effective underground groups financing themselves by trafficking in the drug and using it for bribing officials. It is still uncertain whether heroin was actually parachuted into occupied Europe by the Allied air forces to help the Resistance movements.

* * *

After 1943 some Allied forces were issued with emergency supplies of morphine which could be used even by non-medical personnel giving first and terminal aid to wounded air crew. It was a revolutionary and merciful innovation. When supplies did run out it was because they were being channelled to the black market that emerged almost at once in liberated lands. It was no less malignant though less famous than the trade in adulterated penicillin portrayed in Graham Greene and Carol Reed's *The Third Man* (1949). In a scene in Joseph Heller's *Catch-22* Yossarian, the anti-hero bomber pilot, tries to relieve the suffering of his wounded mate, Snowden. 'I am cold . . .' Snowden whimpers, not surprisingly since his guts are spilling out of his abdomen. Yossarian reaches for the emergency box of morphine. It is empty. All ten syrettes have been removed for sale by the business entrepreneur of the squad. In their place is a share certificate and a note saying: 'What's good for M & M Enterprises is good for our country. Milo Minderbinder'.

But at last the world emerged from the slaughter and began to contemplate peace. Or peace of sorts. Picasso's dove fluttered on the banners but the olive branch in its beak looked increasingly like a Kalashnikov. Then the chilly peace turned into the Cold War. The change was profoundly unwelcome to ordinary people but heralded the resurrection of the international drug trade.

* * *

One of the unforeseen but inevitable consequences of the Harrison Act of 1914 was the emergence in America of a new breed of criminal. He went by the name of 'gangster' and made his debut in New York City. From there the species spread almost as fast as their bullets. By the late 1920s the term had passed into most civilised languages and had become part of the folklore of modern cities. Ethnic layerings existed from the start. The notion that Sicilian gangsters, also known as the Mafia, at first deliberately eschewed narcotics as evil has been a myth fostered by Hollywood; but the morphine and heroin trade was for a time largely in the hands of Jewish gangs led by such charmless characters as Waxey Gordon, Meyer Lansky, 'Dutch' Schultz and Benjamin 'Bugs' Siegel. Their strength lay in their international connections stretching from Marseilles to Shanghai. While the Mafia concentrated on the beatific windfall of bootleg liquor created by Prohibition competition was muted; but such happy divisions of labour rarely last. In 1929 Salvatore C. Luciano, better known as Lucky Luciano – as a young man he had survived a beating that was supposed to kill him but left him only with a permanent squint – had most of his Mafia rivals murdered and allied himself with the Jewish gang of Meyer Lansky. The unorthodox alliance unexpectedly flourished and built up what would be, until the international rise of the Triads, the most powerful criminal organisation the world had ever seen. Their scope embraced every criminal activity in the book but from the start dealing and distributing heroin was a market leader. Their first peak lasted about three years until Lucky's luck temporarily deserted him.

To save his skin one of Luciano's lieutenants turned informer and introduced the expression 'singing like a canary' into police vocabulary. In 1938 the boss was arrested, convicted and imprisoned. To add insult to injury the Second World War almost stamped out the illicit trade. Drugs were still moved in large quantities but mostly by army personnel or under their aegis; illicit routes were blocked or were becoming dangerous. The end of the war was probably a moment in history when organised drug crime – or at least its pre-war form – might have been snuffed out. But when they least expected it, the mobsters found an improbable ally.

* * *

Contacts between the Mafia and American Intelligence dated back to 1943 and were not fully revealed till thirty years later. In 1972 Alfred W. McCoy showed and documented that in 1943, preparing for the landing in Sicily, United States Naval Intelligence spun a close and secret network with local mafiosi.[2] Having provided useful advance information, after the invasion the Palermo Mafia continued to cooperate, hunting down pockets of Fascist resistance. There had never been any love lost between the Fascists and the Mafia: indeed, Mussolini had come as close as anyone in rooting out Cosa Nostra. (It needed a gangster to root out the gangsters.) But the last laugh was with Cosa Nostra. With the American Army turning a blind eye it established a well-organised black market for food, cigarettes, nylons and, above all, drugs. They moved easily to mainland Italy in the wake of the Allied troops; and, at a safe distance, followed the armies in their painful progress north.

In 1946 Lucky was paroled on the grounds of having significantly assisted the war effort, and deported to Sicily. It was not a happy homecoming: like most of his fellow New World mafiosi Lucky had been born in New York and considered himself a fully paid up member of the American dream. But having returned willy-nilly to his ancestral patch, he soon recruited a partly local, partly American-Italian team. With the help of Corsican gangsters who controlled a clutch of heroin chemists the pre-war worldwide narcotics syndicate began to rise from the ashes. Distribution centres were established in European cities – Nice was again favoured – and soon consignments were on their way across the Atlantic. Potentially the United States was still the largest and the richest market and Cuba became a convenient staging post for transshipment. There the brutal pre-Castro regime with its impeccably capitalist credentials welcomed the windfall. It was a beginning but it was still a fragile and perilous business. Then the scenario changed.

In the increasingly tense atmosphere of the Cold War the United States' Central Intelligence Agency (CIA), bedrock since 1947 of the free world's defence against Communism, found its anti-red allies wherever it could. Traces of the old links between the United States Naval Intelligence and the Sicilian Mafia still existed. Now they beckoned. Whatever the Mafia's transgressions in other fields – and perhaps they had been exaggerated – their anti-Communist pedigree was solid. Conversely, staunch capitalists to a man, many stalwarts of the Mafia had developed a sneaking admiration for the CIA. Were they not brothers under the skin? Contacts were reinvigorated and from about 1949 Lucky's operations were transformed. Protected by the Agency and under its anti-Communist banner he closed his homely Sicilian drug kitchens and transferred refining operations to laboratories in Marseilles run by expert chemists known as 'cooks'. Even in the unperfumed air of Marseilles's *Vieux Port* the smell of acetic anhydride was a give-away, but outside the centre disused wine cellars, *faïenceries* and stables could be refurbished with centrifuges, refrigerators and other laboratory paraphernalia. Moving equipment was no problem. Raw opium came from or by way of Turkey unimpeded. At peak production in the 1960s about twenty-five laboratories were operating in the countryside around Marseilles round the clock producing heroin of 98 per cent purity. This was the 'French connection' in all its glory, only slightly romanticised by Hollywood.

After 1949 the Mafia–CIA alliance was additionally protected by the French government. The powerful communist unions bankrolled by Moscow had organised a series of damaging strikes in Marseilles; and the unions were cheered on by the communist City Council. The Communist Party no longer shared in the government in Paris but they were still hoping to become the country's elected majority. To the other side in the Cold War this was a terrible spectre. Its prevention assumed overriding importance. The Mafia–CIA alliance in Marseilles was a thorn in the flesh of the communist City Council. All such thorns had to be lovingly nurtured. Washington and Paris agreed that to turn a blind eye to or even to facilitate the movement of a few consignments of heroin was a ludicrously low price to pay to weaken World Communism. The French government stipulated only that all high-quality heroin manufactured in and around the port should be exported. To this the Mafia bosses, French patriots to a man, readily agreed. Despite such agreements consumption in France rose steadily after 1960.

* * *

The trade still depended on expert conveyancers; but the techniques had been perfected before the war and were easy to bring up to date. Before cheap air travel most of the contraband travelled by sea. Ocean liners and cargo boats offered a wide choice of hiding places. Apart from air shafts, drainage systems, false bottoms to oil tanks and other built-in and self-recommending bolt-holes heroin could (and did) travel as cargo disguised as shrimps, soap, salted peanuts, duck eggs packed along real eggs, Bologna sausages, loaves of bread, barrels of cement, tins of fish and jars of pickled cabbage. Opium could be embedded in ornamental candles or Roman busts as well as in dead kittens. It was in the late 1940s that it began to be carried from Mexico by peasants locally known as 'mules'.

During the early years of the French connection merchant seamen acted as regular carriers direct to New York. When United States Customs cracked down on the route, the trans-shipment was moved first to Canada and then to Latin America. The latter was organised by one of the Mafia's most notorious godfathers, Santo Trafficante Junior. A soft-spoken fan of Maria Callas, to whom he proposed marriage, he was, even by Mafia standards, extraordinarily ruthless. He owned Florida and the Caribbean and his couriers, an elite force, brought heroin to Miami by way of Guadeloupe, Martinique and Puerto Rico. The route was littered with unidentifiable corpses.

* * *

International drug traffickers have been both demonised and romanticised. Amateurs when caught still tend to hit the headlines. They are sometimes seventeen-year-old pretty backpackers on a gap year from High Barnet. But they are a minority. The backbone are professionals. William Burroughs wrote in *The Naked Lunch* (1959) that drug merchants are different from other entrepreneurs only in their illegality:

> Junk to them is neither good nor evil. It is the ideal and ultimate merchandise. No advertising or sales talk is necessary. Overheads are minimal. There is never any haggling. Properly primed the client will crawl through a sewer and beg to buy.[3]

Nor has supply ever been a problem. Even before Afghanistan became the untouchable supplier of 90 per cent of the world's illicit needs, poppy growing had been legal or semi-legal in many countries. Turkey did not ban it until temporarily bribed to do so by the Nixon administration. Pakistan in the 1960s produced legally only 6 tonnes of opium a year but another 200 tonnes regularly disappeared into thin air. A shifting scene was anchored by two constants. First, however miserable the returns at village level, poppy was still the safest subsistence crop for millions of peasants. And second, higher up in the trade profit margins were exemplary. Traffickers and intermediaries were always adequately funded to bribe those who needed bribing. Often anything as crude as a bribe was unnecessary. The narcotics trade commanded (as it still does) the best legal outfits and expert witnesses of impeccable provenance. When in some countries 'morphine salts' as well as 'morphine' were banned, plugging an old legal loophole, it was still possible to trade in benzoyl morphine, not a salt, as a Nobel-Prize-winning professor of chemistry testified in 1976, but a compound. It just happened to be a compound from which morphine was easily extracted.

A permanent boon to the traffic is the value of the merchandise in relation to its volume. A kilo of pure heroin, the size of the latest thriller, can provide after cutting 200,000 fixes. This makes smuggling even small quantities a paying proposition. The rewards on the horizon will always tempt amateurs in financial straits or those being blackmailed. Pregnant women have been found to be podgy not with foetuses but packs of heroin; and quarter-kilo lumps have been surgically implanted into buttocks, the incisions artfully hidden by natural creases. For some years the 'classic' method was condoms tied with dental floss, more tensile than cotton, resistant to gastric acid and unstretched by moisture. They could easily be swallowed after being smeared with syrup. The method seems to date back to 1945 when, acting on information, a Mrs Chowning of Mexico was examined fluoroscopically at Laredo Airport, Texas. The condoms were of a cheap variety and showed up as ghostly foreign objects. Since then techniques have advanced. High-quality appliances are radio-translucent though detectable by more sophisticated screening techniques. Few of those are available in airport customs offices. Dosed with laxatives Mrs Chowning excreted thirty-one packages of heroin. Over the next twenty years the number swallowed and detected has risen, the average being eighty and the record on the files of British Customs 261. But the method can go wrong even without detection. In 1982, forty-two condoms each containing 4 grammes of heroin were recovered from the stomach and small intestine at autopsy of an American student: one condom had burst causing intestinal obstruction and a painful – or perhaps a painless – death.

Such methods are of course regarded as kindergarten stuff by regular couriers, who are rarely caught.

* * *

Next to its clandestine alliance with the CIA, the biggest boost of the narcotics trade came with the Vietnam War. After the defeat of the French in what had been French Indochina an international conference in Geneva in 1954 divided the former colony into Laos, Cambodia, and North and South Vietnam. The divisions proved to be doodles on the map. South East Asia remained a cauldron of independent states, fiefdoms, former colonies now striving for independence, territories ruled by warlords (some more efficiently than internationally recognised states), and remnants of Chiang Kai-shek's defeated Kuomintang army.[4] Over it all hovered what experts in Washington perceived as the menace of World Communism. After the take-over of China by Mao Zedong South East Asia seemed the logical direction for its inexorable progress. Inexorable, that is, unless dependable anti-Communist bulwarks could be erected to stop it. North Vietnam had fallen under Communist rule; and South Vietnam was going to be the next target. It was both desperately poor and potentially rich. Corrupt politics, ethnic divisions and a too quickly growing population were the causes of the poverty. The source of its potential riches was the poppy.

Almost from the start of the ten-year conflict opium and its derivatives were the props of increasingly venal regimes in Saigon. For the first critical years the trade was handled by Corsican gangsters in close alliance with the CIA. The Corsicans contributed the know-how in organising refining laboratories, local networks of dealers, pushers, hit men, corrupt officials and politicians. The CIA provided armed back-up when needed and, most critically, transport from the poppy-growing areas to the refining laboratories. Their fleet of nimble small aircraft expertly piloted, variously known as Air America and Air Opium, was to become a key element in supplying first the native market and then the United States Army.

For the conflict between North and South escalated. In the North the Communist regime drew support from Communist China and later from Russia. In the South what were initially a few hundred and then a few thousand American advisers and military training personnel grew into an army of half a million. After a Christian grandee named Diem was murdered by his disaffected staff impatient with his highminded utterances, the presidency was occupied by a succession of increasingly improbable puppets of the United States. As casualties mounted upbeat pronouncements by the American commanders (each in turn hailed as the saviour) became more and more fatuous. The real victor in the unwinnable conflict was heroin.

When the narcotics market became too complex for the Corsicans the trade was taken over by Chinese Dragon Syndicates or Triads based in Hong Kong. They dispatched trained local managers to cater specifically for the demand by GIs for high-quality 'No. 4' heroin. By 1972 twenty-nine high-capacity refineries were operating in what was soon to be called the Golden Triangle.[5] They were producing thousands of kilos of the drug, including the famous 'Double Globe' brand. Across the top of the neat little packages displaying the logo inscriptions in Chinese and Thai characters in purple (the Chinese lucky colour) advised buyers 'Beware of counterfeits' and 'Travel safely by sea', the Chinese equivalent of 'Bon voyage'. By then American medical officers reckoned that 37,000 of the troops on the ground were heroin addicts. It was almost certainly an underestimate, perhaps a gross one. At least 85 per cent of military personnel irrespective of rank were offered heroin within a day of arrival with an acceptance rate estimated at 80 per cent. The trafficking was entirely open and under the benevolent but eagle eye of General Dang Van Quang, the President's intelligence chief. Many soldiers had their Zippo lighters, part of the kit issued to conscripts (or 'grunts') engraved with drug-related poems or maxims. 'Say Hi! if you're high' was a favourite. Many GIs smoked rather than injected the dope, their pipes sometimes made from cartridge cases or human bones. Black humour flourished in what was perhaps the nastiest war in modern times.[6]

The heroin habit in Vietnam soon had repercussions back home. Thousands of kilos were smuggled out of South East Asia in letters and parcels sent by GIs. The fact was well known but with half a million troops in the theatre of war the United States Army Post Office, always understaffed, was overwhelmed. Not more than one in a hundred consignments could be examined; and when heroin was detected President Lyndon Johnson or Greta Garbo seemed unlikely senders. Soon new and ingenious ways of transporting more substantial amounts were devised. At least half a kilo of heroin could be hidden inside a body bag, sometimes buried in the wounds of the corpse. The method was not detected for a year and then only because the price of heroin back in California nose-dived. To many there this plummeting was welcome. They had long been known as 'junkies'.

CHAPTER 27

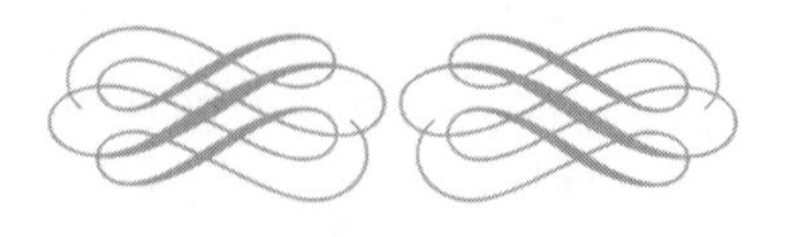

Junkies

THE TERM 'JUNKY' in its dope sense originated in New York City in the early 1920s. For much longer it had meant a junk-man or rag-and-bone man who walked the streets with his cart and cat buying and selling scrap metal, wood and clothing. Aware that garbage sometimes contained saleable items he occasionally rummaged through bins. So did a generation of young people, mostly poor, white and male, who picked through the city dumps to find among the junk something that would pay for their next fix. Hence junkies. The first generation were also dubbed the heroin boys and described in 1916 by a sympathetic neuropsychiatrist, Pearce Bailey:

> They are generally healthy and able to work to start with and are fairly intelligent. Many are engaging personalities; but, as often happens with engaging personalities, they are unstable, suggestible and easily led ... They are mostly of poor, urban, immigrant stock, scrambling for work in an insecure and uncongenial job market, supplementing their income with occasional petty larceny and drawn into neighbourhood gangs which provide a romantic flavour of adventure and camaraderie.[1]

Italian, Jewish, Irish, Scandinavian, Central European, they were the ethnic second generation; and heroin use was part of their American identity. Once initiated, most stayed hooked. Attachment to home became tenuous or snapped altogether. Their new bonding was with other gang members. In New York and later in other American cities female junkies were nearly all prostitutes. Which came first, the addiction or the prostitution, was difficult to tell, as it still is.

What made heroin the standard dope during the early years of the twentieth century was an oversight: it was not included in the first anti-narcotics legislation; it did not become illegal until an amendment to the Harrison Act was hurriedly pushed through Congress in 1924. But it was for several years almost exclusively a New York habit: as late as 1926, 90 per cent of heroin addicts lived within 180 miles of Manhattan. By then most of the drug was injected, first subcutaneously but soon intravenously. Mainlining had the inestimable advantage of giving an instant thrill even when the dope was crudely adulterated. It was this immediate fleeting but overwhelming bliss that junkies craved. The shortening of the period of happiness could always be corrected with another injection. Always, that is, provided the dosh and the dope were available. To junkies their acquisition soon became an overriding priority. A few free clinics were set up by the New York City Health Department in the mid-1920s. They were a failure: the heroin boys suspected, probably correctly, that some of the staff worked hand in glove with the police, the enemy. In reality, as the journalist David Phear commented, 'there was no help for the living dead'.[2] The last two words became another heroin cliché to enrich the language.

* * *

By the end of the decade heroin had spread across the United States and the pattern of the addict population was changing. During the 1930s blacks who had migrated north while the economy was booming became the largest group of junkies. The Wall Street Crash of 1929 plunged the country into the deepest economic depression in its history: the number of unemployed rocketed into the millions and black migrants were at the bottom of the heap. To counter the cold, the hunger and the degradation of the Northern black ghettos there was nothing to do but to play jazz, sing the blues and inject heroin. Procuring the last often added crime to the basics of existence. The peddling by Sicilian traffickers was ruthlessly efficient. Initial samples were often distributed free as loss-leaders until demand had been firmly established. Then the price crept up. What would come first: murder or suicide? Gradually from the black slums heroin escaped and moved west. By the mid-1930s the term 'junky' was understood not only everywhere in the United States but everywhere where English was the lingua franca.

* * *

After the Second World War heroin consumption in the United States continued to increase, at first slowly but then explosively. The timing of the explosion, the late 1950s and early 1960s, was no accident. The post-war baby boomers were coming of age: within one decade the country's population in the 15–24-years age bracket doubled. With the advent of the new generation old inhibitions and prohibitions crumbled. Suddenly the junkies were no longer the forsaken heroin boys of Manhattan or desperate blacks. They were the white middle class from the suburbs with sprinklers and double garages, many of them bright if a little starry-eyed. They met on university campuses, at roaring music festivals and at protest marches. The last were numerous: there was much to protest against. Segregation in the South was one; a colonial war in a far-off corner of the world another. And the corner was not so far off if you or your friends were being drafted or had been drafted already. The world in any case was shrinking. The young backpacked their way through Europe and Asia returning spiritually reborn. And fun though it was to travel, the rave scenes were at home. New York's Greenwich Village and the reincarnated gold-rush town of San Francisco were alive with happenings. Those who made happenings happen laughed at the prejudices of their parents and their grandparents. So much stodge and so many delightful ways of poking fun at it. The young thrilled to new rhythms. They rocked to rock and roll. No portentous speeches; flag waving or even the guillotine were necessary. As the Beatles would immortally bellow a few years later, all you needed was love. Or almost. As flower power bloomed 'Be cool' became the young nation's catchphrase. Almost like a benign signal from heaven (insofar as such a place existed, and why not?), there descended the pill, just a tiny white object but what a boon! Soon there was dope-chic to add to the porno-chic, the two sometimes linked.[3] The dope was not only opium and its derivatives. To help the cooling process a whole range of new substances could be tried. Yet partly because of its horrendous reputation but also because it was still the best, heroin was king. Between 1960 and 1970 the number of users in the United States rose from about 50,000 to 500,000. Nothing like it had ever happened before.

* * *

New social scenes need new heroes. The illiterates found them in pop stars and sporting greats. The literate found him in Jean-Louis (call-me-Jack) Kerouac. The prophet of the 'beat' culture – a term which he himself claimed to have invented but disliked or professed to – was born of French-Canadian parents in Lowell, Massachusetts. Later, like many leading revolutionaries, he oscillated between searching for noble Breton ancestors and tracing his lineage to American-Indian warriors. He wrote French and English with equally slapdash ease, often incomprehensibly but sometimes soaring. At Columbia where he won a scholarship he edited a student newspaper; but when his football career soured after an injury, he dropped out. On an impulse he joined the Merchant Marine after Pearl Harbor but was 'invalided out' after a few weeks. He told his medical examiner that he unutterably loathed everything he had seen of the military: the mere sight of a uniform gave him a 'rhythmic headache'. It was said that the bottom rating for mental fitness to serve was invented for him.

After the war he tried a variety of humdrum jobs but also started to write and scribbled piles of unpublished and unpublishable though in retrospect patchily interesting material. Then, in 1951, while living in New York on the earnings of his second wife, he cut sheets of tracing paper into long strips wide enough for a typewriter and taped them together into 35-metre-long rolls. Feeding those into his machine allowed him to type without interruption even to change paper, without chapters or paragraph breaks and more 'explicitly' than was eventually published. But that was not before he collected a pile of outraged rejection slips often coupled with warnings that most of what he had submitted was actionable. Even today the disjointed but graphic and allegedly autobiographical descriptions of drug use and sexual practices, both homo- and hetero-, in *On the Road* (1957) are eyebrow-raising. Of course, like Coleridge with whom the kinship is remote but clear, he was a liar and a *fantaisiste*, but what he wrote was often spellbinding, and the work has become one of the four or five most influential books of the century.[4] Many have suggested that much of it was written (or 'typed, a big difference', according to Allen Ginsberg) under the influence of heroin; but it is unlike any other opium literature. Kerouac then became interested in Buddhism, as did many of the beat generation, but was demoralised when 'deflated' by Buddhist scholars. He wrote poetry and film scripts as well as several other autobiographical novels. None is as riveting as *On the Road*. During his last ten years he was both a celebrity enjoying fame and a recluse pouring contempt on his fans. He was increasingly soaked in alcohol and under the influence of heroin. He was forty-seven when he died in 1969 from an oesophageal haemorrhage, a complication of alcoholic cirrhosis. Thirty years later his home-town University of Massachusetts bestowed on him the degree of Doctor of Letters.

* * *

Many others of greater or lesser talent gave voice to the beat beliefs and disbeliefs. Many of their songs and writings embody a shocking and yet oddly gentle world-view. Were they like the early Christians as some claimed? Or hell-bent on destroying Western civilisation? Perhaps both.[5] Herbert Huncke, a pioneer, was a petty criminal and heroin addict who met William S. Burroughs in 1946 and introduced the core of New York beats to the junky lifestyle and lingo. To many it was a revelation. Burroughs came from a rich family – his grandfather had invented the Burroughs Adding Machine – and he was a Harvard graduate. He soon became the embodiment of the beat scene and its branches. (There were many.) His first book, *Junkie* (1953), was edited and promoted by his friend Allen Ginsberg whose own seminal poem, *Howl*, sounded like a bugle call. Echoes of Voltaire's *Ecraser l'infame*! there, though the *infame* was no longer the Church but capitalist society. The work also provoked a much-publicised lawsuit for obscenity. Did he win or lose? It mattered not at all. Burroughs's opium-soaked novel, *The Naked Lunch* (1959), also went to court in 1962. Like the *Lady Chatterley* trial in London, such proceedings provided both publicity and entertainment and helped to put beat culture on the map. The much-quoted first stanza of Allen Ginsberg's *Howl*, supposedly addressing a generation of brilliant minds destroyed by 'madness' and their own wild intoxication, set the tone. The exact message was never entirely clear but the poem, a sustained scream rather than a howl, was undoubtedly subversive, inflammatory, and in flashes soaringly inspired. When conceiving the lines the author could credibly have been roaming the dirty 'negro' streets at dawn desperately in search of a fix. Both the desperation and the anger are still scorching.[6]

Burroughs later coined the phrase 'Just say no to drug hysteria' in reply to Nancy Reagan's 'Just say no to drugs'.[7] A less subtle riposte was 'Just say yes to drugs' scrawled on countless lavatory walls. Drugs were at the heart of the beat-hippie-flower-power culture, benzedrine, amphetamines, marihuana and later psychedelics being almost as important as opiates. The last generally came in the form of morphine 'syrettes', little squeeze tubes with a hypodermic needle attached to the tip. Very hip, easy to transport. As the beat movement spread and metamorphosed into beatniks, hippies, flower people and a host of other subspecies, the drug scene too changed from time to time and place to place. The Summer of Love of 1967 was also the Summer of Dope: the beards, the sandals, the caftans, the long hair, the guitars, the no-make-up and no-bra complementing the hoards of multicoloured little pills and capsules consumed by the palmful with gay abandon. But to many it was the Woodstock Festival two years later which embodied the hippie counterculture and its ideals of love and life. Those who provided the entertainment were the brightest stars of the 1960s;[8] and to those who listened to them it was the experience of a lifetime. Sadly, it was also unrepeatable, as such events always are. The next fixture, in Altamont in California, was less benign: eighteen-year-old Meredith Hunter was stabbed during a performance by the Rolling Stones and died. But wasn't death part of life too?

Some of the luminaries of the beat culture were eventually embraced by the literary establishment and elected to academies. Others remained hippies to the end, often an early one.[9]

* * *

Inevitably and in an attenuated form (though forcefully enough to kindle nostalgic past-middle-aged memories) the beat culture hit Britain and the continent of Europe only a few years after conquering young America. But in Western Europe the pre-war background had been different. Auden called the 1930s a 'low dishonest decade', and so it was. Fascism rose. Mass unemployment was an unspeakable blight. But no black ghettos existed in Paris or London to act as incubators of the heroin habit; and the drug culture was kept alive mainly in bohemia. Among its victims was Christopher Wood, rising star of the English artistic renaissance of the '30s which shamed the timorous politicians and boot-licking journalists. He was twenty-nine when in a daze of opium he fell from or jumped to his death from a moving train. In Montmartre experimenting with heroin was de rigueur though many of the district's leading lights – Picasso and Giacometti among them – hated it. So did their rival genius in Montparnasse, Matisse. But Raymond Radiguet, Francis Poulenc and Georges Auric became habituated for a time and dope sellers were part of the street scene on Pigalle. They also became part of its mythology, Maurice Chevalier inimitably stage-whispering *héroïne* though far too sensible himself to become hooked. But Jean Cocteau regarded himself as a prophet even after lapsing into a stupor in the lift ascending to Picasso's flat. Picasso, disturbed in his work (or whatever), was not amused. But Cocteau's line drawing of an addict during withdrawal remains one of the searing images of the world of opium. Guillaume Apollinaire and Louis Leroy too were proudly professing experts on the drug, the latter even publishing an elegant and modestly funny best-seller entitled *Livre de la fumée*. But the explosion of the drug scene in Europe did not come till the 1960s.

To provide for that gay decade the drug trade modernised. Shipments were adapted to containerisation and to specially built fleets of lorries with double floors, mock spare tyres, extra-hollow cylinders and other ingeniously hidden spaces. Britain began to receive most of her requirement by road, as she still does.[10] But the dope trade was not alone in expanding. With only occasional dips international passenger travel too was rising at an unprecedented rate. It was all to the good, social scientists explained: meeting Germans in their *Bierkeller*, savouring French cheese in the Dordogne and importing toys from China were the best ways of counteracting the Dr Strangeloves of this world. That the changes also made effective countermeasures against smuggling difficult was a trivial side effect. Professional traffickers were becoming bold. They accepted that a tiny proportion of their goods and a few of their number would be caught, and budgeted accordingly. Small fry were compensated. The families of middling fry received generous support while the breadwinner was on an enforced sabbatical. The trade in any case flourished in prisons. The international hiring rates fluctuated, but at no time did a shortage of personnel threaten the even flow of the merchandise. Apart from professionals the carriers now included girl guides, international athletes, pop stars and their entourage, church choirs and ambassadors and their wives. Because they were so plentiful they were beginning to be called ants rather than mules.

* * *

Mules, ants or plain traffickers, they were only the front. Behind them and moving them by remote control loomed sober middle-ranking businessmen and above them a few feudal drug barons. The organisations generated sufficient wealth for the last to build any kind of front they fancied. Some became philanthropists, patrons of the arts and socially concerned entrepreneurs. Through donations to political parties their wives' yearning for ladyhood could be satisfied. Honorary fellowships of this, that or the other also had their price tags. As Paul Simonet, a comfortably retired mobster, observed, the business was profitable enough for the barons to adopt any style 'to satisfy ministers, courtiers, police chiefs and cardinals'. It tended to remain ethnically more homogeneous than other multinationals. The traffic to the United States stayed largely in the hands of Italians, to France of Corsicans, to Germany of Turks and to many parts of Asia of Chinese. As the century and millennium drew to a close Nigerians were still comparative newcomers but making headway, and Russians, Ukrainians, Romanians and Albanians were lining up. Hungarians were said to hold the drug trade in contempt unless profits on a deal were mind-boggling. Bonding by language and kinship continued to outweigh the supposed disadvantages of inbreeding.

* * *

Though in Britain the drug scene never achieved the flamboyance or the social menace (as it was perceived by some) of the beat-hippie-flower-power generation of the United States, it was for a time lively – and occasionally deadly. A silly prosecution against the Rolling Stones, the police having been tipped off by the *News of the World* about a 'drugs party' hosted by one of Mick Jagger's friends, set the scene in 1967, *The Times* promoting the non-event to 'a symbol of the conflict between traditional values and the new hedonism'. The future Sir Mick was sentenced to three months' imprisonment, overturned on appeal. The Home Office minister, Lord Stonham, voiced the concern of a more sober generation: pop idols were seducing the nation's young. So they were. For a decade or two pop was drenched in drugs. Who could resist Bob Dylan's mellifluously insinuating '*Rainy Day Women*' with its incendiary chorus that everyone must get stoned? Or the Rolling Stones with their '*Mother's Little Helper*' and '*Connection*'? Or the Beatles with their '*Lucy in the Sky with Diamonds*' and '*With a Little Help from my Friends*'? Or Frank Zappa and the Mothers of Invention with their rousing '*Hungry Freaks, Daddy*' and '*Help I'm a Rock*'? Or Jefferson Airplane with '*White Rabbit*'? Or Procol Harum with '*A Whiter Shade of Pale*'? Or Velvet Underground with '*Heroin*'? Or a hundred other hymns of praise only slightly less durable? It was innocent but not all innocence. After the drug death of Kurt Cobain, the inventor of grunge music, at the age of twenty-five his mother wept: 'Now the silly boy has joined that stupid club.' She meant the club of young geniuses he had always boasted about and who had already died a drug-related death. They included the fabulous Brian Jones of the Stones, the electrifying Jimi Hendrix, Janis Joplin, Jim Morrison, Gram Parsons, Keith Moon of the Who and Sid Vicious of the Sex Pistols; and their number would grow. It was an impressive roll call to celebrate and to weep over.

* * *

As in times past, there were memorable contributions to the opium literature at a more literary level too. Researching for his book *The Quiet American* in Saigon, Graham Greene penned a classic description of the opium dream:

> After two pipes I felt a certain drowsiness, after four my mind felt alert and calm – unhappiness and fear of the future became like something dimly remembered which strangely I thought important once. I, who feel shy at exhibiting the grossness of my French, found myself reciting a poem by Baudelaire to my companion. When I got home that night I experienced for the first time the white night of opium. One lies relaxed and wakeful, not desiring sleep. We dread wakefulness when our thoughts are disturbed, but in this state one is calm – it would be wrong to say one is happy – happiness disturbs the pulse. And then suddenly, without warning one sleeps. Never has one slept so deeply a whole night-long sleep and then on waking the luminous dial of the clock showing that only twenty minutes of so-called real time have gone by.[11]

CHAPTER 28

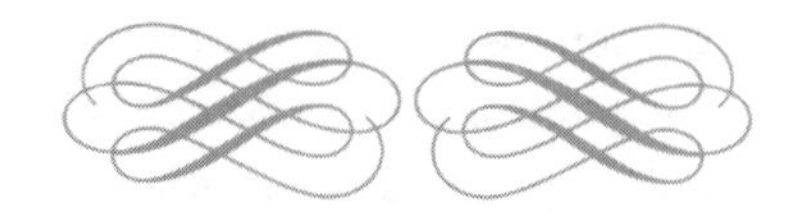

Guardians of the law

POISED TO POUNCE on junkies, flower children, users, habitués, addicts and their suppliers is the vast, and still expanding, army of the national and international anti-narcotics law enforcement agencies. Their model and parent body, the Federal Narcotics Control Board of the United States, dates back to 1928 and was the outcome of the murder of a top New York gangster Arnold Rothstein. Shot but not killed outright, he was questioned about his murderer. He remained shtum. 'But don't you want your murderer caught, Mr Rothstein?' the detective pleaded. 'Yes. But you stick to your job, policeman, and I stick to mine.'[1]

The murderer was never caught, but among Rothstein's papers were documents incriminating top-ranking members of the New York City administration and law enforcement agencies in the heroin trade. The Grand Jury investigation elicited that the dead man had employed the son of Colonel Levi Nutt, director of the Antinarcotics Division of the New York Narcotics Prohibition Unit, as his personal attorney and Nutt's son-in-law as his personal accountant. The Grand Jury also heard evidence that narcotics agents under Nutt's command had routinely taken bribes, lost evidence, padded their accounts with arrests made by other agencies and lied on oath. Witnesses hinted that much of the heroin seized by the police was later sold through special channels at a bargain price. Nutt personally was responsible for grossly inflated figures of narcotic seizures and for narcotics agents under his command regularly arresting the same small dealers but just as regularly letting off the big boys. The case spelt the decline of Tammany Hall and the rise of New York's rumbustious reforming mayor, Fiorello La Guardia. Less happily it also resulted in the shake-up of the old municipal anti-narcotics apparatus. The new Federal Bureau of Narcotics was placed under the direction of the Treasury. The Grand Jury also recommended that the Bureau be given more money and that penalties for drug-related crimes be stiffened. Surrounding the appointment of the director circumstances remain murky but instead of the post going to a retired admiral already nominated by the Treasury, it went to an obscure consular official, Harry J. Anslinger. He was to rule the organisation for thirty years and would leave his mark on anti-narcotics policy in the United States and around the world for many more.

* * *

Anslinger was born in 1892 into a modestly prosperous family of Swiss-German immigrants in small-town Pennsylvania and took pride in remaining a solid middle American. He claimed to have developed his 'ineradicable hatred of narcotics' when his boyhood hero, Jimmy Munger, the best pool player in town, died of an overdose of heroin. His grossly inflated linguistic skills – he learnt a smattering of German from his parents – helped him to a job in the United States Consular Service; and his success in reducing rum running from the Bahamas landed him the directorship of what was not yet but would eventually become the most important subdivision of the Prohibition Bureau. His further rise was not impeded by his marriage to a niece of Andrew W. Mellon, Secretary of the Treasury. Within a few years he was director of the Federal Bureau of Narcotics (FBN), in fact though not yet in name the world's first drug tsar.

Anslinger was not an endearing personality and made no effort to appear one. He was wily rather than clever, suspicious of anybody of superior education or intelligence and unscrupulous in getting his way. When a scapegoat was needed his most loyal assistants found themselves disposable. He relished the title of drug tsar, the invention of a sloppy journalist. The appellation suggested an autocratic solution to a problem to which no such solution has ever existed or ever will. But one doubts if solutions of any kind much preoccupied Anslinger. His credo was simple:

> Drug addiction is an evil which needs not a cure but to be rooted out and be destroyed utterly. We [the United States for which Anslinger regarded himself as an anointed spokesman] must behave like a well-coached football team, crisp in its blocking, sharp in its tackling and well-drilled in all its fundamentals. We need stiffer penalty laws.[2]

He professed to be as resolved to stamp out marihuana as he was to destroy opiates, a counterproductive stance. Since his tales of marihuana were patently exaggerated, his better-founded warnings about heroin lost credibility.[3] Or would have done if people had been inclined to disbelieve him. Few were. He was a persuasive though long-winded speaker, switching on a characteristic choking sound of barely containable emotion at well-spaced intervals. He addressed parent organisations, school speech days, conferences of educators, lawyers, doctors, police, businessmen and, above all, politicians. His charges were modest: he was not financially greedy. His rhetoric was recognised by many as blatantly mendacious but it was effective. Describing the harrowing deaths of addicts or their suppliers his usual peroration was a single sentence thrice repeated: 'How many such murders shall we condone . . . indeed promote?'[4]

As Nutt had done before him and many of his successors and imitators would do later, Anslinger concentrated on apprehending junkies and small-time pushers but leaving the top of the trade untroubled. He never indicted or even investigated organised crime. In return the bosses would stage spectacular pseudo-hauls from time to time which were widely trumpeted in the media, Anslinger being photographed standing in a Napoleonic pose next to crates labelled 'heroin'. Such crates never existed: even the dimmest drug boss was not *that* dim. To the trade such seizures, even when genuine, were loss-leaders.

Apart from 'missionary' work in the streets of big cities Anslinger reserved his special venom for stars of showbiz. To him they were guilty not only of addiction but also of insufferable provocation. Abuse he relished but not ridicule; and he was the butt of New York theatreland. The scum flaunted their friendship with the great and the good, even with the family of successive presidents. Their wealth was obscene. All this, in their own estimation and sometimes in reality, made them untouchable. Untouchable, that is, so long as they remained stars of Hollywood and Broadway. When they were down, there was no mercy. When in 1959 the singer Billie Holiday lay dying of liver failure in a New York hospital Anslinger's thugs raided her hospital room and, ignoring medical and nursing protests – two doctors and a nursing sister had to be treated for injuries after the scuffle – confiscated her heroin, her radio, record player, magazines, flowers and chocolates. It was the heroin which mattered: its loss made her last days a torture. A professional press photographer snapped her in her state of disarray, the pictures later being published with Anslinger's caption: 'A vicious social leper gets her deserts'. President Eisenhower, not an unkind man, sent his congratulations.

But such catches were rare. Most of Anslinger's victims were young blacks and Hispanics of no social consequence. Not only was their apprehension easy, it also fed his contempt for races he regarded as subhuman. The Oklahoma musician Ralph Ellison published his novel, *Invisible Man*, in 1952.

The narrator's grandfather on his deathbed advises his son how to treat the whites. 'Overcome them with yeses, undermine them with grins, agree 'em to death, let 'em swoller you until they vomit ... For your own personal safety let them recognise your danger potential.'[5] The book paints a harrowing but unsentimental picture of the young blacks excluded from the American Dream taking up drugs partly to spite white authorities but also to handle the stresses of early adulthood in the slums. The author becomes one of the tens of thousands of Anslinger's victims. But one need not rely on fiction, however close to reality. Statistics from the Lexington prison hospital confirm the trend. Of 2,943 admissions in 1946, 7.3 per cent were black; in 1950 of 4,534 admissions 42.2 per cent were black; and in 1956 of 5,875 admissions 72 per cent were black. Eighty-one per cent were admitted for drug-related crimes. Drugs were freely available in jail – though at a price. Even youths first admitted for non-drug-related crimes left as addicts. More than two-thirds of all admissions – some estimates are as high as 85 per cent – were readmissions.

Anslinger's own figures never added up. They served his objective of the moment. On the one hand 'the drug war was being won and would be over by now if penalties would be commensurate to the evil'. On the other in 1955 he declared that there were still 45,000 heroin users in the United States, the highest ever, 'and their number is growing'. It depended on which set of figures was more likely to achieve his purpose. His purpose, power, was constant. Generally he was believed, as one narcotics expert commented, 'because his audience wanted to believe him'. He reached his personal peak with the 1956 Narcotics Control Act, one of the most oppressive pieces of legislation ever passed by Congress. Penalties for drug-related offences were again hiked. They included the unprecedented provision in federal law of allowing arrest without a warrant 'in the belief' that a heroin-related offence had been committed. At state level, in some states citizens could now be prosecuted for offences against narcotics law even when the authorities had no proof of either purchase or possession. Anslinger flourished in this atmosphere of lies and fear and he found like-minded allies. Among them was the Junior Senator from Wisconsin, making a name for himself.

The ostensible seal of the Anslinger–McCarthy alliance was their common appreciation of the Chinese Communist menace, but what formed the underlying bond was McCarthy's addiction to morphine. Recognising his friend's vulnerability to blackmail Anslinger arranged for Joe to pick up the best-quality morphine he needed in a Capitol Hill pharmacy which functioned under Anslinger's protection. As Anslinger later related in his thinly disguised autobiographical novel, *The Murderer*, published after his retirement:

> One of the most influential members of Congress at the time and one of my most dependable supporters was a confirmed morphine addict. He was an amiable man but would do nothing to help himself to get rid of his addiction . . . He refused medical advice and insisted that no-one would ever be permitted to interfere with him or with whatever habit he wished to indulge in . . . He was also a heavy drinker but it was his addiction to morphine which was the gravest threat to himself and his country even though, in the national interest, his uninterrupted supply of the drug was guaranteed by my Bureau. On the day he died I mourned him deeply as a friend but also thanked God for relieving me of a great burden and a certain danger.[6]

Anslinger had much in common with Joe but, unlike Joe, whose ascendancy lasted four years, Anslinger reigned for three decades. To his own surprise he was reappointed by newly elected President John F. Kennedy in 1961. Full of honours, he retired a year later.

CHAPTER 29

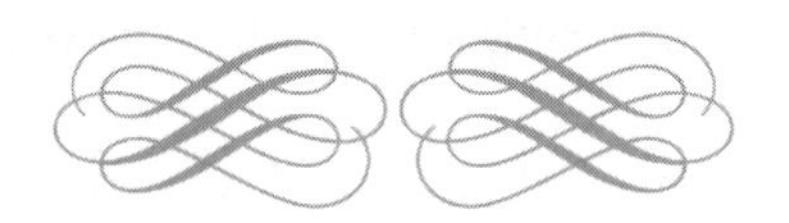

Zero tolerance

IN 1975 DISTRICT JUDGE Whitman Knapp was asked to study explosive allegations made by a dismissed police officer. In his report the judge concluded that 'attempts so far to legislate drug addiction out of existence seem to have resulted in *more* addiction, *more* business for organised crime and *more* evidence pointing to police corruption'.

> Corruption has grown in recent years to the point where high-ranking police officers acknowledge it to be the most serious problem facing the Department. In the course of our investigation we became familiar with . . . practices including:–
> Keeping money and/or drugs narcotics confiscated at the time of an arrest or raid;
> Selling narcotics to addict informants in exchange for stolen goods;
> Passing on confiscated drugs to police informants for sale to addicts;
> 'Flaking' or planting narcotics on an arrested person in order to obtain evidence of law violation;
> Storing narcotics, needles and other drug paraphernalia in police lockers;
> Accepting and demanding money or narcotics from suspected narcotics law violators in payment for the disclosure of official information; and
> Financing heroin transactions.[1]

But even as the Knapp Commission was issuing its revelations, President Richard M. Nixon was driven to declare a 'War on Drugs', the first such dramatic declaration by an American president. His alarm was not unfounded. Since American involvement in South East Asia the country's addict population had been rising; and there had been a roughly proportional rise in addiction-related crime, overcrowding in prisons, stagnation in the law courts and deaths from needle-transmitted diseases and overdose. In a much-heralded television broadcast on 17 June 1971 the President declared: 'The country should stop looking for so-called root causes of drug crime and put its money instead into increasing the number of police and the severity of the law.' He was gripped by the fanatical zeal of the clandestine binge drinker:

> America's leadership class will be remembered for the role it played in helping to lose two wars: the war in Vietnam and, at least so far, the war on drugs. That class is made up of highly educated people in the arts, the media, the academic community, the government bureaucracies and even business. It is characterised by intellectual arrogance, a hollow obsession with style, fashion and class and above all, by a permissive attitude to drugs.[2]

The Woodstock peace-and-love music festival, admittedly no tea party on the vicarage lawn, provoked his especial fury:

> To erase the grim legacy of Woodstock we need total and implacable war against drugs. That means war on all fronts. It will be our second civil war.[3]

But claiming the mantle of Abraham Lincoln was a mantle too far. The first civil war was hard enough. Winning wars against drugs, crime, terror and other abstractions would prove harder. The Nixon administration bribed Turkey into banning poppy growing, an achievement hailed as a historic breakthrough. The ban lasted less than two years, was much resented in Turkey and accelerated the shift of poppy cultivation to Iran, Afghanistan and Pakistan. Heroin consumption in the United States did not fall. Nixon, like many before and after him, was fighting the law enforcers as well as the trade; neither in truth wanted the war to end. Gore Vidal wrote:

> The antinarcotics bureaucratic machine has a vested interest in playing cops and robbers . . . Both enforcement agencies and criminals want strong laws against the sale and use of drugs because if drugs were sold at cost there would be neither money nor jobs in it for anyone . . . If drugs were cheaply available addicts would not commit crimes to pay for their next fix and the Bureau of Narcotics would also wither away, something they are not going to allow without a struggle.[4]

Nixon departed, but other presidential drug wars followed. Both Presidents Reagan and Bush Senior were strong on rhetoric 'against the scourge of drugs' and allocated billions of dollars to campaigns against it. United Nations surveys and academic studies suggested that money spent on treatment and rehabilitation was three to four times more cost-effective than money spent on enforcing prohibition and sending users to prison; but facts had little effect on policy. Three-quarters of new funds were always spent on law enforcement.

* * *

Many had hoped for a change once the fight against Communism had ceased to override all other objectives, a more scientific, rational, even human approach. It did not happen. After 1989 the War on Drugs became the overarching moral imperative. The development was not without irony. Unacknowledged either by President Reagan or by his friend, Prime Minister Margaret Thatcher of Great Britain, drug lords were the most committed upholders of the market forces and of entrepreneurial freedom they both extolled. As Philippe Bourgeois of the University of Pennsylvania wrote:

> Like most other people only perhaps more so drug traders in the United States are scrambling to obtain their piece of the pie as fast as possible . . . They are aggressively pursuing careers as the much lauded private entrepreneurs, they take risks, they work hard and they ask for no public handouts. They pray devoutly for good luck. They are the ultimate rugged individualists braving an unpredictable frontier where fortune, fame but also possible destruction are all just round the corner and where the enemy must be ruthlessly hunted down.[5]

Task forces, drug initiatives, control strategies and threatening acronyms proliferated. The Anti-Drug Abuse Act passed with a large majority by Congress in 1986 allocated an extra $6 billion to drug eradication and made prison sentences for dealers mandatory. William Bennett's reign as drug tsar surpassed some of the absurdities of the Anslinger era. One effect was obvious. Over the ten-year period from President Reagan's election to 1994 the United States prison population, already the highest per capita in the Western world, increased threefold. It was, as before, a selective rise. By the end of the period more black men aged between fifteen and twenty-five were in prison than attending an educational establishment. Black and Hispanic women were nine and four times respectively more likely to receive a prison sentence for a drug-related offence than white women. The average prison sentence in federal courts for low-level heroin dealing and first-time offenders was ten years six months, 59 per cent longer than for rape or manslaughter. As the Justice Department explained, the 'three strikes and you're out' policy meant that many with a history of untreated addiction and a small tally of drug-related offences, mostly young blacks, would be imprisoned for life. And the Higher Education Act of 1998 denied eligibility for student loans to all those convicted of a drug-related though to no other type of offence. There was more replay of past attempts at 'crushing drugs in one fell swoop' (in William Bennett's phrase). 'This was not a war on drugs', the first openly HIV-positive heroin addict to address the International AIDS conference told the meeting in 1993. 'It was, as it has always been, a war on the victims of drugs.'[6]

* * *

Britain, for long regarded as culpably soft on drugs by American hardliners, started to catch up in the 1980s. After United States troops pulled out of Vietnam the big suppliers of South East Asia began to concentrate on Europe. Some Iranian exiles tried to save their fortune in the form of heroin from the grasp of the ayatollahs. But most important, the trade, never short of material rewards, was getting more lucrative by the day. In millennium year, one kilo of heroin bought in the form of opium from a peasant farmer in Afghanistan or Pakistan for the equivalent of £500 was worth about £10,000 by the time it left distribution centres in Kabul or Islamabad and twice as much on arriving in London or Paris. It was worth £40,000 once it was broken into smaller units. It was sold on the streets in bags costing £5 or £10. The aggregate value of the original kilogram had by then risen to £100,000.

After all the cutting, much of the contents of the bags was too unpleasant to inject; and 'chasing the dragon' – heating the powder on aluminium foil and inhaling the fumes – caught on in British cities.[7] Dale Beckett, a psychiatrist and level-headed observer of the drug scene, usefully distinguished 'seed', 'soil' and 'climate' in the spread of addiction.[8] If heroin was 'seed', the 'soil' was the miasma of despair that settled over dismantled industrial communities. There the heroin business at street level became the preserve of young unemployed working-class men, one of the few opportunities open to them in Mrs Thatcher's Enterprise Britain. In the jargon of the trade a distinction was soon drawn between 'smack-heads' who dealt to raise enough money for their next fix, the fodder of police raids, and 'bread-heads' who dealt with an eye to making it in a free-enterprise society. Not a few did, one or two ending up as financial grandees in the City of London. In most towns the 'drug scene' became localised to a circumscribed area. After the collapse of coal mining in the Barnsley region of Yorkshire a few streets in Grimthorpe were virtually taken over by drug gangs. It made civic sense: law-abiding citizens could avoid gangland and most of the time gangland was left unmolested.

Among the underlying 'climatic' changes, faith in existing approaches to treatment was faltering. General practitioners disliked being 'heroin dealers by appointment to H.M. Government'. Their grouses reflected dissatisfaction with the National Health Service. By the 1980s its novelty and pioneering spirit were becoming frayed and an expanding private sector was thought to cream off rich addicts for expensive and sometimes successful therapy. One well-known psychiatrist who refused to treat addicts on the National Health Service 'on moral grounds' was later revealed to be treating twenty private patients at £1,000 per head per week. The National Health Service clinics too tended to run into difficulties. Management of addiction was becoming more complicated. Crack, amphetamines, barbiturates and a host of new drugs were now part of the narcotics scene. Addictions to different drugs have certain features in common but each has its own particular problems; and clinic staff were rarely trained to deal with them all.

The political climate too changed. Like Presidents Reagan and Bush across the Atlantic, Conservative governments in Britain had little patience with investigating social causes, promoting rehabilitation or spending money on hopheads. Addicts are often accused of being focused on obtaining their next fix to the exclusion of all else, and so they are; but they are also sensitive to 'atmosphere'. Many noted how attitudes in Whitehall were changing. In 1970 Arthur Hawes expressed his 'very deep gratitude to the Drugs Branch of the Home Office . . . Addicts would often turn up for help and advice; and their reception was always helpful, never hostile.'[9] Bing Spear, appointed head of the branch in 1977, saw himself as the guardian of the 'Rolleston tradition': he knew every registered heroin addict in London personally and minded what happened to them. By the mid-1980s he and his kind were fighting a losing battle. Attitudes were almost palpably stiffening, 'penalisation' was no longer a dirty word. Indeed, it was emerging as the most effective course. As Home Secretary Douglas Hurd was comparing the influx of heroin to a 'medieval plague we are visiting on ourselves', Margaret Thatcher vowed that peddlers of heroin, however small time, 'will find no safe haven in Britain'. 'Zero tolerance' was to be the guiding principle. The fateful label 'wet' began to hover over those who expressed reservations.

The picture was muddied when cocaine became part of the polydrug scene whose primary ingredient was still heroin. A new kind of addict combined the two, injecting them serially or mixed together in concoctions known as 'speedballs'. It was not an easy addiction to deal with; and it created its own mythology. The nickname 'Crack City' was bestowed by press and police on a housing estate in Lewisham in south London where the core problem was in fact intravenous heroin. The 'Yardies', a small-time gang of young

Jamaican immigrants were picked up and demonised by politicians and the media.

> They are bound by their origin and reggae culture to travel from one location to the other; and such is their nomadic life-style that serious offences such as murders are and will continue to be committed wherever their cultural bandwagon happens to stop.[10]

As late as 1997 the National Criminal Intelligence Service described them as 'threatening the basic security and stability of the nation'. Much of the original misinformation was supplied to the police and the press by white south London drug barons who resented the intrusion of lumpen blacks on their patch. The links between the barons and the police were later found to be uncomfortably close.

While paranoia ruled, heroin spread. Between 1978 and 1982 annual registrations increased by 250 per cent. The total number of addicts notified to the Home Office rose from 2,657 in 1970, to 6,107 in 1980, to 14,688 in 1985 and 19,987 in 1990. As always, notification represented a gross underestimate. By the end of the Thatcher years the illicit drug reception and distribution centres in Britain were as police-proof as they had become in many parts of the United States.

* * *

An additional and terrible cloud began to spread over the drug scene in the 1980s. In contrast to the sonorous but wildly exaggerated messages of national decline as a result of heroin issued by bodies like the National Criminal Intelligence Service, the emergence of Acquired Immunodeficiency Syndrome (AIDS) was a development of biblical proportions. It had long been recognised that shared needles among addicts was a health risk: the spread of hepatitis and malaria had been a serious concern of the authorities in Egypt in the 1920s and among immigrants to New York since the First World War. None compared to the pestilence which at first seemed to strike the gay communities selectively. Its emerging link with needles shared by addicts added an extra coating of prejudice. In fact and in retrospect the connection had a faint silver lining. The growing prevalence of AIDS among drug users of all sexual orientations was crucial to the emerging British strategy which tried to present the virus as a threat to the general population rather than as a by-product of the gay lifestyle. Yet, though the strategy was partly adopted to reduce bigotry and violence against gays, it could backfire. Roy Robertson, a general practitioner on the Muirhouse estate of Edinburgh had for many years treated addicts since most consultants in the city's hospitals refused to run a drug-dependence clinic. Now police descended on his surgery and confiscated the syringes. His patients started to frequent 'shooting galleries' where twenty or thirty would share a syringe. When HIV testing became available in 1985 more than two-thirds of the 164 heroin users from the practice tested positive.[11] Efforts by general practitioners elsewhere to introduce a viable needle-exchange programme could cost them their livelihood.[12]

* * *

By the early 1990s the gravity of the AIDS–drugs connection was clear. The reported frequency of infection among European drug users in 1993 – already out of date by the time the reports were published – ranged from 16.4 per cent in England and 18 per cent in West Germany to 32 per cent in Italy, 45 per cent in Switzerland and 68 per cent in Austria, Scotland and Spain. The differences made little sense: the true figures were probably everywhere higher. Yet in some quarters denial prevailed. A committee of the Royal College of Psychiatrists stated in that year that 'it would certainly be wrong to take AIDS as an excuse for promiscuous prescribing'.[13] A more alert report by a committee chaired by Ruth, Lady Runciman urging an expanded clean-syringe service was summarily rejected by ministers as 'blatantly condoning' drug use.[14]

Another aspect of Reaganite and Thatcherite drugs policy has received comparatively little attention. By the late 1980s the value of the global turnover of drugs exceeded that of the alcohol and wine industries. Huge transactions in the former, probably worth between $300 and $500 billion annually, escaped taxation. But the secrecy raised problems beyond the loss of revenue. In the past the black economy had formed the confluence of tens of thousands of small transactions in a vast range of goods and services. To a large extent their fluctuations in both directions were self-balancing. Now the movements of a single massive block of black merchandise, heroin, had come to dominate the market and those movements remained wholly uncharted. In a modern society predicting the financial future, the main purpose of existence of chancellors of the exchequer and presidents of the Bank of England, depends on the flow of accurate information. Rises and falls in the price of oil, cereals, gold or coffee might be troublesome but they are known. Rises and falls in the global illicit heroin trade remain guesswork. As one Treasury official observed, 'few admit it but the criminalisation of drugs has made financial forecasting a farce'. In the more weighty words of Sir Malcolm Delevingne more than half a century earlier, 'the noxious trade, by remaining hidden as well as illegal, represents a fundamental challenge to the capacity of the state to govern'. He was right.

CHAPTER 30

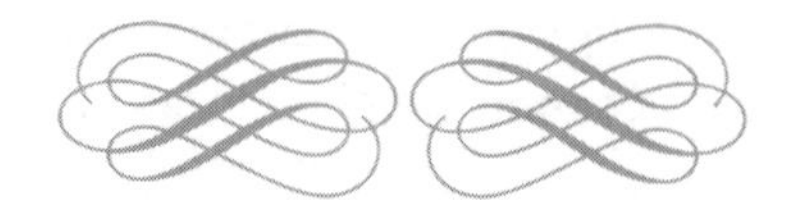

God's own medicine

THE TRIBULATIONS OF governments and the noxious trade always represented only one side of the opium story. As the second half of the twentieth century advanced developments on the other side were no less momentous.

* * *

Even birds and animals look after their wounded and dying – some do so with extraordinary devotion – and special institutions dedicated to terminally ill humans have existed since medical records began. In modern times Mme Jeanne Garnier, a wealthy and devout widow of Lyon, is sometimes remembered as the pioneer. Having lost her two sons to a slowly disabling and painful neurological ailment, in 1875 she and a few friends opened the first of a series of refuges for the untreatable called somewhat despairingly 'Calvaires'. Twenty years later the Irish Sisters of Charity chose the name 'hospice' for their home for the 'passing' in Dublin; and in 1905 the Order founded a similar institution in the East End of London. Protestant churches were not far behind; and it was St Luke's Hospital for the Dying in east London, a Methodist foundation, which, on 20 April 1948, received a new recruit.

Cicely Saunders was then thirty, unmarried, tall, gawky, short-sighted and suffering from chronic backache. But charitable homes had been desperately short of nurses since the end of the Second World War[1] and, despite her bad back, Miss Saunders seemed keen and capable. But it is a fair guess that nobody on the appointment board expected her to change the face of modern medicine. Her past history was unremarkable. She came from a well-to-do but dysfunctional north London family. She had been sickly and friendless at an exclusive boarding school where she had been packed off at the age of ten and not much happier in wartime Oxford where she read politics, philosophy and economics.[2] Towards the end of her undergraduate years she converted from a vague agnosticism to a then unfashionable brand of evangelical Christianity; and, instead of setting out on a career as a personal secretary to a respectable but not too important Tory politician, she enrolled in that most traditional of nursing schools, the Florence Nightingale School of Nursing at St Thomas's Hospital, London. She did well but remained friendless; and once again her career was derailed. A back injury compelled her to move from nursing to train as a 'lady almoner', a somewhat obscure 'auxiliary' profession tucked away in the bowels of large hospitals. It was a far from auspicious beginning but her education did endow her with that intangible and wholly unselfconscious air of 'class' which in mid-twentieth-century England was still an asset when taking on the establishment.[3] Later she would successfully cajole, persuade and chide medical grandees, ministers, senior civil servants, archbishops, newspaper editors and minor royalty with the same kindly but firm assurance with which she approached unprivileged patients and their families.[4] Her work at St Thomas's Hospital also gave her an insight into the glaring need of the terminally ill. Or rather, the need was glaring to her but not, it seemed, to anybody else.

This was the time when in the acute wards of the famous teaching hospitals of London those requiring long-term palliative care quickly acquired the label of 'bed blockers' (or to callous medical students 'grot'). It was a sentence of expulsion. Teaching students and training future hospital consultants – no other career in medicine was regarded worthy of consideration – was also deemed to require a quick turnover of patients with 'good' (that is bad) clinical signs. Such patients were expected to get better or to die. With this overriding objective the 'chronics' were transferred to other institutions designated for that purpose. In truth, they were designated for that purpose only in the sense that they were unsuitable for anything else. Some were Victorian workhouses with hermetically separated wings for male and female inmates. Many were former tuberculosis sanatoria made unexpectedly redundant by the advent of streptomycin. The latter were recognisable by their open-air loggias but even more by being tucked away in remote parts of the country deliberately inaccessible to friends and family. Remoteness also made recruitment of medical and nursing staff problematic, sometimes impossible. In some of these long-stay hospitals – for dying could take a long time – local general practitioners took turns to call at irregular intervals. Or not.

Varied as they were, these establishments shared one creed. A period of severe physical pain was a sad but inevitable – and even perhaps a desirable – prelude to dying. This was not for want of humanity nor was it ever spelt out. But that death should come (in the late Victorian phrase) as a 'blessed relief' was not perhaps an evil. The medical and nursing catchphrase parroted by textbooks was 'painkillers on demand'. It was first heard in the 1880s and it meant – or seemed to mean – that patients were given opiates, the only effective painkillers for most of them, when they 'needed' and therefore 'demanded' them. In reality it meant that they got them late or not at all. The view universally held was that opiates were addictive, dangerous and probably sinful. Dying or not dying, they therefore had to be used sparingly or, preferably, not at all. This was made clear to the patients. The implication was that they should 'demand' opiates only when the pain was intolerable. Dying was no excuse for whining. Underpinning the moral precept was the shortage of hands. Opiates meant an injection requiring a doctor or a qualified nurse. Often neither was available. For patients it was often a choice of pain or humiliation. To be spared homilies, reproving looks and loaded silences many preferred the former.

One fact generally recognised today was not appreciated or, when appreciated, ignored. It is far easier to prevent pain than to abolish it. In semi-quantitative terms, the *extinction* of pain may need ten times more opiates than its *pre-emption*. Why this should be so is still far from clear though a variety of more or less unconvincing explanations has been advanced.[5] Caring families often discharged relatives in mental and physical distress to be nursed at home. Without expert guidance, support and easy access to opiates, that too could cause much suffering. The persistently evil reputation of cancer as an inevitably painful end dates from this period rather than from the more distant past.

To teach herself the basics of what was not yet called palliative medicine, Cicely Saunders searched the literature. 'That did not take long,' she later told the present writer, 'since there was none.' Standard medical texts devoted at most a paragraph or two to the terminally ill; and nursing manuals were similarly lackadaisical.[6] Catholic authors tended to emphasise the power of prayer; and the *bona mors* or 'good death' had a large devotional literature. But religious texts often exalted pain rather than suggesting ways of relieving it.[7]

* * *

Despite such pervasive lack of interest (at least as reflected in medical publications) a few Victorian doctors did publish articles calling for fresh thinking. In 1896 Dr John Snow of the Cancer (now the Royal Marsden) Hospital in London wrote a paper on pain relief in cancerous disease and concluded that available analgesics other than opiates were 'less than satisfactory' and opiates were 'difficult to handle'.[8] Moral support was all-important. He praised – rightly – the hospital chaplaincy service, then a comparatively recent innovation. Most terminally ill patients, he concluded, were happier at home than in hospital; but home care even among the affluent was often taxing emotionally as well as physically. In 1905 Sir William Osler, recently arrived to adorn the Regius chair of medicine in Oxford, reviewed the case histories of 500 dying patients. He found that '390 suffered great bodily pain or distress of one sort or another; 210 showed profound mental apprehension, half of those positive terror'.[9] He deplored this 'deeply unsatisfactory state of affairs' but had few practical suggestions for how it might be improved. But he insisted (much to the consternation of some of his colleagues) that opiates, especially morphine, should be prescribed more readily. There simply was no substitute. But the zeitgeist was against him.

Perhaps the most remarkable though now barely remembered forerunner of modern palliative care was a New England family doctor. In 1935, at the age of eighty, Alfred Worcester published the text of three 'non-compulsory' lectures which he had delivered to Boston medical students under the title *Care of the Aged, the Dying and the Dead.* It was the second category which preoccupied him most. He pointed to the growing pressures of modern urban life which made the nursing of the terminally ill in the home difficult. Yet institutional provisions even in 'our rich and ancient city' were often inadequate or non-existent. Fear of opiates was irrational yet all-pervasive. But in dying patients relief of pain and anguish should surely take precedence over most other needs.[10] Like Osler, he was ahead of his time and found few disciples; but in 1945 four hospital social workers, no longer called lady almoners in the United States, Ruth Abrams, Gertrude Jameson, Mary Poehlman and Sylvia Snyder, made an impassioned plea in the prestigious *New England Journal of Medicine*:

> There should be competent medical supervision uninterrupted throughout the course of the last illness, a good nursing service to supplement medical supervision, the services of medical social workers to aid doctors, nurses and, above all, the patient's family and carefully controlled but readily available opiates ... No patient in the twentieth century should be allowed to suffer prolonged and severe pain.[11]

It must have sounded like a pipe dream then, and more than sixty years later in many parts of the world it remains one.

* * *

Determined to change all this, Cicely Saunders consulted friends. One eminent medical consultant told her that she stood no chance of changing clinical practice since she was not a qualified doctor. This was meant as a kindly warning off; but she re-entered St Thomas's Hospital as a medical student and qualified as a doctor in 1957 at the age of thirty-eight. Her historic paper, 'Dying of cancer', was published in the *St Thomas's Hospital Gazette* a few months later. It was probably read by few at the time but her report on 900 patients treated at St Joseph's Hospice aroused wide interest though little immediate action. Yet it changed the science and practice of pain relief, the use of opiates and the care of the terminally ill.

> Our study clearly shows that opiates are not addictive for patients with advanced cancer; that the regular giving of opiates does not cause a major problem of tolerance; that giving oral morphine works and that it works by relieving pain and not by inducing indifference to pain. Set alongside prevailing myths and fears of tolerance and addiction that a patient arriving at St Joseph's should say 'The pain in the other hospital was so bad that if anyone came near my bed I would scream: "Don't touch me! Don't come near me!" With regular *prophylactic* treatment with morphine balanced to her need . . . she became alert and cheerful and she retained her composure until her death two months later.[12]

In the first place 'prophylactic treatment' meant establishing the 'rhythm and time-line' of the pain and focusing on preventing its recurrence rather then on trying to cope with it after it had struck. This diminished rather than increased the required dose of analgesics. From this simple discovery flowed a body of complex palliative measures, physical and psychological, designed to preserve incurable patients' mental, physical and spiritual dignity. The measures eventually grew into a new specialty.

The outcome of Cicely Saunders's efforts in mobilising ministers, local charities, the churches, the press but, above all, thousands of ordinary folk living in the south London conurbation of Sydenham, was the opening in June 1967 of St Christopher's Hospice where her principles were put into practice.[13] The number of patients admitted was fifty-seven. A home-care programme was added two years later. Within five years hospices were recognised as a basic medical need. Though still underfunded, today they and their home-care extensions look after tens of thousands in Britain alone.[14]

* * *

The new specialty brought about a wider change. To most doctors 'symptomatic treatment' had been for centuries a thinly disguised synonym for second-class medicine. Merely to treat symptoms – the headaches, the dysuria, the cough, the sleeplessness – however successfully, was a confession of failure. Getting at the root cause of illnesses, even if nothing could be done about it, was a matter for hearty congratulations. Now suddenly symptomatic treatment became a science-based skill with dedicated journals, textbooks and a medical hierarchy. The synchronicities were ironical. In many countries the second half of the twentieth century saw opium recognised as a menace requiring presidential wars, drug tsars and a colossal international police force. At the same time palliative medicine revealed it as an incomparable blessing, unique in combating not just 'pain' but what Cicely Saunders called 'total pain'. The last term was daunting in its implications. It covered the sensation of hurt conventionally referred to by the word but also its often hidden and infinitely complex ramifications. It meant patients regaining a measure of control over their own lives and being relieved from incapacitating feelings of guilt. (Cicely Saunders recognised how many felt mortified by the thought of being a burden.) That the essence of poppy juice should often provide relief from such fears would be almost unbelievable if developments in the laboratories had not revealed even more astonishing but experimentally verifiable facts.

* * *

Moments of scientific revelation do exist but they are rarely appreciated as such at the time or indeed for a long time afterwards. Claude Bernard came from a family of poor labourers in the vineyards of Saint-Julien in Burgundy, attended the village school, excelled not at all, was apprenticed to the local pharmacist and dreamt of becoming a writer. In 1838 when he was twenty-five he walked to Paris with the manuscripts of some of his plays in his knap-sack and showed them to the well-known theatre critic and fellow Burgundian, Saint-Marc Girardin. Girardin thought the plays were dispensable but made the inspired suggestion that the young man who showed 'remarkable powers of observation' should enrol in the Faculté de Médicine. Four years later Bernard qualified near the bottom of his class but then his fortunes changed. He became assistant to François Magendie, France's most controversial – some would call him bloody-minded – clinician, who loudly proclaimed that 90 per cent of what students were taught in medical schools was garbage and who admired what he described as Bernard's 'resistant mind'. The recognition of genius was mutual and in 1855, when Magendie died, Bernard took his place as professor at the Collège de France. His work there rejuvenated every aspect of medicine. In 1864 (between discovering the chemistry of diabetes and establishing the revolutionary concept of the constancy of the internal milieu) he carried out a series of experiments with the newly isolated South

American arrow poison, curare. As every biology student knows today but did not know then, the application of an electric current to the nerve which supplies the muscles of a frog's isolated hind leg makes the hind leg twitch. Claude Bernard found that if the frog had been injected with curare before being destroyed the response was abolished. Either, he argued with impeccable Gallic logic, curare made the nerve incapable of conducting the impulse or it rendered the muscle unable to respond to it. To his annoyance, a series of experiments showed beyond doubt that the nerve still conducted the impulse and, directly stimulated, the curarised muscle still twitched. Gallic brilliance then took over from Gallic logic. If curare acted neither on the nerve nor on the muscle its impact had to be on a hitherto unknown structure between the two, a special site where the nerve made contact with the muscle. Without actually being able to demonstrate it, he called that site the *jonction neuromusculaire*. Modern neuroscience was born.

Or would have been if the discovery and hypothesis had been followed up. But Bernard had his plate full with more urgent projects and for almost fifty years nobody took any notice of his discovery. Then J.N. Langley of Cambridge, studying the inhibitory action of nicotine on curare in 1909, came to the conclusion that a minute amount of a chemical is released when a nerve impulse reaches a neuromuscular junction and it is this chemical which transmits the command from one to the other. After doing so, the transmitter either returns to its original receptacle or, more likely, it is destroyed and resynthesised in its original site.[15]

But another twenty years passed before this bizarre notion was taken up by the young H.H. (later Sir Henry) Dale at University College London and the Austrian pharmacologist and later Dale's friend, Otto Loewi. In a simple and lucid series of experiments they showed that nerve impulses to the heart and involuntary muscles were mediated by the comparatively simple molecule acetylcholine, the first neurotransmitter to be identified. This at last started ripples and in 1936 earned the discoverers the Nobel Prize in medicine, but the real revolution came another thirty years later. Then, over a comparatively short period, an extraordinary range of chemicals were shown to transmit impulses not only from nerve to muscle but also from nerve to nerve. They were in fact the links in the immense complexity of brain circuitry and between the brain and its peripheral extensions, the nerves. And it was soon recognised that one of their most striking and generally shared characteristics was their specificity or, more precisely, the specificity of their 'fit'.

Provided it is not taken too literally, the model of lock and key will serve. For a neurotransmitter to transmit its message from one nerve cell to another (or from nerve to target organ) it must not only be released at the right moment in the right amount in the right place but it must also fit into a specific receptor site. This can be compared to the right key and no other key fitting into the right keyhole and no other. It is this specificity which brings the story back to opium. Contrary to what might be expected from the analogy, in the case of the body's own morphine and morphine-like substances it was not the key which was first discovered but the keyhole. In 1973 Dr Candice Pert and Dr Sol Snyder at Johns Hopkins Hospital in Baltimore showed that specific morphine receptor sites – that is the keyholes – existed in the pig's brain; and since it seemed improbable that they would have been created in anticipation of the pig receiving extracts of the juice of the white poppy, the researchers adopted Sherlock Holmes's perennially sound principle and put forward a marginally less improbable notion: that the receptor sites existed for the pig's own morphine or morphine-like product. Pert and Snyder called these still somewhat enigmatic mediators endorphins (from endogenous morphines).[16]

There is always a right moment (as well as a plenitude of wrong ones) for a fundamental discovery in science; and since the time these receptors were identified progress has been fast and furious.[17] Dr John Hughes of Edinburgh and his colleagues in Scotland and Cambridge established that morphine-like neurotransmitters existed in the human brain and other tissues as well and called the body's own opiates 'encephalins'. It is now known that there are at least three distinct endorphins and several subtly different encephalins, all with their specific morphine-like properties. Similarly, there is not one receptor site but at least three as well as some 'orphaned' ones with so far no clear function.[18] Some of the body's own morphines appear to be chemically different from poppy-opium morphine but others are virtually identical and they are present or inducible in a variety of organs.[19] The field is advancing so fast at both the clinical and the experimental levels that any examples can only be a random selection.

One condition which emerged early as a possible manifestation of a burst of endogenous morphines was the 'jogger's high'. The phenomenon is of no great clinical importance but is known by sight to most modern city dwellers and remains at many levels a puzzle. Could endorphins or encephalins enable bodily functions to go over their normal physical limits? They are not the only candidates for the role; but recent imaging studies of athletes' brains provide evidence in support of the idea.[20] Endorphines may be important in modulating stress response, helping to cope with situations which might otherwise cause mental and physical collapse and death. The occasionally impressive effect of acupuncture has also been investigated. Could the procedure trigger a burst of endorphins? Still somewhat inconclusive evidence suggests that the pain-relieving effect of this ancient treatment can be blocked by the morphine inhibitor naloxone.[21] After the third month of pregnancy the foetal component of the placenta seems to secrete endorphin-like substances into the maternal blood. If these make expectant mothers mildly endorphin – that is morphine – dependent, then untoward post-partum manifestations like postnatal depression could be the result of sudden withdrawal. In animal experiments encephalins and endorphins have a powerful antidepressant action.[22] White blood cells too can elaborate endorphins and *in vitro* release is enhanced by nicotine, alcohol and cocaine.[23] Both experimental and clinical work suggest that the often abnormal craving for sweet and fatty foods, and the consequent obesity, may be partly mediated by endorphins.[24]

But most important perhaps, the new endogenous substances may be changing attitudes to opiate addiction. Could some cases be a kind of empirical replacement therapy, correcting a lack of endogenous substances, as 'physiological' or at least as therapeutic as insulin is in diabetes? One recalls the recurring figure of the long-term addict on a steady dose of opiates not only surviving but living a long and creative life. The species is not extinct.[25] While this is pure and perhaps reprehensible speculation, it seems not unreasonable to hope that current advances may one day lead to a better understanding of addiction and to comparatively painless ways of treating it.

CHAPTER 31

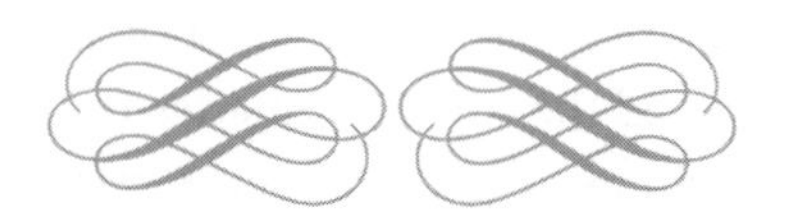

Treatments and cures

ONE DAY – BUT it would be a bold person who predicted when. Failure to take advantage of advances in the basic sciences has never been for want of trying. During the early years of the twentieth century the antitoxin hypothesis found an enthusiastic following. The idea evolved from the great discoveries of the immune response. In healthy organisms bacterial toxins trigger the elaboration of highly specific antitoxins; and, though the molecular mechanism is still imperfectly understood, in some previously fatal diseases – among them rabies and diphtheria – antitoxins proved triumphantly successful in treatment. Such advances sometimes prompt purely verbal and therefore misleading analogies. Dr Lorenzo Gioffredi, a fashionable practitioner in Rome, argued that morphine and heroin were poisons or 'toxins' and therefore 'had to be' capable of generating antitoxins. He postulated that the primary action of the drugs was on the stomach lining and that it was the lining cells which could therefore be expected to produce these brave molecular warriors. Dogs were made addicted with oral morphine and their blood serum which supposedly contained the antitoxin was tested on patients. The patients survived; and in wishful therapeutics such a modest result is sometimes interpreted as a success. But there was more wishful evidence to support the theory. If morphine was constipating, the antitoxin was predicted to cause diarrhoea; if morphine made the patients feel good, the antitoxin would make them feel bad. Both predictions came true, especially when the antitoxin was supplemented with small doses of belladonna, strychnine and galvanic and faradic stimulation. Belladonna containing atropine was an ancient antidote to opium, replacing the euphoria of morphine with babbling drowsiness. (This prevented patients from even thinking of sneaking out of the ward in search of their accustomed fix.) Strychnine served to further stimulate the bowels; and, on no evidence whatever, electrical stimulation was expected to invigorate the so-called antitoxin.

During the first years of the century the regime was adopted in some of the best – that is most expensive – private detoxification clinics in Western Europe and led to claims of significant successes. The bilious faeces in particular, one of the most unpleasant features of the treatment, became the sign of a cure. The fad then crossed the Atlantic. Charles B. Towns, a Georgia insurance salesman, forged a lucrative career for himself treating addicts and searching their bedpans for the 'green stool of hope'. In 1906 Towns's approach was endorsed by no less an authority than Alexander Lambert, personal physician to President Theodore Roosevelt; and, joining forces, the two men designed a horrendous 'hyoscine and antitoxin regime'. Towns demonstrated the depth of America's friendship for China by treating a hundred coolies at an 'auxiliary clinic' set up during the Shanghai Opium Conference; and received an official citation from the State Department recognising his contribution to Sino-American brotherhood. It described the Towns–Lambert Method as 'the most successful treatment of opium and morphine addiction on record'. Soon a four-floor Charles B. Towns Hospital rose on Central Park West in New York City where rich addicts paid $300 a day for a private room and the less affluent $70 for a bed in a public ward. The institution folded ten years later but left Towns a rich man and Lambert an influential Congressional lobbyist for strong anti-narcotics legislation.

One craze led to another and both the interwar and post-Second-World-War years saw a proliferation of new regimes of varying complexity, absurdity and expense. All aimed at weaning addicts painlessly and permanently off their servitude to morphine or heroin. Carbon dioxide treatment consisted of inducing unconsciousness first with nitrous oxide and then with a mixture of 30 per cent carbon dioxide and 70 per cent oxygen. As late as 1972 Dr Albert LaVerne lectured on the efficacy of the method; but trials were abandoned when one patient never recovered from her coma. Heavy doses – 300–500 micrograms – of lysergic acid diethylamide (LSD) were tried on volunteers in prisons in Maryland, combined with psychotherapy. A third of the treated group was said to have been off heroin six months after their release. A course of therapy based on a full-scale Freudian interpretation of addiction gained advocates in the 1920s and '30s both in Europe and in the United States. The craving for morphine was identified as a pharmacological sex thrill; and other features were given complex Freudian explanations. To provide a cure was simple: one only had to 'irradiate and sublimate the libido which the addict was wantonly wasting on the fetish of drug addiction'. The concept was welcomed in principle by the Federal Narcotics Bureau but its implementation was judged premature. Psychoanalysis and other forms of intensive psychotherapy were nevertheless used extensively in private practice.

Not all trials were as benign. During the 1930s and '40s a host of new drugs were tried in special clinics, most notoriously at the Lexington Antinarcotics Farm, to which patients were referred by the courts and various anti-narcotics agencies for 'mandatory detoxification'. 'Why do you call this torture chamber a farm?' one inmate asked. 'Do you actually grow the weeds here?' Some of the programmes were pursued with malign zeal. Most inmates/patients tried and tried again to escape and the successful suicide rate, never actually disclosed, was said to exceed 25 per cent. All trials demonstrated that after any pharmacological 'cure' patients had a strong tendency to lapse into deep depression or to go back on the needle within six months. But negative results seemed to do nothing to douse the enthusiasm of the teams of experimenters.

* * *

At the Nuremberg Trials of Nazi war criminals in 1945 many wondered why the United States representatives refused to sign the part of the final protocol which prohibited (and still prohibits) the performance of medical experiments on prisoners in jails, even after 'full explanation and consent'. Of all revelations of horrors during the court proceedings nothing shocked the world more deeply than the use of concentration camp inmates for so-called medical experiments. The reason for the refusal was that the signing of the protocol coincided with the endowment of the Lexington Addiction Research Center with a handsome new budget by the National Institute of Mental Health for new anti-narcotic drug tests. Although the worst outrages had by then been stamped out, 'voluntary prisoners from jail' were still the largest prospective group of guinea pigs; and elaborate detoxification programmes with such drugs of evil reputation as nalorphine and cyclazocine, both toxic to the liver, bone marrow and other organs in a significant proportion of cases, continued.

But it was general attitudes which most effectively frustrated progress; and they were not confined to Lexington. In 1971 Dr David Smith wrote about the doctors and nurses at one emergency hospital of high reputation:

> Most public health representatives, like most policemen, look on young drug abusers as sub-human. When adolescents come to Park Emergency for help, the doctors frequently assail them with sermons, report them clandestinely to the police or submit them to complicated and long-drawn-out referral procedures. The nurses sometimes tell prospective patients to take their

> problems elsewhere. The ambulance drivers 'forget' calls for emergency assistance. They and other staff members apparently believe that the best way to stamp out addiction in the districts is to let their younger residents destroy themselves.[1]

The first anti-addiction drug of merit and still in use was synthesised during the years before the Second World War in Germany but then put on the back burner. Only when war in the Balkans threatened to cut off Turkish opium were some of Germany's leading chemists set the task of finding a substitute. Pre-war efforts were redoubled and the outcome was a drug known today as methadone but originally called *dolophine*.[2] The medication was used in analgesic mixtures during the Second World War; but in the post-war bonanza of 'liberated' German patents it was forgotten. Twenty years after the war Vincent P. Dole, professor of psychiatry at the Rockefeller University in New York City, and his colleague and future wife, Marie Nyswander, were searching the literature for possible lines of detoxification. They had been increasingly disturbed by the treatment of addicts as 'social lepers' in various 'rehabilitation' and 'detoxification' centres with a relapse rate approaching 100 per cent. Eventually they came across forgotten experiments with dolophine as a possible substitute for heroin. Fortunately the drug, renamed methadone, was at the time sufficiently unknown not to be on the legally prohibited list. It was soon shown to be an effective though far from perfect detoxifier. Its main action was to block the euphoriant effect of heroin and therefore the craving for it. Given in large but gradually reduced doses it abolished or at least mitigated 'abstinence sickness'. This chimed with the astonishing experimental finding that, though the molecular shape of the substitute is unlike that of morphine, it seemed to fit one of the endorphin receptor sites.

Soon methadone allowed some addicts to resume a normal life, learn trades, hold down jobs and enjoy social functions. This did not stop it from being fiercely opposed by the Federal Bureau of Narcotics and other official bodies as well as by individual politicians and policemen. What the opponents objected to was not so much the drug as Dole and Nyswander's attitude to addiction. But the opposition could only slow down progress. Dole remained a believer and wrote after thirty years' experience that the treatment had helped thousands of patients who otherwise would have died or would have ended up incarcerated:

> For me the most educational experience of the past three decades [he wrote] was to learn that the traditional image of the narcotic addict – weak character, hedonistic, unreliable, depraved, dangerous – is totally false . . . The

> myth, believed by the majority of the medical profession as well as by the general public, had distorted and corrupted public policy for seventy years . . . My experience of thousands of patients from widely different cultures has shown the typical heroin addict is a gentle person trapped in a chemical slavery, pathetically grateful for understanding and effective treatment . . . But discussions about maintenance programmes have been hopelessly muddled by political interests, ideological posturings and disinformation in the media and by public figures aspiring to public office.[3]

No current detoxification programme was or is perfect. Methadone converts one addiction to another, and none is desirable. In some respects it is the addict's family rather than the addict who benefits: after three years on a methadone programme addicts are said to be 90 per cent less likely than before to be arrested for theft, assault, drug sales, dud cheques and other petty fraud, vagrancy, fencing stolen goods, public intoxication, prostitution, attempted suicide or disorderly conduct. On the debit side many addicts detest the treatment, describing its effect as making them 'zonked'. They may also find that withdrawing from it is as hard as withdrawing from heroin. In every country where the method is used the drop-out rate is high but a significant proportion of dropouts return to the programme. The hopeful and sometimes realised objective is that once the craving for heroin has been extinguished addicts will find pursuits more worthwhile than chasing their incubus. They may even discover the mundane pleasures of home life.

One of the incidental consequences of successful or relatively successful rehabilitation programmes was the diminished likelihood of becoming infected with AIDS. As the epidemic spread and claimed more and more young lives, addicts seeking a cure and the use of methadone rose. Or did where it was allowed to. Mention of even a limited programme still provokes outrage in Russia. But in most countries the search for other and better drugs and 'pro-drugs' – precursors which are converted in the body to the active principle – continues.

* * *

Drugs are of course not the only road to salvation. Some addicts have been weaned off heroin by acupuncture and other holistic methods. Religious faith too can move mountains, as it did for Jacqueline Pullinger. In 1966 this twenty-six-year-old English music teacher travelled to Hong Kong on a cargo boat: more precisely, she travelled east in search of a mission and alighted in Hong Kong on an impulse.[4] She found work as a primary school teacher in Kowloon Walled City against no competition. The 'City' was a wedge of buildings traversed by alleyways into which the heat of the sun beat down but light barely penetrated. It had acquired the reputation of being the world's most dangerous square third of a mile; and the reputation was deserved. In the 1960s the ground below was still a warren of cellars and subterranean passages extending to a depth of three or four floors. Above ground the tenements were divided into tiny, teeming apartments and one-room factories making everything from fishcakes to sex toys. But above all Kowloon Walled City was the world's premier centre of heroin manufacture. The white powder, exceptionally pure, could be purchased by the kilo by Triad-accredited traffickers. It could also be purchased by freelancers provided they were ready to risk having a knife slipped under their ribs. But profits could be luminous. Few residents of any age were not heavily addicted. Half-mummified ancients – but perhaps young in calendar years – crouched in doorways, rag-doll symbols of death. Most were immobile except for jerkily taking a puff from a pipe every few minutes. But simply inhaling the air was enough to get the inhaler hooked – or at least give him a headache. By the 1930s officialdom had tacitly given up policing the quarter, allowing it to be ruled by the Triads. Their sway was absolute; their tolerance zero.

It was in this lower circle of hell that Jackie (as she became known) settled to teach and to establish her evangelical prayer meetings. To these addicts flocked, travelling from Australia, the Americas and Europe. During the meetings some went into a trance, swaying, muttering and falling to the ground, and some came out of their trance exhausted but cured. Jackie and her few acolytes continued to work unharmed, protected by the Triads. (Later publications do not mention this but they could not have survived otherwise.) Tourists to Hong Kong were warned off, in fact forbidden to visit the quarter unless accompanied by Jackie or one of her circle. In their company the visitor was safe. Eventually the group, renamed St Stephen's Society, gained the support of the Hong Kong government who provided it with a plot on which to build a hostel. As a religiously sceptical historian of opium remarked: 'it was almost as if God, feeling guilty at having made His own medicine a curse, was offering His own relief from it and it is perhaps not just divine inspiration but divine irony that He should exercise His love in China, at the core of the opium story'.[5] The Walled City was demolished in the early 1990s to be replaced by a boring patch of greenery; but the Society continues to function and now has branches outside Hong Kong. Religious foundations on a smaller scale exist elsewhere; and unorganised and unsung individuals use faith as the universal healer in probably every country.

* * *

Faith, not necessarily religious, or at least a desire to be cured, remains essential for any kind of detoxification. The desire tends to be taken for granted by non-addicts, especially those committed to help. The belief is not always justified. The world of addicts looking in is different from the world of addicts looking out. Looking in there is the degradation, the disease and the prospect of ruin. There is also the anguish of friends and family. Looking out there is sometimes nothing but the folly, the greed and the vanity of the non-addicted world. That looking-out image has to change before a cure can be expected. No drug on its own can accomplish that.

Clinics under expert psychiatric guidance dedicated to permanent detoxification now exist in most Western European countries, relying on a combination of psychotherapy and non-addictive tranquillisers; and some achieve respectable results. But the task remains labour-intensive and therefore expensive. It can also be argued that those rescued are mainly patients who have become addicted through peer pressure or temporary adverse circumstance rather than 'natural' addicts. Does the second category exist? Many authorities deny it as an article of faith; but in the light of developments in the basic sciences the concept is difficult to rule out. Free clinics also offer help, many staffed with a dedicated workforce, often past addicts. It is possible that detoxification regimes based on combining a substitute drug with carefully controlled small and slowly diminishing doses of the drug of addiction will prove to be a significant advance.

CHAPTER 32

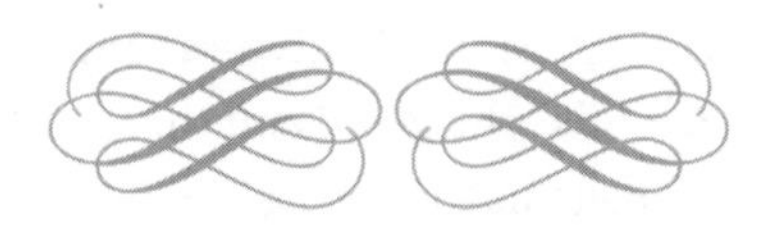

Life and death of a drug lord

THE DRUG LITERATURE of the past fifty years is littered with geometrical shapes – triangles, crescents and most recently a square – usually described as golden.[1] The shapes refer to areas on the map which have successively supplied the world with its illicit drugs. The 'golden' presumably hints at the profits of a few criminals living thousands of miles away.[2] For most ordinary inhabitants of the shapes the decades have been anything but golden.

Chronologically the Triangle rose to fame first, about half a million square miles of South East Asia, the mountainous and beautiful northern parts of Myanmar (Burma), Vietnam, Laos and Thailand.[3] The triangularity ignores man-made frontiers: they have always been figments. It is traversed by majestic rivers interrupted by magnificent waterfalls and wild mountain ranges much of them covered by jungle. Contrary to what is sometimes stated, the region is not one of the ancient homelands of the poppy. Chinese fleeing south after the opium wars and Taiping Rebellion established the plant; and large-scale cultivation did not begin till after the Second World War. Apart from the displaced Chinese and their descendants, the land is inhabited by several nations or tribes, most numerously the Shan.[4] In the closing years of the Second World War they may have been promised independence by shadowy uniformed characters dripping with gold braid; but no Balfour Declaration supports their claim.

The war and Japanese occupation had blasted the credibility of colonial rule but had left much of South East Asia in turmoil. The remnants of the Kuomintang Army, defeated but still equipped with smart American weapons, invaded the region from the north. What used to be French Indochina erupted in post-colonial wars in the south. Neighbouring Thailand, ruled by a small, self-serving elite, was deeply corrupt.[5] Newly independent Laos was a close rival.[6] Burma's fledgling democracy was threatened by ethnic, religious and ideological divisions. Confusion may favour new developments but the fight for Shan independence still needed material resources; and of those the movement possessed only one. To exploit it there emerged a charismatic leader, perhaps the hero of yet unwritten chronicles, perhaps the most wicked drug lord of all times.

* * *

Khun Sa, his *nom de guerre* roughly meaning Prince of Prosperity, was born Zhang Qifu in 1934 in Loi Maw in the Mong Yai state in what was then still the north of the British colony of Burma. His father was Chinese, his mother Shan. His father died when he was three and his mother then married a landowner of some consequence. In a hierarchical society this mattered: Khun Sa never lost his aristocratic airs. While his three stepbrothers were sent to a missionary school and later christened Oscar, Billy and Morgan, Zhang was raised by his learned Chinese grandfather. His only outside education was a few years as a temple boy in a Buddhist monastery; his correspondence later had to be read to him and his own letters were dictated. But he was no country bumpkin. He had a cultivated taste in Chinese art, preferring T'ang terracottas to nineteenth-century kitsch, a good knowledge of history beyond South East Asia, and a command of mathematics sufficient to sustain a position in a global trade.

At about the age of sixteen he was forcibly recruited into the Kuomintang Army which maintained itself in 'temporary' sanctuaries. It was fitfully supported by the Kuomintang government in Taiwan and more regularly by the CIA. Though a rabble, they had gold-plated anti-Communist credentials and were proclaiming their intention of reconquering China any day now. For the local population they were a disaster, collecting extortionate taxes, forcibly enlisting local youth and encouraging poppy cultivation to finance themselves. But Khun Sa learnt from them the techniques of guerrilla war and the subtleties of South East Asian poppy politics. He soon formed his own armed band to fight his former masters; and he did so successfully enough to be recognised as the leader of a 'home guard' unit by the government in Yangon, formerly Rangoon. With government money he purchased French and American military equipment in neighbouring Thailand and Laos and at thirty-three challenged the Myanmar regime. He was soundly defeated and taken prisoner, but his army melted away into the jungle. The story of his eventual release, allegedly in exchange for two Russian doctors kidnapped for that purpose by his loyal deputy, is embedded in the literature but even by poppy standards extravagantly implausible. Where did the Russian doctors come from? What were they doing in xenophobic Myanmar? And why should the government in Yangon release a prize captive in exchange for them? The truth, like most of that commodity in modern South East Asia, must have been more brittle, involving convoluted transactions never uttered, let alone written down. (Records in the opium trade, whether in Myanmar or more recently in Afghanistan were and are means of obfuscation rather than of illumination.)

Whatever the details, between 1976 and 1990 as Prince of Prosperity he ruled over a state roughly the size of England and commanded a well-equipped force known as the Shan United Army. The main purpose of that army was to protect the opium traffic to and from refineries, guard over the new opium-processing laboratories, the first to be built in the region, and to escort the finished product to distribution points in Thailand and elsewhere. It also put down unauthorised trade: villages which were tempted by rival offers to sell their produce to interlopers – that is past or would-be drug lords – were destroyed. Boys and young men were recruited into the Shan United Army. Annual opium production in Khun Sa's kingdom started with about 100 tonnes in 1976, rose to 4–600 tonnes by 1980, to over 1,000 tonnes by 1985 and to over 3,000 tonnes by 1995. Satellite pictures show how the area under poppy cultivation increased from 34,000 hectares in 1987 to over 200,000 hectares in 1991. The harvest was converted locally to heroin of exceptional purity and all of it was exported. Though every stage of the business had to be buttressed by payouts to political bosses and lubricated by sweeteners at lower levels, the product on the streets of New York, Los Angeles, Toronto, Sydney, London and Paris, known as 'China white', was still good value; and the profits at base paid for the construction of roads, hospitals and schools and a vaccination programme as well as for trifling personal luxuries for the prince. Other warlords in South East Asia who had had an interest in the trade, some older and better established than Khun Sa, were ruined, extinguished or brought into the fold. Assassination was a last resort. The illegal Burmese Communist Party which had controlled a swathe of poppy fields on the Burmese–Chinese border suddenly found that their refineries stopped working. Nobody came to repair them

Khun Sa's links with the Dragon Syndicates, also known as the Triads, remain opaque; but they were probably close. He ran his fiefdom efficiently but he was in no position to organise a worldwide sales network. The Dragon Syndicates, on the other hand, had been involved in the opium trade since the 1880s and by the 1970s controlled drug transit through Bangkok and Hong Kong. Rooted in the Chinese diaspora of 40–60 million, by far the largest in the world, it had developed its own banking system based on traditional family ties and a seamless money laundering apparatus. It owned men of influence in most countries and the heads of state in some. But, unlike the Sicilian Mafia with their sleek yachts and private jets, the Syndicates shunned publicity; and Khun Sa was the perfect front. He could be demonised in the media as the linchpin of the global narcotics trade. He was also a skilled negotiator with the legitimate governments of Myanmar, Thailand, Laos and Vietnam.

But even with support from the Syndicates, temporary setbacks were unavoidable. Under American pressure the Thai Army moved against the Shan United Army in 1982. What ensued depends on which of many lies one chooses to believe. It is possible that the Thais defeated the Shans in a pitched battle lasting ten days and resulting in considerable bloodshed. More likely, an amicable agreement was negotiated after a virtual battle, the Thai Army retiring covered in martial glory and Khun Sa moving his capital across the frontier to Hmong in Myanmar. Here his career reached its peak.

Though sometimes referred to as the 'jungle hideout of bandits', Hmong was, according to one visitor, 'the nearest thing in South East Asia to Milton Keynes'. It had an efficiently laid out grid of streets and a mall lined with well-stocked shops. All necessities of Western life, from laptop computers to fake Rolex watches, could be bought at affordable prices. Payment was in dollars. Between 20,000 and 30,000 souls lived in timber or concrete houses surrounded by fruit trees, clipped hedges and gardens in bloom with bougainvillea and marigolds. The scent in October was intoxicating.

> Clear signs indicated where one could obtain travel permits to Thailand and hopefully elsewhere, even to the United States of America. The town boasted two schools and a Buddhist monastery. Though the Shan are devout Buddhists, visitors from the Reverend Moon's church in South Korea were welcome ... A hospital with an operating theatre and X-ray facilities was manned by courteous and competent Chinese doctors. There was a video hall, a karaoke bar, a hotel, a disco and a small park with benches for tired feet. Overseas calls could be placed from a commercially run telephone booth. Local artefacts and atrocious historical paintings in the European style were on display in a one-room cultural museum ... Among projects under construction in 1989 were a hydroelectric plant to replace the diesel-powered generators and an 18-hole golf course. Khun Sa was said to be adept at the game; but the facility was to be mainly for visitors from Thailand, Taiwan, Singapore, Hong Kong, South Korea and Japan. Some came under the pretext of buying precious stones from the gem centre which the Prince ran as a hobby. But if visitors expected a drug-paradise they were disappointed. Drug dealing locally was forbidden and, in contrast to Bangkok, San Francisco or Manchester, the prohibition was enforced.[7]

The prince himself lived in a bungalow distinguishable from others only by its more extensive grounds, planted with Norfolk pines and beds of orchids and wild strawberries, his favourite fruit. It was also ringed with concrete bunkers sheltering 50-calibre anti-aircraft machine guns. Bursts of ack-ack were said to reassure the populace that they were being guarded and protected. But Hmong was also a garrison town. Soldiers lounged in the cafes and screeching armoured cars threatened life and limb. Most tellingly, people in the streets were uncannily well behaved (as they were in Stalin's Moscow and Saddam's Baghdad): no drunks, no yobs, no antisocial behaviour. 'I have an army, so I am free,' Khun Sa said to the journalist and writer Bertil Lintner. 'Poor Aung San Suu Kyi has no army and therefore she is under house arrest.' This in his opinion was the only difference between himself and the Nobel-Prize-winning democratically elected leader of the country. According to a visitor

> he was not on first meeting endearing. He had jerky movements, a hoarse voice, a malarial complexion and a bad breath; but when relaxed his face could light up in a childish and engaging grin. And he was, in rare and sudden moments, not without a sense of humour. 'I am not a rich man, as you can see. Of course I could be a millionaire if I turned myself in to the police in New York.'[8]

While some of this may sound cosy, by 1990 Khun Sa's state with the commercial help of the Triads was providing the bulk of the illicit heroin consumed in the United States and Australia and a significant proportion of what was imported to South America and Europe. In 1989 a Washington State Department spokeswoman publicly branded him the Prince of Death and put him on a par with the most evil of Mafia dons. In 1992 a reward of $2 million was offered by the Drug Enforcement Administration for his capture, *Time* magazine named him the world's most wanted gangster, and William Brown, United States ambassador to Thailand, described him as 'the worst enemy the world has today'. But that was not the whole picture. Lady Brackett, a former model and now a New York socialite, boasted of having met him and having found him a pet. She showed her friends a gem-studded pair of slippers, a present from the fiend. More significantly (and perhaps more truthfully) Jean-Louis Destruelle, a retired French army colonel, had devoted years to trying to trace the graves of (and, just conceivably, any still alive) French prisoners from the first or 'French' Vietnam war. He had battled against insistent and malicious obstruction from authorities in Hue, Saigon and Yangon (including a brief spell in jail as a spy) until he sought the help of Khun Sa.

The prince was sympathetic and his sympathy, as Destruelle later recounted, was 'like the waving of a wand. I did not have to mention his name or show documents. From that moment doors were opened not only by officials but also by village elders hundreds of miles away.'[9] A similar experience gained Khun Sa the friendship or at least the gratitude of James 'Bo' Gritz, a highly decorated American Vietnam veteran, whose search for prisoners of War and for those missing in action after the Second or 'American' Vietnam War was helped by the warlord. Gritz's efforts were partly financed by Ross Perot, oil tycoon and United States presidential candidate, who endorsed Khun Sa's help. Shirley D. Sac, a New York gem dealer, proposed to raise funds for the Shan Liberation Army under Khun Sa's leadership. Khun Sa also had high-level contacts among the military junta in Yangon and friends in high places in both Bangkok and Singapore; he could not have survived otherwise.

* * *

Early in his career Khun Sa met Joseph Nellis, aide to Congressman Lester Wolff, then chairman of the United States House Select Committee on Narcotics, and presented a 'six-year plan' to eradicate opium from the Golden Triangle in return for an 'undisclosed but substantial' sum. The Carter administration declined. Some years later through the mediation of Stephen Rice, an Australian journalist, the prince offered to sell his entire opium crop to the Australian government for about A$800 million a year for eight years. Whether the arrangement would have worked remains problematic; but the Australian government refused. Senator Gareth Evans declared that 'this Government is not in the business of paying criminals to refrain from criminal activity'. Yet, after a comparative lull of a hundred years and with a coastline difficult to police, his country was again struggling with an escalating drug plague.[10]

Khun Sa's next surprise move came in 1996 when, filmed by television crews, he surrendered to the Myanmar army. He was arrested and transported to Yangon. The move allegedly transformed him in the eyes of his Shan followers from liberator into traitor; but the facts were probably infinitely more complicated. The United States was rumoured to be raising the reward for his capture to $5 million; and against such an incentive even the best-guarded target may begin to experience frissons of apprehension. His five sons and three daughters were ready to enter higher education or the world of business. There was little scope for either in the Shan States, however booming. Perhaps his health was no longer robust. And he surrendered on his own terms. What concessions if any he achieved for his United Shan States is unclear but personally he was assured an important position in the Myanmar economy and of the safety of his family. The ruling junta of Myanmar, however repressive, live by their military code and their promises were kept. His old contacts in the financial citadels of Singapore and Bangkok were probably useful for laundering millions of government drugs profits. Three of his children were sent abroad to university, one to an ancient establishment in the United Kingdom. In 2010 his favourite son was running a luxury hotel and casino in the picturesque town of Tachilek near the Thai border. He himself lived in a well-appointed colonial villa in one of the suburbs of Yangon where, in 2003, a distinguished British physician attending a medical conference in Bangkok was invited for consultation. The doctor found the prince (as the patient was still addressed by all) slightly despondent. His complaint (which also embittered the last years of President Nasser of Egypt) was diabetic neuropathy. 'It is ironical,' Khun Sa remarked, 'that even the blessed heroin has no effect on my suffering.' Little could be done to ease his pain but he pressed a ruby tiepin in the shape of a dragon on his visitor.[11]

* * *

Khun Sa died peacefully in October 2007. Belying rumours that his desertion had outraged his Shan followers the current leader of the Shan United Army told Reuters news agency that the prince would be deeply mourned by his people. 'He was a man of lofty ideas. Circumstances had conspired against him but time will vindicate his aims.' His departure also caused or, more likely, coincided with a geographical shift in the international drug trade. The amount of heroin originating from the Golden Triangle declined, the region becoming instead the world's main supplier of methamphetamine.[12] Dealing in madness would prove as lucrative as dealing in dreams, and just as lethal.[13] Opium cultivation drifted from the Golden Triangle to the Golden Crescent or, in more prosaic terms, from South East Asia to Afghanistan.

CHAPTER 33

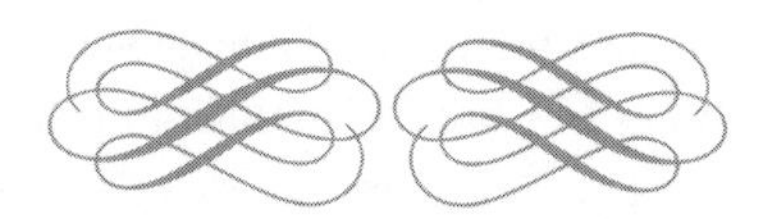

The making of a modern narco-state

UNLIKE THE GOLDEN Triangle, the area of Central Asia known as the Golden Crescent has been home to the poppy for centuries. Straddling the Great Silk Road between China and the Mediterranean the region has always sent some of its produce to the West; but in global terms the trade has been small. The last League of Nations narcotics survey reported that Afghanistan at the heart of the Crescent contributed less than 2 per cent to the world's illicit heroin.[1] What has made it a modern narco-state in the first decade of the twenty-first century is politics.

For a century before the First World War the country was the focus of an undeclared but fiercely contested war known as the Great Game. The contestants were the British and the Russian empires and the mainspring of their conflict was paranoia. Pan-Slav circles in St Petersburg watched with beady eyes the expansion of British India into the Himalayas, ostensibly threatening the largely empty spaces of Central Asia. On the other side spokesmen of the 'White Man's Burden' in London, no less mentally deranged, perceived in Russian interest dark designs on India 'by the back door'.[2] Nothing less than occupying these deserts could ensure the safety of the Raj and the survival of the Manchester textile industry. At the epicentre of this particular mayhem Afghanistan, a poor but not yet a desolate country, had a long history of near-total devastations by nomadic hordes alternating with periods of glittering achievements in the arts, crafts and literature.[3] By a combination of luck, valour and the blunders of would-be conquerors it preserved its independence and Islam remained a dominant though also a fragmenting force.

* * *

It might have been expected that after remaining neutral in the Second World War the country would at last settle into a period of calm and prosperity; but centuries of psychotic suspicion are not easily stilled. In 1972 the Soviet Union, allegedly anticipating Western designs, invaded the country and imposed on it a succession of notably inefficient Communist governments. At least a million Afghans fled; but the invasion galvanised Islamic sentiments at home and abroad. Young Muslims from all round the world came to join the anti-Russian resistance, among them the rich Saudi businessman Osama bin Laden. In the West the mantle of lunacy had by then descended on the United States and the mujahideen (from the Arabic *mujāhidīn*, literally those who engage in jihad), the main resistance group, were lavishly supplied with American weaponry and intelligence. Despite such support few at the time expected them to prevail: the purpose of the ground-to-air missiles which became the nemesis of Russian helicopter gunships was to inflict as much damage on the Soviets as possible, not to defeat them. But the mujahideen did prevail; or rather, under Mikhail Gorbachev Russia experienced one of those moments of sanity which occur at widely spaced intervals in the history of most great powers. After years of sustaining heavy losses in men and materiel – as all invaders of Afghanistan had done since time immemorial – Moscow unexpectedly pulled out its troops. What remained were burnt-out Soviet tanks scattered over the arid countryside like a herd of prehistoric monsters caught in a gale of asteroids. They might have acted as a warning to the next invaders thirteen years later but did not.

* * *

Despite Russian withdrawal and the subsequent break-up of the Soviet Union war in Afghanistan continued between tribal lords, pockets of mujahideen and criminal gangs. All sides exacted crippling taxes from the peasantry and promoted poppy cultivation to finance their rampage. It was in response to this that the Taliban emerged from religious schools in Pakistan – the term means student in Persian – and soon they were earning hurrahs in the West. They defeated some warlords and conciliated others, sacked incompetent officials, appeased the mujahideen and promised to establish lucrative gas and oil pipelines to Pakistan. Most important, they issued a fiercely worded ban on poppy cultivation. By 2000 and to most observers' surprise they had reunited their devastated land.

But devastation remained and became the root cause of subsequent events. Bridges, roads, wayside shelters and warehouses had been destroyed. Without them hopes for a rapid agricultural revival faded. So did the West's love-in with the Taliban. The imposition of a legal and ethical code which had been humane in the seventh century but seemed barbaric in the twentieth horrified many. What goodwill lingered exploded with 9/11. Claiming ignorance of his whereabouts, the government in Kabul refused to extradite Osama bin Laden, mastermind of the atrocity. The West declared War on Terror, whatever that meant: in effect war on the Taliban in Afghanistan. In time-honoured fashion anti-Taliban allies were found among a scattering of disaffected tribes and were elevated to the status of a 'Northern Alliance'. Kabul was conquered by a BBC television crew under the command of Mr John Simpson, the Corporation's world affairs editor. Under the concerted onslaught of other American, British and European Union forces on the rest of the country the Taliban crumbled within weeks. Or so it seemed.[4] An anti-Taliban government was installed in Kabul. The newly elected president, Hamid Karzai, was invited to address both houses of Congress in Washington. As an honoured guest he attended a cabinet meeting in 10 Downing Street. In strong language he condemned terrorism, corruption, opium and sin. The new Afghanistan was promised generous aid and would rise from the ashes like a phoenix or, better still, an Asiatic Switzerland.

* * *

This was the background to the transformation of the country into a twenty-first-century narco-state. However low the economy had sunk under Russian occupation, it sank even lower under the subsequent mujahideen and Taliban regimes. Under the latter's harsh social policies Afghans with technical skills and business expertise fled. But it was the countryside which suffered most. No reliable census had been conducted for fifty years; but an estimated three-quarters of the population, currently about 18–20 million, have always lived off the land. Wheat had been the main agricultural product followed by maize, rice and barley. In most provinces the system of irrigation on which these staples depend – less than 20 per cent of cereal production in Afghanistan has ever been rain fed – now lay in ruins. Until the Russian invasion the country had been known for the excellence of its citrus fruits, figs, dates and pistachios (the best in the world). The commercial infrastructure necessary to sustain their cultivation no longer existed. Before the 1970s animal husbandry provided not only meat and dairy products but also skins and wool, including that of the famous karakul sheep.[5] The industry had been devastated by a combination of drought and war. Much of the country's timber had been exported to Russia or used for fuel. Deforestation would take decades to reverse. The litany could be continued. For millions of farmers only one lifeline remained.

Official documents describe the main cause of the upsurge in poppy cultivation as 'food insecurity'. In ordinary language that means famine, farmers watching first their livestock and then their families 'perish by inches'. (The biblical phrase refers to a process unimaginable unless experienced or at least witnessed.) When traders emerged in the late 1990s offering *advance* payment for opium, the so-called *salaam* system of credit, considerations other than survival did not arise.[6] Poppy cultivation is highly labour-intensive: David Mansfield has estimated that in Afghanistan, depending on the composition of the workforce, 250–400 person-days are required for cultivating one hectare of opium compared to approximately 45 person-days for a hectare of wheat and 57 person-days for a hectare of black cumin. But labour was never in short supply. In peasant families every man, woman and child is a 'person'; and their number was swelled by displaced and despoilt kin. Hundreds of thousands had lost their homes and had their land laid waste. Their sweat cost nothing. The skills already existed. Even in the barely remembered days of peace and plenty life without the poppy would have been harsh indeed. In terms of soil and even of water the blessed plant was less demanding than wheat or barley. If destroyed it could be replanted. Its seeds were wondrously plentiful and could be stored forever.

Official reports published under the auspices of the United States Antinarcotic Agency, concerned branches of the United Nations, the Afghan government and various charities have suggested that part of the opium produced in the country is a lucrative sideline of greedy farmers and a few prosperous landowners. Both species exist but they are a tiny minority. Most peasants eke out a bare living cultivating the poppy. Of course it would not even be a bare living if the opium were not converted to heroin and then exported. Afghans could not consume more than 0.1 per cent of what they grow. But after the opium has been collected, dried and packaged – well preserved it will survive for at least a year – its further fate is out of the growers' hands.

A few figures (not always reliable) may illustrate the speed and scale of the transformation. In the 1970s, before peacetime agriculture fell victim to the war, Afghanistan produced on average 700 tonnes of opium a year. In 1995 the harvest was an estimated 4,600 tonnes. When in 2000 the Taliban made their bid for international respectability the figure was slashed (as testified by United Nations observers) to 185 tonnes, though how much of the rest was hoarded for a future change in policy is uncertain. It could still have been a landmark event: not only was it a 95 per cent drop on the previous year but the lowest in living memory. The trend was not maintained. After Afghanistan was declared a terrorist state and the Taliban stopped courting the West the 2001 bumper harvest yielded 5,800 tonnes of opium. This was convertible to 580 tonnes of heroin, on the wholesale markets of Pakistan worth roughly

$2.5 billion. On the streets of London and New York, processed into white powder, it would fetch many times that amount.

* * *

In World-Bank-speak the opium problem in Afghanistan is 'multidimensional'. Beyond the first dimension of peasant families wanting to survive, there is the dimension of moving the opium to the refineries and then beyond the frontiers. The trade is brisk and profitable but condemned by the country's president and in more muted tones by visiting politicians, the United Nations, the European Union and most of the world's media. It continues to flourish for three reasons. First, because there is a bottomless demand for heroin outside Afghanistan. Second, because directly or indirectly it provides most Afghans with their livelihood. And third, because law enforcement, like every other aspect of public life, is deeply corrupt.

Corruption can mean many things:[7] and examples of the Afghan variety can be plucked from a cornucopia of absurdities. Posts of commanders at border checkpoints are effectively for sale by the ministry charged with frontier control. The buyer recoups the purchase price and a little extra from handouts from the drug trade. The drugs are then let through and a small percentage is added to their future sales price. The money collected by the ministry is distributed to the needy, including cuts to parties who might otherwise make a fuss; that is, talk to foreign media. Those who still do, however high ranking and respected, are sacked or shifted to safer pastures. For foreign consumption periodic small to moderate seizures of 'contraband' are staged and, with the television cameras humming, the culprits are arrested and sent for trial. They are duly sentenced but immediately released on a variety of grounds or none. Arresting officer and villain may meet in the marketplace a few days later. Hearty greetings are exchanged. The family gifts and legacies distributed to the judiciary are welcome supplement to their inadequate salaries. Early in 2010 about 80 per cent of the police and about 60 per cent of the new Afghan Army were reckoned to be on the take.[8] So were – and are – most ordinary people. It is not a crime, not even a sin. It is a way of life.

The occupying powers are themselves the third and last dimension. For a few years after the Allied invasion, with the Taliban apparently collapsing, eradication of the source of heroin was a war aim second only to the rooting out of terrorism. The end of poppy cultivation would be part of the country's political and moral renewal. In official doctrine it was the Taliban who had been forcing the peasantry to cultivate the plant to finance terrorism worldwide. Anti-narcotics agents, journalists and academics crowded in behind the military; and international conferences in Kabul were upbeat. The mood lasted for about two years. Mostly in response to military commanders on the ground the anti-poppy clamour then virtually ceased. What the soldiers objected to was the double-talk. Winning hearts and minds and destroying livelihoods were incompatible. No alternative survival strategy existed.

This is not to say that alternative survival strategies have failed. Many experts believe that they have never been tried.[9] Until the Soviet invasion and the subsequent waves of civil war Afghanistan survived on a non-opium-based economy. It could do so again if the material basis for such an economy existed. It is fashionable in the West to say that 'material basis' means democracy and the end of corruption. However desirable, this is questionable. What matters is the eradication of hunger – or the fear of it. A non-opium-based economy could achieve this but would require long-term investment and the rebuilding of the country's commercial infrastructure. This might be almost as costly as the war. It would also be more complex than some reach-me-down solutions suggest. One of many difficulties, as explained by David Mansfield, is the diversity of the problem. 'Treating opium poppy farmers as a homogeneous group will always prove ineffective.'[10] But not to do so is complicated and time-consuming. Important theoretical advances in crop-substitution have been made over the past ten years (though sometimes couched in a barely comprehensible language) but at the time of writing the prospect of any of those being translated into practice is slim.

Inexpensive alternatives exist. Forced eradication from the air or at ground level has been tried on a small scale in other countries. The approach was widely publicised and apparently successful in Thailand. Sporadic attempts have been made in Afghanistan too. In 2005 an eradication force clashed with about two thousand villagers in the southern district of Maiwand. At least one soldier was killed and a 'large number' of villagers were injured. Protests in Kabul were predictably shrill. Aerial spraying was also discussed and perhaps tried. Reports of unauthorised action in Nangarhar province led to a fevered investigation by the Afghan Ministries of Agriculture and Health. They reported that non-narcotic as well as narcotic crops had been destroyed by 'unidentified but poisonous chemicals'. An increase in 'related illnesses' – the symptoms and signs non-specific but grave, calling for commensurate compensation – occurred among inhabitants of many nearby villages. Zalmay Khalilzad, United States Ambassador at the time, issued a statement that 'some drug-associated people' must have sprayed the crops 'in order to create a sort of distrust and problem between the Afghan people and its allies'.[11] The idea of aerial eradication was then vetoed by President Karzai, who cited public health and environmental concerns. Mixed-crop cultivation still exists and serious damage to legitimate crops would terminally silence the mantra of hearts and minds.

Yet ignoring the heavy scent of opium in the air at poppy-harvest time as Western troops wend their way along heavily mined country roads is not easy. Some of the troops may have had their lives crossed by heroin before enlisting. 'My mum was a junkie,' says Guardsman R. 'It did not make for a happy childhood. Now here we are babying the people who make the stuff. It makes me sick.' Ending the drug trade had been extolled by speakers in the training camps. What then was being done about the acres of white poppies swaying gently in the breeze? Was it to preserve those that the troops were there and were from time to time being blown up? Such questions tend to be sidestepped by visiting chummy politicians in shirtsleeves. But in 2011 some of the men and women in uniform may ponder the news item that the country whose precarious resurgence they are there to promote has increased its lead as a heroin supplier to the free world from 91 to 93.5 per cent.

CHAPTER 34

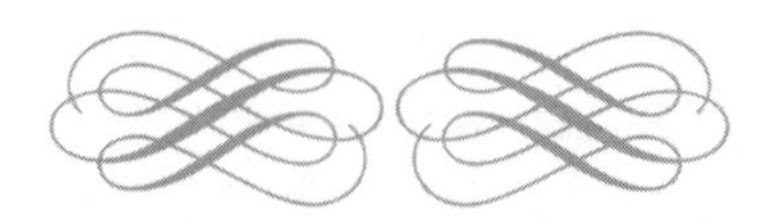

New trails: old tribulations

WHATEVER THE OUTCOME of the war in Afghanistan, it will not make heroin disappear. In 1995, twenty years after Richard Nixon declared a 'War on Drugs', Thomas Constantine, head of the United States Drug Enforcement Administration, admitted to Congress that the 'availability and purity of . . . heroin is at an all-time high'.[1] And the trend is maintained. Rough estimates put the number of heavily dependent addicts worldwide at 16–20 million and the number of occasional and experimental users at five times as many. Even when the drug is not hitting the headlines in one country it is penetrating others never exposed to it before.

* * *

Historically the people of Russia have been partial to vodka rather than to the poppy.[2] Today Siberia probably has more heroin addicts than Britain. Students in the lively new universities are proving especially receptive, as were young people in the United States and Western Europe in the 1960s. This is confirmed by thunderous notes addressed by Moscow to the new Central Asian republics which separate Afghanistan from Russia. 'The trade in heroin is contrary to international law. Governments must not provide safe passage and even safe havens for criminal traffickers.' Such démarches are reminiscent of those addressed by the Celestial Kingdom to the Red Barbarians in the early nineteenth century. The parallel does not raise high hopes; nor does a glance at the map. Frontiers snaking thousands of miles over steppe and mountain are impossible to patrol, let alone seal.

Inevitably the people of Uzbekistan, Kazakhstan, Tajikistan, Turkmenistan and Kyrgyzstan have become consumers as well as traffickers. The drug locally is cheap: in Osh in south-west Kyrgyzstan, a dollar and a minute's stroll at any time of the day or night will secure a top-quality fix. On the surface much money is swilling around; but the affluence is paper thin. Underneath, poverty is deep and despairing. In most regions youth unemployment hovers around 60–80 per cent. Tajikistan, said to be the worst affected by the drug plague, is also the poorest; and its 1,400-kilometre frontier with Afghanistan is probably the most porous. On paper penalties for trafficking are severe but only tadpoles get caught: the sharks continue to frolic. Street corner traffickers tend to be – as always and everywhere – paid in kind: the trade therefore leaves behind a trail of hard-core addicts. A quarter of the half-million in Uzbekistan are said to be children, some under ten. Corruption is rife: the livelihood of the police is as dependent on the narcotics trade as that of the traffickers. Family codes condemn inebriants but old taboos are crumbling.

The transit business of Central Asia is now worth about £15 billion; and it will continue to grow so long as people live on the edge of starvation. It will also lead – as it has always done – to bloodshed. The violence which followed the presidential elections in Kyrgyzstan in the summer of 2010 and which resulted in the forcible removal of the second elected president since independence was largely fuelled by drug rivalries, as were the more recent clashes between Kyrgyzs and Uzbeks which have claimed thousands of lives. In their inhospitable lands these people have lived side by side for centuries, united in their resistance to the heavy-handed rule of St Petersburg and later of Moscow. Cut-throat competition over the spoils of the heroin trade has transformed shared aspirations into corrosive hatred.

* * *

But the new Central Asian republics and even Siberia are only transit stops. Like the Great Silk Road in the Middle Ages the heroin trail moves west and divides into a southern and a northern branch. The northern route traverses Russia to the Baltic, Poland and the Czech Republic. Until recently much of the cargo was carried in rusting Soviet-era railway carriages nobody would suspect of transporting merchandise worth millions. From distribution centres in Poland and the Czech Republic – provincial towns are generally preferred to capital cities – the journey continues to Scandinavia, Germany and points west. Belarus and Ukraine with their comparatively good road and rail networks are becoming cheaper and shorter alternatives. Most seizures, occasionally genuine but more often staged, are made on the Polish–German frontier. The sums into which the hoards are translated for the benefit of Western television audiences sound like lottery jackpots. The southern or Balkan route crosses the Caspian Sea to the Caucasus and Turkey. From there the heroin is shipped to Greece and Italy or transported in trucks through

Macedonia, Serbia and Hungary to Austria. The formerly sedate university city of Graz where Anton Bruckner played the organ a hundred years ago was for some years the East European hub of the trade; but in Hungary Pécs too has proved to be conveniently situated, just 150 miles from Budapest. The trade there is making headway among school leavers. The juvenile prisons are full but so are the cafes where addicts and their suppliers meet. The business is efficiently managed by mainly East European syndicates and still largely organised on an ethnic basis. Their occasional turf war provides one of the few openings for the anti-narcotics police; but infiltration remains a deadly risk. The syndicates do not like making mistakes and take no prisoners. The Albanians enjoy a particularly chilling reputation: their trail is signposted with the mutilated bodies of suspected double-crossers. Not all the profits go to buy luxury yachts for the bosses: some will procure weapons and explosives for far-flung corners of the world.

* * *

Near the western end of the trails Britain has now been for some years in the lead among consumers; and her addict population has been getting younger. Of the five young women murdered by the serial killer in Ipswich between October and December 2006 three had started to inject themselves before their sixteenth birthday, the other two before their eighteenth.[3] All had become prostitutes to finance their habit. Ipswich is a prosperous county town: it is hard to see it as a den of iniquity. Nor is it: the tragedy could have happened in a hundred 'red-light districts', a grandiose name for a stretch of grubby pavement near the main shopping mall or the cathedral close. Death rates from accidental overdose and suicide are rising. Explanations for the trend in Britain (in contrast to the comparative stability in most other Western European countries) abound. None is entirely convincing but a parallelism between drug use and prison population undoubtedly exists. In the latter, Britain tops the European league table.[4] In some prisons inmates become routinely initiated even during short remand periods. The official answer to the overcrowding is more prisons.

The transfer of Hong Kong to China in 1997 has contributed to the criminal landscape. Before the handover a Parliamentary Commission declared that 'we are satisfied that nothing suggests a move of Triad bases from Hong Kong to Britain. The word Triad can usefully be dropped from the police vocabulary.' Few such declarations have ever been so monumentally wrong. By 1990 English and Scottish police officers were being urgently dispatched to Hong Kong for training courses. Not only did the Triads establish themselves in London, Glasgow and Manchester (declared a 'Dragon City', one of only four outside China) but they have also found new and profitable fields of enterprise to combine with the narcotics trade. Some destitute inner-city squats housing thirty or more human cargo per room, sleeping on the floor in turn, have become heirs to the coolie tenements of nineteenth-century Hong Kong: as in old Hong Kong, the hold of heroin on most of the labour fodder is absolute.

* * *

Thanks to the abundant flow of Afghan heroin the cost of a fix in London or Glasgow today is roughly that of a cheap cinema ticket. Yet British addicts spend on average £20,000 a year on their habit. Most of this is derived from crime, mainly mugging, burglary, robbery, shoplifting and cartheft. This is in addition to domestic pilfering, a terrible but private predicament. Most patients in expensive detoxification clinics in Spain and Switzerland are young Britons. Economic fluctuations like the recession following the banking crisis of 2008 tend to confirm memories of the past; drug use tends to rise in the economically most depressed areas.

* * *

As addiction increases, the argument over decriminalisation continues and occasionally erupts in verbal and sometimes physical violence. No topic of social policy is so divisive: even historians committed to impartiality cannot approach it with total detachment. To parents whose child has been broken on the wheel of narcotism, prostitution, crime, suicide and murder the idea of 'legalising' drugs will remain an obscenity. But would the tragedy have happened if drugs had already been decriminalised? In the sense that criminalisation has almost certainly not stemmed the tide of addiction, legal constraints have failed. The only exception has been Communist China where the methods used would be unacceptable even in China today. Historically anti-narcotics legislation has usually been counterproductive; but controlled experiments have never been possible nor can they be. The failure of alcohol prohibition in the United States in the 1920s remains the reformers' flagship argument; but in truth alcohol and drugs pose different problems. Prohibition was introduced after centuries of freedom to indulge in alcoholic drinks, a habit more deeply embedded than drugs. And when some reformers suggest that the 'criminalisation of drugs has been an unmitigated catastrophe' they are overstating their case. Children in Western Europe are no longer dosed with opium to keep them quiet. No infant death from opiates has been reported in the United Kingdom for eighty-four years.

* * *

In Europe decriminalisation – or depenalisation – has been pioneered in Portugal and is currently judged to be a success. Before the change drug-related crimes were paralysing the law courts. More than half of those newly infected with HIV were drug addicts. HIV-positive diagnoses among addicts were running at 3,000 a year. Today the corresponding figures are 20 per cent and less than 2,000. The improvement may seem undramatic but should be seen against predictions of doom. In most Western European countries it is still axiomatic to forecast that, whatever the long-term consequences, the novelty of depenalisation must lead to a temporary increase in the number of experimental users. But it did not happen in Portugal and would not happen anywhere where the ground had been properly prepared. What had been recognised in Portugal is that imprisonment has a promoting or at best no effect on the prevalence of addiction. What might – just might – lead to a decline is some form of treatment. Arrest for being in possession of heroin for personal use now means not a court appearance but an invitation to a Dissuasion Board. The board sits in a modern office with no wig or police uniform in sight. It can request – but cannot compel – attendance at facilities for assessment and therapy. Organisers have been overwhelmed, by the uptake rate, by the readiness of addicts to talk about their burden and by the willingness of many to try a cure. Even more surprising has been society's acceptance of addicts as a social rather than as a criminal problem, a sea change in a Catholic and traditionalist country. But most remarkable has been the conscious turning on its head of the hallowed principle that all those caught in a crime must be treated as equals before the law. 'On the contrary,' a magistrate told the present writer, 'we now do our best to apply the law to fit the individual.' While recreational users, for example, may be fined, a financial penalty tends to be avoided in cases where the risk of it leading to crime is high. Not everybody has been converted. Some even among early enthusiasts now fear that a humane approach takes too much pressure off addicts. 'They now have clean clothes, adequate food and moral support. Why should they change?' But there never have been social advances without temporary setbacks.

* * *

Disputes are simmering in many other countries. Chile decriminalised the possession of small amounts of drugs for personal use in 2005 following the unpublicised lead of Paraguay (1988) and Uruguay (1998). Brazil changed its policy in 2006 and Argentina in the summer of 2010. Martin Acuna, the Buenos Aires judge and eloquent advocate of reform, has ruled that prosecuting and jailing people for the mere possession of illicit substances is unconstitutional. In 2009 the presidents of Colombia and Brazil called for a new global strategy based on public health considerations rather than on penalisation. Few doubt that the incidence of HIV infection has fallen significantly in every country where a more liberal approach has been tried. But the profits of the illegal trade are too high. And incorruptibles brave enough to face assassination are few. In Mexico the war between government and rival drug cartels (dealing mainly in cocaine but more recently also in heroin) has so far claimed the lives of more than 30,000 people, the majority innocent bystanders. Even in the bloody annals of narcotics wars this is an unprecedented toll.

European responses have been fluctuating. In the traditionally permissive Netherlands cannabis in any form has always been a soft drug and its smoking is allowed in public as well as in private; but trafficking in hard drugs is illegal. Many impeccably liberal citizens dismayed by drug tourism – too many bloated bodies floating in the Herengracht – want existing regulations tightened. In Switzerland in Zurich's Platzspitzpark the experiment of free clean syringes started auspiciously in 1987. Addicts flocked to the park, bringing their own heroin and injecting it openly. By 1992 the numbers attending rose from a few hundred to an unmanageable 20,000; and the experiment in its original form had to be discontinued. In the United Kingdom commitment to reform has been the staple of rising and retiring politicians but not of politicians in sight of power.

Some countries seem to be frozen in old attitudes. In the presence of President Medvedev and Prime Minister Putin, Tatyana Dmitrieva, deputy chair of the International Narcotics Control Board, recently declared in Moscow that 'We are not for harm reduction . . . We are for supply reduction.' And the minister of health promised that Russia will always remain 'categorically against providing substitution treatment . . . Even the distribution of sterile needles and syringes only stimulates social tolerance.' Methadone and other opiate analogues are illegal, syringe exchange services are few and detoxification programmes are non-existent. The policy is leading to a continuing rise in HIV infection both among addicts and in the general population. In a shrinking world it is a disquieting picture.[5]

* * *

Two cross-fire victims of the controversy have been common sense and scientific rigour. Opponents of reform tend to suggest that 'we should learn from the 5 million victims a year of other drugs like tobacco.' This is in the same league as the assertion by a top-ranking scientist that horseback riding is more dangerous than Ecstasy. Cigarette smoking is an aetiological factor but not the cause of any known disease except nicotine allergy. That kills about one person every ten years. And how does one measure the dangers of either horseback riding or of Ecstasy, let alone compare the two? But on the subject of drugs, sound-bites tend to replace argument. 'To do so-and-so,' politicians intone, 'would send out the wrong signal.' What does this nonsense mean? The notion that signals from on high will guide citizens in deciding what to think is a legacy of another age. From Mr Gladstone such hints might have swayed multitudes; but clones of Mr Gladstone at present are few.

The most disabling obstacle to rational debate is the non-recognition of the fact that the real choice may be not between good and bad but between bad and not-so-bad. Relative costs must also be considered. Despite the widely recognised failure of the War on Drugs the United States still spends $40 billion a year on trying to eliminate the supply of heroin. European countries are less transparent about their finances but in proportion to their size may not lag far behind. The United States also arrests five million of its citizens each year for drug offences, locking up half a million. Drug laws are the main reason why one in five black American men spends some time in jail. The country's current president could have suffered such a fate for an unguarded sniff in his youth. With such a record he would not now be president. The cost of legal procedures arising from drug-related crimes and the cost of imprisonment of those found guilty is incalculable. So is the cost in ruined lives.

Criminalisation is justified if it deters potential delinquents and protects the innocent. Little if any evidence suggests that current legislation does either. The crimes associated with illicit drugs, moreover, have always been among the most vicious. Al Capone and his cronies of Prohibition days were no softies; but they could take lessons in brutality from the quiet men in suits who now run the drug business in Britain. And much blood is shed in forgotten countries like Guinea-Bissau as well as in Mexico; and toddlers as well as presidents are being gunned down.

In a field beset with special pleading one looks for hard evidence. Casting doubt on claims by law-enforcement agencies that in the United States and Western Europe more than half of all illegal drugs are now seized and destroyed is the fact that the street price of heroin hardly changes.[6] Slightly more credible is the unofficial but well-informed estimate that 88 per cent of heroin sent from any point on the globe to any other has a good chance of arriving, a considerable advance on the arrival rate of Christmas cards. In support of official claims it has been suggested that even if the price remains stable, the purity of the powder sold on street corners declines. Since purity equals safety, this is not a matter for rejoicing. The disastrous poppy harvests in Afghanistan in 2010 have led to a shortage of heroin on the streets of Britain for the first time in decades. New cheap and toxic adulterants are making intravenous injections dangerous. But the risks have not affected the rate of addiction. Most important perhaps, criminalisation and education remain incompatible: one cannot instil the moral sanctity of private property into the mind of a burglar while imprisoning his body. Yet education, however slow and ponderous, must surely be the best prospect for change.

Against decriminalisation the most powerful argument is that in global terms it would favour the rich and damn the poor. Even at a time of economic difficulties Portugal could and can afford to substitute education for punishment. So could Britain and other Western governments if they put their minds to it (as they should). But the rehabilitation of addicts would come a long way down the list of health priorities in countries struggling to survive. As legalised heroin became cheaper, it would compete with tobacco and alcohol as an escape route from desperation. Heroin would probably win. Even in a cauldron of evil uncertainties the possibility cannot be ignored.

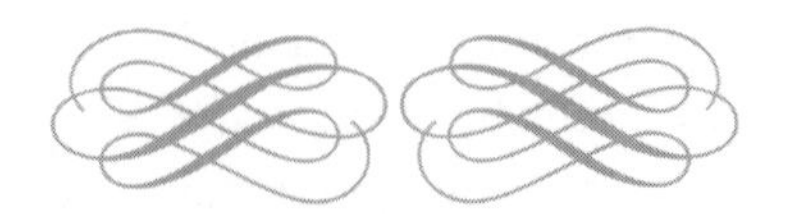

Epilogue

IT WOULD BE pleasant to end this narrative on an upbeat note. The image of the pipe-smoking, bushy-bearded Afghan farmer surveying his field of poppies at nightfall, knowing that it will save him and his family from starvation is real. The young poppy bursting into bloom remains an enchantment. The relief provided by opiates to suffering all around the world is an incalculable blessing. The praise heaped on the poppy by poets, musicians and writers of the past has been much dwelt on in the present book. But such an ending would border on the fraudulent. Opium and its derivatives remain a deadly threat. No other drug can so insidiously and captivatingly distort reality. No other will gain so irresistible a hold. None is so contemptuous of race, class and virtue. None destroys so completely. The arguments over punishment versus depenalisation reverberate and seem irreconcilable. The controversy has never been political or economic. It touches on health but it is not predominantly medical. It is not and never has been a straight choice between right and wrong. Yet it arouses elemental responses. To millions love of the drug remains overwhelming. Hatred of it no less so. No argument can extinguish the first. No statistics can tarnish the second. And those in its grip, directly or indirectly, continue to suffer and to die.

But unmitigated gloom would also be unjustified. Significant advances towards effective detoxification are being made. In the light of laboratory findings the nature of addiction is now better understood than it was half a century ago. In practical politics common sense may eventually prevail: the less bad may be allowed to replace the bad. Distant rumbles too early to evaluate predict other possible developments. Microorganisms are being genetically teased to synthesise morphine, as some already synthesise hormones. The technology may spell the end of ten thousand years of history, the back-breaking cultivation of the white poppy, the laborious processing of its juice, the extraction of its magical essence, another step perhaps in man's ascent to the stars.

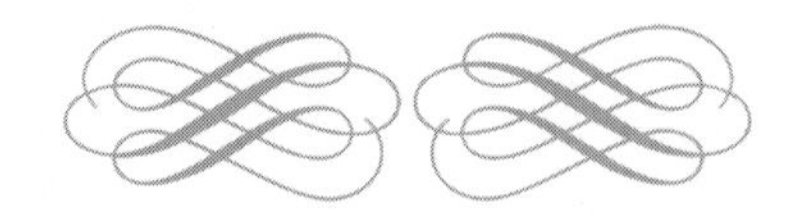

Acknowledgements

IS THERE SUCH a thing as an opium bore? If yes, my family have been putting up with a virulent specimen with angelic patience. Indeed, they have encouraged his obsession in a hundred and one ways. Without Liz there would be no book. Charlotte has guided me through obscure (to me) but rewarding byways of English literature and has shared with me her insights and prejudices. Richard tactfully put me right on scriptural uncertainties. Michael has been my long-suffering source of information on classical lore; and Letty has drawn my attention to several exceedingly boring but important books. She has also tried, sometimes almost successfully, to keep me up to date on twenty-first-century medical advances relating to the drug.

Friends, patients and addicts who have helped me are too numerous to mention individually; but I must record my thanks to a few. Most generously Joe, Mike and Tessa have provided me with personal insights. George Watts imparted much surgical information and wisdom. John Zieger, a cousin several times removed was always close enough when advice, help and constructive criticism were needed. Ian Douglas-Wilson, old friend and a great former editor of the *Lancet*, has been, as always, a great listener and critic. Cicely Saunders and Adrian Tookman taught me all I know about palliative care. I am grateful to Dr Judith Zerkowitz for providing useful information relating to Geza Csath and the notable Hungarian chemist and inventor Janos Kabay. Miklos Ghyczy has sent me valuable material from Germany. Ted Chapman was my main source of information about the pharmaceutical aspects of modern opiate manufacture and pointed me to important developments. I am most grateful to them all.

The great literary art of creative editing is often obituarised today. Happily, it is still alive and in full bloom at Yale University Press in London. Rachael

Lonsdale not only suggested the book in the first place but her wonderful editing has also immensely improved the product. I am deeply grateful to her. My warm thanks also to friend and publisher Heather McCallum for unfailing support, encouragement and always constructive and wonderfully upbeat criticism. The copy-editing of Beth Humphries has been superb and it has been a pleasure to receive editorial guidance from Tami Halliday.

To a technologically challenged writer faced with new computer systems the patience, friendliness and expertise of the Library Staff of the Royal Society of Medicine, the Wellcome Medical Historical Library and the British Library have been priceless assets. Thank you very much.

My beloved siblings and dedicatees of this book, Daisy and John, are jointly and entirely responsible for all the errors, mistakes, misprints and misinterpretation in the text. [The above facetious sentence was written before Daisy's death in March 2011; but she remains so alive in my heart and mind that I let it stand.]

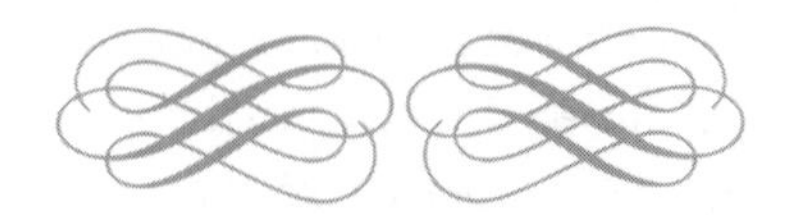

Notes

1 PETRIFIED BUNS

1. Many such villages in mountain lakes have been discovered since, not only in Switzerland but also in regions as far apart as Spain and Slovakia, but Meilen was the first. See F. Keller, *The Lake Dwellings of Switzerland* (London, 1866).
2. The discovery of the lake village gave Swiss national pride a boost. A diorama commissioned by the Federal Government in Berne was the centrepiece of the Swiss Pavilion at the Paris World Exhibition of 1867.
3. *Papaver somniferum* was the name given to the opium poppy by Linnaeus in his great *Genera plantarum*. But the root may also have produced colloquial terms like 'poppy' in English and 'pipacs' in Hungarian. The word 'opium' is derived from ὄπιον meaning poppy and ὀπός meaning vegetable juice in general.
4. The papyri date from about 700 BC but they convey knowledge and information that have been passed down the generations for 2,000 years. They were acquired by Smith in 1862. He did well selling his skilful fabrications to early Cook's tourists travelling up the Nile – many are now the prize exhibits of prestigious collections round the world – and amassed a fortune. On his death in 1906 the Smith papyrus was donated by his daughter to the New York Historical Society. A splendid facsimile edition with transcriptions and translation was published by James Henry Breasted in 1930 with medical notes by Dr A.B. Luckhardt.
5. The Ebers papyrus was claimed to have been discovered between the legs of a mummy in a Theban necropolis. In his lifetime Ebers was often credited with the actual discovery, a flattering misapprehension which he was in no hurry to correct. Towards the end of his academic career he became a successful author of children's adventure stories set in ancient Egypt: *The Bride of the Nile* in which stupor induced by poppy juice is a critical episode is a particularly nail-biting tale. The papyrus is discussed in R. Scholl, *Der Papyrus Ebers. Die grösste Buchrolle zur Heilkunde Altägyptens* (Leipzig, 2002).
6. Opium does not measurably affect the acute pain of surgery; but it can make the perioperative periods less stressful; and it can in some mysterious way make even the operation more bearable. The possibility of the use of opium makes the finding of trephined skulls dating from the Neolithic period a little less gruesome. New bone formation around the edges of the holes shows that the sufferer survived for at least some months. This rather than being eaten at Stone Age tea parties may have been the purpose of the lake dwellers' poppy-seed buns.
7. This was still practised in parts of Central Europe in the early twentieth century.

8. The period was in some respects similar to the intellectual ferment in the nineteenth century AD in Western Europe.

2 THE MAGICAL SEEPAGE

1. Casual visitors breathing in the air over a poppy field at the time of opium harvesting may rapidly become drowsy – as did Dorothy in *The Wizard of Oz* passing near poppy fields – but not the local harvester.

3 A GIFT OF THE GODS

1. Book IV, v. 219 of Homer's *Odyssey*. An opium derivative by that name survived in the British pharmacopeia until the mid-twentieth century.
2. This is Pliny the Elder (uncle of the letter-writing Pliny the Younger), admiral, imperial administrator and polymath. His insatiable curiosity kept him in Pompeii during the eruption of Vesuvius in AD 79 where he perished, probably from suffocation, aged fifty-six.
3. Pindar, *Pythian Odes*, 3.1.
4. Homer, *Iliad*, XI, 847.
5. So is the Lotus-eaters' drink celebrated by Alfred, Lord Tennyson two thousand years later. Tennyson disapproved of the opium habit of many of his friends; yet his poetic reconstruction of the story is tinged with nostalgia.
6. Not least for his former enemies. A monument erected by the Roman emperor Septimus Severus was still standing in the eleventh century and was admired by the first Crusaders.
7. Eventually Nero tried to get rid of her, first by trying to drown her in a leaking boat (but she gamely swam ashore) and then, more effectively, using hired assassins.
8. *The Banquet of Xenophone*, trans. James Wellwood (London, 1750), Book II, 24.
9. Virgil, *Aeneid*, VI.
10. Virgil, *Georgics*, I, 77–78.
11. Virgil, *Georgics*, II, 14–16.
12. Horace, *Fasti*, IV, 661.
13. Horace, *Odes*, III, 21.
14. The paternity of Triptolemus was disputed, various classical authors nominating different mortal heroes. Plato later promoted him to be one of the judges of the dead in the Underworld.
15. But to be on the safe side Louis XV promulgated a law banning the sale of Dutch poppy seeds. The law remained in the *Code* till 1926.
16. He is just off-centre next to Aristotle in Raphael's great *Philosophy* in the Vatican.
17. Theophrastus' *De historia plantarum* gives a description of 550 species of plants, gathered from Spain to India. He died in 287 BC.
18. The description of the actions of opium was given in Theophrastus' second great work, *De causis plantarum*. This was neglected in the Middle Ages; but, when rediscovered during the Renaissance, it inspired the first botanical gardens in Italy and France. Poppies were Lucrezia Borgia's favourite flower.
19. Lucian, *True Story* in Vol. I of *Works*, p. 337 with an English translation by A.M. Harmon (Cambridge, Mass. and London, 1961).
20. Though little read during the Middle Ages, in 1528 under the title *De re medicina* it was one of the first medical books to be printed and quickly went into many editions.
21. Few other textbook writers could have summarised the complexities of acute inflammation in four rolling words: *dolor* (pain), *rubor* (redness), *tumour* (swelling) and *calor* (heat). *Functio laesa* (impaired function) was added by Thomas Sydenham 1,600 years later.
22. Celsus, *De medicina,* 3 vols with translations by W.G. Spenser.Vol. II, p. 230.

23. Unlike the work of many of his contemporaries, a significant part of his writings survived the collapse of the Western Roman Empire and was available for translation by Islamic scholars in the ninth and tenth centuries.
24. Quoted by V. Nutton (ed.), *Galen: Problems and Prospects* (London, 1981), p. 56.
25. The term derives from the Greek καροῦν, to stupefy, and καροτίδεζ carotid arteries of the neck. Rufus of Ephesus says: 'The ancients called the arteries of the neck carotids because they believed that when they are pressed hard the animal becomes drowsy.' The ancients were right and their observations apply to humans as well.
26. Quoted in Nutton, *Galen*, p. 63.

4 RIVAL BREWS

1. For some reason his enormous influence was not immediate and he died in relative obscurity. The *Urtext* of his work is the beautifully illustrated *Codex Anicia*, presented to Juliana Anicia, daughter of Anicius Olibrius, Emperor of the West in AD 442.
2. They were standard constituents of 'premedication' before general anaesthesia since they can help to dry up secretions in the mouth and respiratory tract. Atropine (or analogue) is still widely used as a dilator of the pupil in ophthalmology.
3. Now widely attributed on exceedingly flimsy evidence to acute intermittent porphyria.
4. Quoted by V. Robinson, *Victory over Pain* (London, 1947), p. 17.
5. I. Benedek, *Mandragora*, 2 vols (Budapest, 1962).
6. Today such a statement relating to an authoritative medical textbook might raise an eyebrow.
7. C.J.S. Thomson, *The Mystic Mandrake* (London, 1973), p. 71.
8. Quoted by A. Maday, *Romai orvostudomány* [Roman Medicine] (Budapest, 1936), p. 98.

5 *AFFYON*

1. It replaced Damascus as the Abbasid capital and eclipsed Ctesiphon, former capital of the Persian Empire, 30 km to the south-east.

   Harun al-Rashid died in 809 aged forty-three. Charlemagne died in 814 aged seventy-two. They exchanged greetings and presents. The Caliph sent two elephants, which completed the journey to Aachen in fourteen months.
2. Not till the eighth century did religious houses begin to preserve classical scrolls; and their interest then was mainly in theological texts.
3. They were better paid and held in higher esteem than translators have ever been before or since.
4. Local palm leaves carefully folded created a packaging almost impossible to reproduce elsewhere.
5. In the West throughout this dark period the only international currency were holy relics.
6. They included camphor, nutmeg, senna, mace, tamarind, manna and ginger. Many are no longer easy to identify.
7. In his own words he lived his life 'in breadth, not in length'.
8. In this Circle Dante espied great men who, like his guide, Virgil, 'are innocent of sin; / yet lacking in Baptism, they could not claim / its saving grace, and thus are doomed forever'. In this sad place the poet recognises Dioscorides, 'the scientist of herbal essences', as well as Hippocrates, Galen and Avicenna. The doctors shared their abode with non-medical worthies like Socrates, Plato, Diogenes, Hector, Aeneas, Seneca, Orpheus, the Elder Brutus, Homer, Horace, Ovid and Julius Caesar. An odd assemblage: apart from sighing together, what did they talk about?
9. Quoted by S.M. Soheil in *Avicenna, his Life and Work* (London, 1958).
10. O. Cameron Gruner, *A Treatise on the Canon of Avicenna* (London, 1930).
11. Ibid., p. 342.

12. In these potboilers he did not aim at high-flying poetry. Arab scholars criticise his style as being 'cumbersome with new coinages' but praise his precision. In his scientific works language had no magic for Aristotle either.
13. One English 'Docteur of Physic' entertained his fellow pilgrims to Canterbury by boasting of having read all the works of the great Persian.
14. Quoted by W.H. Grohlman, *The Life of Ibn Sina* (London, 1930), p. 56.
15. He was born in 1135 and died aged sixty-nine.
16. Known in Islamic lands as the Magnanimous, Saladin overthrew the Fatimids, the ineffective ruling dynasty in Egypt, and eventually became sultan. He died in Damascus in 1193, aged fifty-six.
17. Maimonides, *Correspondence*, ed. I. Twersy (New York, 1972), p. 357.
18. Quoted by R. Weiss and C. Butterworth (eds), *The Ethical Concepts of Maimonides* (New York, 1975), p. 193. In a cynical vein, this still makes the public manifestations of grief in Muslim countries better television than comparable disasters in Anglo-Saxon settings.
19. As recently as 1979 the Ayatollah Khomeini pronounced that wine and other alcoholic drinks were impure under all circumstances but opium when used for medical purposes was not.
20. They were called *maristans*. Officers in Napoleon's army during their occupation of Egypt marvelled at the *tendresse* with which the insane were treated by their jailers. (General J.-P. Bertrand, *Mémoires*, Paris, 1836, p. 47.)
21. Quoted by D. Latimer and J. Goldberg, *Flowers in the Blood: The Story of Opium* (New York, 1981), p. 42.
22. Quoted ibid., p. 38.

6 THE SLEEPY SPONGE

1. Even the greediest in the city felt that some of this treasure should be channelled into good causes. The outcome was the *Grande Scuole*, charitable, not educational institutions.
2. The final assault was led by the doge, Enrico Dandolo, ninety and blind.
3. This was the moment in Venetian history which caught the imagination of whoever thought up the story of Antonio the merchant and Shylock and his pound of flesh. By the time Shakespeare worked it over the moment had passed.
4. As it has remained ever since.
5. This act was frequently illustrated in medieval codices. The superstition probably had its origin in the vaguely anthropomorphic shape of the root, reminiscent of the little man who nowadays directs seekers of relief to the male lavatories.
6. Shakespeare refers to it seven times, to the poppy only twice.
7. Artful and effective packaging was one of the generally recognised accomplishments of Venetian merchant houses.
8. N.G. Siraisi, *Medieval and Early Renaissance Medicine: An Introduction to Knowledge and Practice* (Chicago, 1990); also Roy Porter's *The Greatest Benefit to Mankind* (London, 1997).
9. B. Lawn, *The Salernitan Questions: An Introduction to the History of Medical and Renaissance Problem Literature* (Oxford, 1963); and P.W. Cummins, *A Critical Edition of 'Le Regime Tresutile et Toututilisable pour Conserver et Garder la Santé du Corps Humaine* (Chapel Hill, NC, 1976).
10. In English the most popular translation was by Sir John Harington, courtier to James I and inventor of the water closet. The exhortations of his splendid creations, Drs Quietly, Diet and Merriman were a store of common sense. Moderation was the portal to a happy life; melancholy and constipation were the bringers of grief. Against the former, poppy juice was the top recommendation, 'whatever the cost'; and indeed the king, suspicious by nature, never travelled without it. The concoction was made palatable with honey and wine.

Harington's translation was reprinted by P.I. Hoeber of New York in 1920 and reprinted again by A.M. Kelley in 1970 under the title *The School of Salernum, Regimen Sanitatis Salernitanum. The English Version by Sir John Harington with a History of the School of Salernum by Francis R. Packard, M.D. and a Note on the Prehistory of the Regimen Sanitatis by Fielding H. Garrison, M.D.*

11. Theodoric, *Chirurgia*, Vol. IV, quoted in A.S. Ellis, *Ancient Anodynes* (London, 1936). The device would come into its own seven centuries later. On 30 March 1842 Dr Crawford Williamson Long, town physician of Jefferson, Georgia in the USA sprinkled sulphuric ether on his bunched-up handkerchief, a direct descendant of the medieval sponge, and held it to the nose of Mr James Venables who needed to have a tumour removed from his neck. Mr Venables promptly fell asleep and woke up having felt nothing of the operation. General anaesthesia was born.
12. Vinegar or acetic acid may have been occasionally effective. It generates carbon dioxide in the body which stimulates the respiratory centre in the brain.
13. The acetyl group, -COOH, in which fruit juices are rich, is relatively easily transferred from one substance to another. Although acetylation may appear a slight change on paper, it can drastically affect the biological properties of a substance.
14. This was the chemical reaction by which Heinrich Dreser in 1898 would convert morphine to heroin. See Chapter 22.
15. Sections of the book were still reproduced almost verbatim in the *Captain's Handbook* issued by Their Lordships of the Admiralty to captains of the Royal Navy in Nelson's day.
16. E. Nicase, ed. and trans., *La grande Chirurgie de Guy de Chauliac, compose en l'an 1363* (Paris, 1890), p. 436. The lavish original copy is in the Bibliothèque Nationale in Paris.
17. He died in 1370 aged sixty-three.
18. The anxiety (*angor animi*) which accompanies ischaemic pain (*angina pectoris*) generated in the heart muscle may itself aggravate the condition. Morphine can interrupt the vicious circle.
19. See T. Dormandy, *The Worst of Evils* (London, 2006).

7 GREATER THAN CELSUS?

1. The term may have come from the Greek *ladanon* (ά̓δανον), an oriental shrub, rather than from the Latin *laudo*, I praise. The oriental shrub was known to Pliny and may at one time have been thought to be the source of opium. Michael Scott in 1200 used the word, referring to a heavy and soporific scent, perhaps of the poppy, but the name never caught on until invented or reinvented by Paracelsus.
2. Such adopted names were a humanistic fad, no sillier than other such fads were and are.
3. Quoted in W. Pagel, *Paracelsus. An Introduction to Philosophical Medicine in the Era of the Renaissance* (London, 1982), p. 134.
4. Ibid., p. 187.
5. *Johannistag* was celebrated in Basle as in Wagner's contemporary Nurenberg.
6. J. Jacobi (ed.), *Paracelsus, Selected Writings* (London, 1951), p. 267.
7. Ibid., p. 325.
8. He did not realise that one day they would manage to combine the two.
9. Pagel, *Paracelsus*, p. 252.
10. Ibid., p. 342.
11. Jacobi (ed.), *Paracelsus*, p. 452.
12. Ibid., p. 468.
13. See A.G. Debus, *The English Paracelsians* (New York, 1966).
14. In 1982 Charles, Prince of Wales told an increasingly glassy-eyed audience assembled for a British Medical Association conference that they 'could do worse than look at the principles Paracelsus believed in, for they have a message for our time when science and medicine have become totally estranged from nature . . .' He did not specify the message. For his espousal of 'folk wisdom' Paracelsus was promoted in Nazi Germany too but he can hardly be blamed for these idiocies.

15. Wren was thirty-seven, his architectural career still in the future.
16. See L. Jardine, *Ingenious Pursuits: Building the Scientific Revolution* (London, 1999).
17. See T. Dormandy, *The Worst of Evils* (New Haven and London, 2006), p. 118.
18. He was rarely entirely wrong. Despite advances in the neurosciences, no other site of the soul has yet been identified; nor has any other clear role been assigned to the pineal gland.
19. Thomas Sydenham (see next chapter), an ardent admirer of Cervantes, wrote that *Don Quixote* was the 'best Treatise from which to learn medicine'.
20. Well reviewed by F. Lopez-Munoz, P. Garcia-Garcia and C. Alamo, 'The virtue of that precious balsam: approach to Don Quixote from the psychopharmacological perspective', *Actas Espagnolas Psiciat*, 2007 (35), 19.
21. The Elector liked animals and preferred experiments performed on prisoners in exchange for a partial remission of their sentences.
22. Bethlen died in 1629, aged forty-nine. G. Halasz, *Erdely Tortenete* [History of Transylvania] (Budapest, 1932), p. 68.
23. Sir T. Browne, *Works*, ed. H. Robbins, 2 vols (Oxford, 1986), Vol. II, p. 342.
24. R. Burton, *Anatomy of Melancholy*, in *Works*, ed. T.E. Faulkner, N.K. Kiesling and R.L. Blair, 3 vols (Oxford, 1989–94), Vol. I, p. 343.
25. J. Dryden, in *Selected Poems*, ed. D. Hopkins (London, 1998), p. 256. Even more woundingly (and memorably) Dryden satirised Shadwell as 'the last great prophet of tautology' and 'The rest to some faint meaning made pretense / But Shadwell never deviates into sense'.

8 THE TINCTURE AND THE POWDER

1. But even the poor little girl has her legs in calipers, strongly hinting at congenital syphilis.
2. In the days when rheumatic fever in childhood was common, one of its complications, uncoordinated involuntary movements due to the involvement of the central nervous system, was called Sydenham's chorea.
3. One brother died in battle; another rose to occupy important positions in Cromwell's government.
4. He was on friendly terms with many leading intellectuals of the day, including John Locke, a doctor himself though not a practising one.
5. The first edition of the *Opera omnia* was published in 1683. The Sydenham Society published a translation by Dr Greenhill in two volumes with a sympathetic biographical introduction by Dr R.G. Latham. A short biography with excerpts of the works was published under the editorship of K. Dewhurst in 1966.
6. T. Sydenham, *Collected Works*, trans. Dr Greenhill, 2 vols (London, 1884), Vol. II, p. 34. Laudanum is sometimes referred to as a 'solution'. Opium is in fact insoluble in alcohol as well as in water and Sydenham's famous liquid was a fine *suspension* (like mayonnaise only more dilute). The difficulty in preparing stable and fine suspensions accounts for the variations in the quality and the price of laudanum over the next 200 years.
7. Ibid., Vol. I, p. 345.
8. In essence it ruled that the religion – Catholic or Protestant – of the population of a region or country should be decided by the religion chosen by the ruler: *cuius regio, eius religio*. This applied to the 300 tiny principalities into which Germany was divided as well as to kingdoms like France.
9. Witches as well as heretics were persecuted and burnt during the religious wars on an unprecedented scale and on both sides of the religious divide. Only Zwingli of Zurich decreed that they did not exist.
10. Wars did not cease after the Treaty of Westphalia but, compared to the religious conflagrations, the dynastic campaigns of the eighteenth century were professional, even elegant affairs in which the civilian populations suffered comparatively little.

11. The major arts of the baroque, such as architecture, were magnificent confirmations of religious faith; but they would have been incongruous settings for the rise of such 'minor' acts as porcelain.
12. T. Dover, *An Ancient Physician's Legacy to his Country* (London, 1732), p. 543.
13. They were called privateers when British, pirates when foreign.
14. A few years later Selkirk's four solitary years inspired Daniel Defoe's story (based on the account of the master of the boat) of Robinson Crusoe. Or so it is still generally believed. The source has been disputed by T. Severin in his searching study, *Seeking Robinson Crusoe* (London, 2007).
15. Dover, *Ancient Physician's Legacy*, p. 246.
16. Ibid., p. 248.
17. R.S. de Ropp, *Drugs and the Mind* (London, 1958), p. 56.
18. Sadly it made him even more bewitched.
19. P. Pomet, *L'Histoire générale des drogues, traitant des plantes, des animaux et des mineraux, etc.* (Paris 1694), p. 243. The French edition contains more than 400 illustrations. Translated into English anonymously it was published in a less lavish format in 1712 in London under the title *A Compleat History of Drugs*.
20. Ibid., p. 312.
21. Monsieur Pomet was nevertheless expected to abolish or at least to ease the royal pains arising from an anal fissure.
22. Mme de Sévigné, *Selected Letters*, trans. and ed. with an Introduction by L. Tancock (Penguin Classics, 1982), p. 283. The abbé was a man of letters and an academician.
23. Quakers banned alcohol but not opium, a 'valuable easer of pain', as Joseph Jackson Lister, Joseph Lister's father and a respected scientist himself, described it.
24. J. Jones, *Mysteries of Opium Revealed* (London, 1700), p. 49.
25. Ibid., p. 387.
26. Ibid., p. 412.
27. Ibid., p. 485.
28. J. Boswell, *Life of Johnson* (Everyman edn, London, 1949), p. 347.

9 ON THE BANKS OF THE GANGES

1. Like many decisive military engagements in history the victory of Plassey was due largely to accident and misperceptions. In numbers the British were heavily outgunned; but enemy gunpowder had been inadequately protected against rain and did not explode. Noticing this, Mir Jaffar, one leader of the enemy coalition, and his troops changed sides in mid-battle. Clive nevertheless showed sound judgement and restraint and earned the plaudits of history. As a commander Lord Macauley compared him favourably to Napoleon Bonaparte.
2. The scion of one of the oldest but impoverished families of Shropshire, as a schoolboy Clive was successively expelled from St Mary's Parish School in Market Drayton, the Market Drayton Grammar School and Merchant Taylors' in London. He never qualified as an accountant and never attended a university.
3. J. Auersperg (ed.), *Erinnerungen des Prinzen Joachim von Auersperg, 1740–48* (Vienna, 1932), p. 364. See Bibliography.
4. The first of the joint stock East India companies of Europe, it was granted a Royal Charter by Elizabeth I on the last day of the seventeenth century. Before opium it traded mainly in cotton, silk, indigo dyes, saltpetre (used for gunpowder) and tea.
5. B. Stout and M. Misra (eds), *Reports of the English East India Company* (Bombay/Mumbai, 1932), p. 34.
6. Ibid., p. 153.
7. Retrospective diagnoses are always difficult. Clive's illness occurred before tuberculosis became by far the commonest chronic abdominal illness.
8. Irish peerages like Clive's and later Lord Palmerston's did not automatically admit the holder to the House of Lords but made them eligible for a seat in the House of Commons.

10 THE TURKISH CONNECTION

1. J.-P. de Récamier, *Voyages au Levant* (Paris, 1850), p. 58.
2. It was in Smyrna that Lady Mary Wortley Montagu performed her first experiments with vaccination against smallpox.
3. Punishment for accepting bribes was death by quartering.
4. Venice was given to the Habsburg Empire at the Congress of Vienna in 1815 and, apart from a brief resurgence of independence in 1848, remained an Austrian province till 1859. It then became part of the Kingdom of Italy.
5. D. Latimer and J. Goldberg, Blackwood's Magazine quoted in *Flowers in the Blood* (New York, 1981), p. 134.
6. Ibid., p. 136.
7. J. Hill, *Family Herbal* (Bristol, 1812).
8. G. Young, *A Treatise on Opium* (London, 1765), p. 84.
9. S. Crumpe, *Inquiries into the Nature and Properties of Opium* (London, 1793), p. 65.
10. R. Boyle, *The Skeptical Chymist*, quoted by M. Booth, *Opium* (London, 1996), p. 57.

11 ROMANTIC OPIUM

1. T. Trotter, *A View of the Nervous Temperament* (Edinburgh, 1788). An excellent and illuminating work.
2. The 'Celtic Bard' was for a time more popular in continental Europe than Shakespeare but in most European languages it was Shakespeare who became the greatest German, Polish, Hungarian and Russian dramatist.
3. His superb oratory successfully mobilised France to resist invasion by the First European Coalition. At the 'Battle' of Valmy the doddery Duke of Brunswick was so surprised to encounter any resistance (believing that the revolution would collapse at the mere news of his approach) that he ordered a general retreat. Goethe, on the staff of the duke, perceived the dawn of a new age.
4. Aretaeus, *Opera* (mid-second century AD) trans. F. Adams (London, 1856), p. 78.
5. They received the opium derivatives morphine or heroin. The year 1947 marked a watershed and Orwell was one of the first to be treated with streptomycin. Unfortunately he proved to be allergic to the drug and died soon after.
6. Clark was a believer in horseback riding as the best remedy for phthisis, a fashionable dogma at the time. A kindly man, he was to notch up an impressive list of illustrious victims in his career, including Sir Robert Peel and the Prince Consort.
7. The duke was the son of Napoleon and was created 'King of Rome' at birth. He was also the grandson of Francis I, Emperor of Austria. This did not prevent him catching tuberculosis and dying aged twenty-six.
8. 'This is my death warrant,' Keats exclaimed when he saw the fresh blood on his handkerchief. Shelley drowned in a sailing accident but would have been dead of tuberculosis within a year or two.
9. Dumas *fils*'s first love, Marie Suchet, died of tuberculosis. Murger himself died of the disease aged thirty-four.
10. Quoted by P. Toussaint, *Marie Dupléssis, la vrai Dame aux Camélias* (Paris, 1958), p. 342.
11. After some delay the British government obliged. The drug may have eased the exiled emperor's last days.
12. Quoted in G. Denes, *A Savonai Fogoly* [The Prisoner of Savona] (Budapest, 1942), p. 43.
13. Tube in the singular. Laënnec's stethoscope was a small wooden trumpet. The modern binaural stethoscope was not invented till the 1880s.
14. Wife of the heir presumptive, Charles X's only son.
15. Quoted by R. Kevran, *Laënnec: médecin breton* (Paris, 1955), p. 346
16. J. Keats, *Complete Poems*, ed. John Barnard (London, Penguin Classics, 2003), p. 234.
17. Ibid., p. 266.

18. Richard Cappell's comment on Schubert's *Requiem* or *Allerseelen* in his book *Schubert's Songs* (London, 1954), a judgement with which the present writer humbly agrees.
19. 'Még nyilnak a völgyben a kerti virágok!' ('Garden flowers are blooming still in the valley' – but untranslatable, as is most great poetry.)

12 THE PLEASURE DOME OF XANADU

1. The correspondence is from J. Gillman, *The Life of Samuel Taylor Coleridge* (London, 1834). Further passages in the text are quoted in the Introduction to the sympathetic biographical study of Coleridge, *Samuel Taylor Coleridge, A Bondage of Opium*, by Molly Lefebure (London, 1974).
2. Gillman, *Life*, p. 25.
3. Ibid., 1, p. 27.
4. The friendship is described in an illuminating recent study by A. Sisman, *Wordsworth and Coleridge. The Friendship* (London, 2006).
5. A quart = a quarter of a gallon, approximately 1.2 litres.
6. Yet his dissertation on *Hamlet*, one of the least regarded Shakespeare plays in his day, is still quoted in theatrical programme notes today.
7. The house in the Grove, Highgate and the extension Dr Gillman built to accommodate his lodger still stand.
8. It is an abnormal sensitivity reaction to the haemolytic streptococcus, the cause also of scarlet fever and other ailments; but the aetiology is still not fully understood. The knees, elbows and other medium-sized joints were most commonly affected. The commonest heart structures to be destroyed were the two cusps of the mitral valve.
9. Salicylates, the only effective painkillers in rheumatic fever, were not introduced till the middle of the century (see T. Dormandy, *The Worst of Evils* (New Haven and London, 2006)).
10. S.T. Coleridge, *Collected Works*, Gen. ed. K. Coburn (Princeton, 1971–2001), p. 324.
11. Ibid., 10/I, p. 23. Kendall Black Drops was a popular preparation of opium in vegetable acids, supposed to be four times stronger than conventional (Sydenham's) laudanum. It was invented and manufactured by a Quaker doctor, John Airy Braithwaite and, after his death in 1810, by his widow and then his daughter, Hannah, in Kendal in Westmorland.
12. Blessedly for Coleridge (and to the relief of at least one future doctor) Dr Gillman more than redeemed his profession.
13. N. Fruman, *Coleridge, the Damaged Archangel* (London, 1971).
14. Or, to borrow from T.S. Eliot, 'Immature poets imitate, mature poets steal'.

13 THE OPIUM EATERS

1. St Augustine, *Confessions*, trans. H. Chadwick, 3 vols (Oxford, 1992).
2. Most notably John Bunyan's *Grace Abounding to the Chief of Sinners* (1666), his best-loved work after *Pilgrim's Progress*.
3. He never exceeded five feet in height.
4. T. de Quincey, *Confessions of an English Opium-Eater and Other Writings*, ed. and with an Introduction by Barry Milligan (first published 1822; London, 2003), p. 29.
5. Ibid., p. 34.
6. Ibid., p. 68.
7. M. Booth, *Opium. A History* (New York, 1996), p. 111.
8. De Quincey, *Confessions*, p. 156.
9. Prince Myskin refers to De Quincey in Dostoevsky's most autobiographical work, *The Idiot* (1868).
10. As the inscription on his gravestone testifies: 'Born of humble life, he made himself what he was, breaking through the obscurity of his birth by the force of his genius; yet

he never ceased to feel for the less fortunate, entering into the sorrows and wants of his poorest parishioners . . .' Byron called Crabbe 'Nature's sternest painter, yet the best'. Thomas Hardy was another admirer.

11. 'Peter Grimes,' in *The Borough* (1810), in *The Works of the Rev. George Crabbe* (London, 1828), p. 56.
12. Ibid., p. 64.
13. J. Markus, *Dared and Done: The Marriage of Elizabeth Barrett and Robert Browning* (London, 1995), p. 54.
14. L. Foxcroft, *The Making of Addiction: The Use and Abuse of Opium in Nineteenth-Century Britain* (Aldershot, 2007), p. 24.

14 THE PEOPLE

1. But not in the deeply concerned mind of the Prince Consort, who worried about the excessive use of poppy tincture by ordinary people. The eminent surgeon Sir Benjamin Brodie advised him to 'let sleeping dogs lie'. To the prince this was 'a very English answer'.
2. V. Berridge, 'Fenland opium eating in the nineteenth century', *British Journal of Addiction*, 1977 (72), 275.
3. C. Kingsley, *Alton Locke* (London, 1850), p. 79.
4. Quoted by V. Berridge, *Opium and the People* (London, 1981), p. 38.
5. R.G. Vicars, 'Laudanum drinking in Lincolnshire', *St George's Hospital Gazette*, 1893, (1), 24.
6. The Mitterands, an old Auvergnat family, were famous growers.
7. Berridge, *Opium and the People*, p. 40.
8. *Das irdische Leben*, one of the *Des Knaben Wunderhorn* cycle set to music by Gustav Mahler.

15 A SALVE FOR ALL AILMENTS

1. Enlightened royalty (like the Emperor Joseph II) liked to visit madhouses and foundling hospitals within walking distance of their palaces.
2. T. Dormandy, *The Worst of Evils* (London, 2006), p. 348.
3. W. Bateman, *Magnacopia* (London, 1839).
4. I. Benedek, *Semmelweis Élete és Kora* (Budapest, 1960), p. 47.
5. V. Berridge, *Opium and the People* (London, 1981), p. 68.
6. Ibid., p. 70.
7. At the age of twenty-eight Louis qualified from the University of Paris in 1813 and soon realised the futility of much medical argument because of lack of reliable data. His 'statistical' studies gradually gained him followers, especially in the United States.
8. F. Langhans, *Erinnerungen eines Arztes* (Tübingen, 1852), p. 67.
9. W. Colton, *Turkish Sketches* (Philadelphia, 1836), p. 67.
10. *New York Times*, 15 March 1840, p. 132.
11. Most authorities believe that an inclination to become a morphine or heroin addict is not significantly predisposed genetically. For social reasons the children of addicts may be less rather than more likely to become addicts.
12. This pernicious use of opium and later of morphine persisted well into the twentieth century. The death of the poet Dylan Thomas in 1956 in New York was ascribed by his attending physician, the deplorable Dr Feltenstein, 'specialist in alcohol-related illnesses', to delirium tremens; but it is more likely that on this particular occasion the poet suffered from pneumonia and that the immediate cause of death was a 'generous' injection of morphine by the doctor. Although Thomas was undoubtedly a severe alcoholic, at autopsy the liver showed little evidence of cirrhosis and the brain seemed normal. The lungs were 'congested'.

16 THE SHAPE OF DREAMS

1. Somewhat unfashionable today as judged by the demand for university places – perhaps deceptive.
2. Many of the most notable chemists of this golden age of organic chemistry started their careers as pharmacists. They included Scheele in Sweden, Dumas in France and Liebig in Germany as well as Serturner.
3. M. Seefelder, *Opium. Eine Kulturgeschichte* (Hamburg, 1987), p. 105.
4. T. Dormandy, *The Worst of Evils* (New Haven and London, 2006), p. 207.
5. He later moved to the Sorbonne and eventually to the Jardin des Plantes. He died in 1850, aged seventy-eight.
6. It was one of the lesser prizes but the only one awarded to non-French scientists. Serturner died in 1841, aged fifty-eight.
7. Strychnine, caffeine, nicotine, aconitine, atropine, paraverine and quinine were all characterised during the same decade.
8. Aneurysms of the arch of the aorta (as distinct from the currently far more common aneurysms of the abdominal aorta) were a late complication of syphilis. They could be extremely painful and were almost invariably fatal.
9. J. Pereira, *Elements of Materia Medica and Therapeutics* (London, 1839), p. 246.
10. F. Accum, *A Treatise on the Adulteration of Food and Culinary Products* (London, 1820).
11. V. Berridge, *Opium and the People* (London, 1981), p. 68.
12. He also invented the steam engine to drive ornamental fountains and the water clock.
13. F. Rynd, 'The subcutaneous injection of medicated fluids for neuralgias and other painful ailments', *Dublin Medical Press*, 1845 (5), 564.

17 THE MOST WICKED OF WARS

1. Quoted by J. Beeching, *The Chinese Opium Wars* (San Diego, Calif., 1976), p. 12.
2. M. Booth, *Opium. A History* (New York, 1996), p. 202.
3. R. Newman, 'Opium smoking in late Imperial China: A reconsideration', *Modern Asian Studies*, 1995 (29/4). Revisionist history briefly reviewed in V. Berridge, *Opium and the People* (London, 1981), p. xxvi.
4. They included a state coach, a telescope, a chiming clock, a book of maps, the Bible and a stuffed wild boar. To Macartney's annoyance Sir George Staunton Bart in the resplendent scarlet gown of an Oxford doctor of something or other was regularly mistaken for the king of England. The baronet was accompanied by his twelve-year-old son, a prodigy who had learnt Mandarin on the journey out and exchanged a few friendly words with the emperor. As a grown-up he would become a nuisance but at twelve he was by all accounts delightful.
5. Quoted by Booth, *Opium. A History*, p. 219.
6. For some reason Britain continued to invest this with enormous importance.
7. Quoted by W. Travis Hanes III and F. Sanello, *The Opium Wars* (Naperville, Ill., 2002), p. 58.
8. At least 3,000 years before Charles II the legendary Chinese emperor Shen Nung was visiting a region where an epidemic was making the consumption of unboiled water dangerous – a fact recognised in China thousands of years before Pasteur. A leaf from a wild teabush, a plant native to the Central Provinces, is said to have floated into the imperial cup full of hot water. Despite the alarming discoloration, the beverage – the first cup of tea – not only smelt and tasted divine but gave the dejected ruler the strength and determination to cope with the devastation of the province.
9. M. Collis, *Foreign Mud* (London, 1952), p. 76.
10. Quoted by Travis Hanes III and Sanello, *Opium Wars*, p. 58.
11. Beeching, *Chinese Opium Wars*, p. 134.
12. They were not factories in the modern sense but business and personal accommodation for the 'factors' or managers of the various companies trading openly in tea

but mainly and illicitly in opium. Each of the main trading nations had its own factory.

13. John Napier of Merchistoun (1550–1617), an inventor of logarithms, was also the populariser of the decimal point.
14. Collis, *Foreign Mud*, p. 134.
15. Beeching, *Chinese Opium Wars*, p. 149.
16. W. Turner, *British Opium Policy and its Results in China* (London, 1876), p. 56.
17. The hoisting of the Union Jack was in fact a routine event and had little to do with patriotic provocation. News from Canton still took 4–10 months to reach London and was usually garbled by the time it arrived.
18. Beeching, *Chinese Opium Wars*, p. 157.
19. John Gladstone, his father, had founded his huge fortune partly by investing in West Indian sugar plantations worked by slaves. That background was far removed from the Lancashire country house where William and Helen grew up with their ponies, family prayers and music-making. But William's speech in the opium debate marked a turning point in his career. The unbending young Tory who a few years earlier had voted for the flogging of soldiers and the exclusion of Jews from Parliament began to move at a tangent, eventually to emerge as leader of the Liberal Party.
20. Beeching, *Chinese Opium Wars*, p. 109.
21. Ibid., p. 110.
22. Only exceptionally and to meet an emergency was such an appointment made in China.
23. Most notably, when in 1828 the Sung River flooded and the people rose in protest against imperial taxation, he crossed the river in a boat unaccompanied and resolved the emergency without bloodshed.
24. His poetry is still highly esteemed in China.
25. Quoted in Hanes III and Sanello, *Opium Wars*, p. 42. Lin's first difficulty was to have his letter translated into English. The official interpreter made such a dog's dinner of the text that Lin sought the help of the best linguist in Canton, the American Dr Peter Parker. Parker had started life as a stonemason but, after studying theology at Yale and medicine in Philadelphia, came to China and was running a free eye clinic for poor Chinese. 'With God's help and in a good cause' he did his best. Lin also imagined that Britain was just beyond Calcutta and that his letter would arrive after a journey of about two months. In fact it took nearly ten months for it to reach London by a route still not known and was there delivered to the editor of *The Times*. It was to be frequently quoted in parliamentary debates, eliciting varying degrees of hilarity.
26. Sir C. Elliot, *Memoirs* (London, 1863), p. 453.
27. Quoted in Collis, *Foreign Mud*, p. 92.
28. A.F. Lindley, *Taiping and Tien-kwoh: The History of the Taiping Rebellion*, 2 vols (London, 1866).
29. S.Y. Teng, *The Taiping Rebellion and the Western Powers* (Oxford, 1971).
30. The event is analysed by D. Hurd in *The Arrow Incident* (London, 1967). The real culprit who fanned the flames of an incident into a full-scale war was probably Lord Palmerston's odious appointee as Consul in Hong Kong, Sir John Bowring (see note 35).
31. Lord Palmerston's assessment.
32. J.L. Morison, *The 8th Earl of Elgin* (London, 1928). The 7th Earl in particular had bankrupted his family by paying for the removal of the Parthenon friezes to London and by presenting them to the British Museum as a free gift. He kept only a marble foot as a doorstop.
33. T. Waldrond (ed.), *Letters and Journals of James, 8th Earl of Elgin* (London, 1878), p. 98.
34. He was the son of Antoine-Jean Gros, Napoleon's best battle-scene painter.
35. Sir John Bowring started his career as a literary radical though some evidence suggests that he may have been one of Lord Palmerston's private informers on what was

happening in radical and revolutionary circles. He was made governor of Hong Kong; and, after Lord Elgin's arrival, showered the newcomer with unsolicited advice. Baron Gros too suffered from Bowring's verbal bullying.

36. Marquis de Moges (ed.), *Recollections of Baron Gros's Embassy to China and Japan in 1857–58* (London, 1860), p. 234.
37. F.S. Turner, *British Opium Policy and its Results in China* (London, 1876), p. 46.
38. Morison, *8th Earl of Elgin*, p. 298.
39. *The Times*, 19 September 1859, p. 15.
40. Moges (ed.), *Recollections*, p. 256.
41. L. Oliphant, *Narrative of the Earl of Elgin's Mission to China*, 2 vols (London, 1859).
42. French and Italian Jesuits contributed to the design of several of the buildings, combining Chinese traditional styles with Counter-Reformation baroque at its most exuberant.
43. J.-M. Comte de Tassin de Moligny, *Mémoires de ma jeunesse* (Paris, 1880), p. 368.
44. Thanks to the newly installed telegraph the future Lord Rothschild had his agent there apparently disguised as an American major and acquired several prize lots.
45. It is still part of the Crown Jewels.
46. The breed is still cherished for its loyalty, intelligence and fearlessness.
47. There may be something to be said for the looting of palaces like the Summer Palace. One way or another most treasures have ended up in public collections like the British Museum or in private collections now open to the public like the Frick in New York. They gradually awakened interest in Chinese art (which barely existed at the time of the looting) and have given pleasure to millions who might never have travelled to Peking or even Beijing. But even to the most cynical art lover the torching of the Summer Palace remains an act of total barbarity.
48. G.J. Wolseley, *Narrative of the War with China 1860* (London, 1862), p. 245. This is one of the best accounts of the burning of the Summer Palace. The almost tearful revulsion may seem surprising in a professional soldier but the future field marshal was no ordinary soldier.
49. R. Swinhoe, *Narrative of the North China Campaign of 1860* (London, 1861), p. 168. The author was an official interpreter and an excellent witness.
50. Only Queen Victoria and Prince Albert expressed disapproval or at least reservations. The queen wrote that the 'burning of royal palaces even to set a salutary example is *wrong*'. The prince, whose views of Chinese art were far in advance of his times, was distressed at the destruction of a 'priceless heritage'. By contrast, Lord Elgin explained to the Royal Academicians: 'I have been repeatedly asked whether the interests of art are likely to be in any way promoted by the opening up of China. I don't think so . . . I don't think our artists have anything to learn from that country . . . Nevertheless I am inclined to believe that under the mass of abortions and rubbish there lies some hidden spark which the genius of my countrymen may nurse into a flame.'
51. Morison, *8th Earl of Elgin*, p. 462.
52. He enjoyed this high office only for eighteen months, dying in 1864 from either a 'seizure' or a ruptured aneurysm (accounts differ).

18 THE YELLOW PERIL

1. The peace treaty was to be known as the Peking Convention. The 'lull' did not apply to the ongoing war with the Taiping.
2. William Jardine and James Matheson left China though their firm still flourishes. Jardine invested his fortune in works of art, bought himself the parliamentary seat of Ashburton and entered Parliament as a supporter of Lord Palmerston. He did not enjoy his political career for long, dying in 1843 of an undiagnosed but painful illness. His death created the myth of a curse on all those who had profited from the opium trade, ignoring the fact that his partner was to live to ninety-one. Matheson also entered Parliament and married in 1887. He bought the entire island of Lewis off the

coast of Scotland and built a castle there costing half a million pounds. He also endowed a Chair of Chinese in London University. His heir and nephew, Donald, who came to hate the opium trade and resigned from the firm as a young man, became in old age chairman of the Society for the Suppression of the Opium Trade in Britain.

3. The war poet Siegfried Sassoon was a collateral descendant.
4. W. Mott, *Über Opium in China* (Tübingen, 1880), p. 237.
5. It was Gordon's exploits in China that led to his later appointment to take command of the Sudan.
6. Quoted by D. Latimer and J. Goldberg in *Flowers in the Blood* (New York, 1981), p. 210.
7. C. Dickens, *The Mystery of Edwin Drood* (1870; Penguin edition, London, 1980), p. 342. But Dickens's most brilliant and horrific excursion into opium lore is in Chapter 11 of *Bleak House* where the lawyer, Mr Tulkinghorn, discovers the body of a solitary, forsaken addict, who has died of an opium overdose.
8. With little formal education, Rohmer/Ward was a gifted writer though he could not spell. After the Second World War he moved to New York, where he died in 1959.
9. It was a clever spoof entitled *The Fiendish Plot of Dr Fu Manchu*; but it was never completed or publicly released.
10. Quoted by Latimer and Goldberg, *Flowers in the Blood*, p. 214.
11. Ibid., p. 235.
12. Ibid., p. 237.
13. Ibid., p. 239.
14. Ibid., p. 238.
15. Total prohibition on Chinese immigration to the United States, strictly enforced for twenty years, was lifted in 1943 when China became a gallant ally in the war against Japan.
16. With grim postludes. In 2007 a British citizen resident in Finchley, north London was arrested in China for carrying 5 kilos of heroin and, despite pleas at the highest diplomatic level, was executed by lethal injection in the last week of 2009.

19 DOCTORS RULE

1. Anita Brookner's resonant phrase.
2. Marx's *Communist Manifesto* of 1848 was still a thoroughly Romantic document. In his *Das Kapital* there is little trace of Romantic sentiment.
3. Approximate equivalents:
   1 drop = 1/20 ml
   20 drops = 1 ml
   1 fluid ounce = 29.5 ml
   20 fluid ounces = 1 English pint.

   1 grain of opium = approx. 25 drops of laudanum, according to De Quincey. A grain was the smallest British weight, the average weight of a seed of corn, about one 7,000th of a pound.
   24 drops of laudanum = 1 grain of opium = 6.4 mg morphine
   1 grain of opium = 64 mg opium = 6.4 mg morphine
   1 liquid ounce of laudanum = 590 drops of laudanum = 24 grains opium
   24 grains opium = 1520 mg opium
   1520 mg opium = 152 mg morphine.
4. In *Eminent Victorians* (1918) Lytton Strachey wrote: 'It was not by gentle sweetness and womanly self-abnegation that she had brought order out of chaos . . . It was by strict method, by stern discipline, by rigid attention to detail, by ceaseless labour and by the fixed determination of an indomitable will.' No mention of morphine, more important perhaps than any of these admirable attributes: even to the irrepressible ironist medical needs and private drug habits were not to be raked over by biographers. (All his four 'eminent Victorians' had interesting illnesses, none of them mentioned by the author.)

5. Mortality among new admissions to the hospital in Scutari was almost twice as high as among participants in the charge of the Light Brigade.
6. Estimates of the drugs used in the United States for most of the nineteenth century tend to be based on Customs data relating to imported opium and morphine. The dramatic rise during the war years due to home-grown produce is often missed.
7. Horace Day, *The Opium Habit* (New York, 1868).
8. Except in the Shavian sense according to which all professions are conspiracies against the laity.
9. B.A. Morel, *Traité sur la Degénération* (Paris, 1857). It became the doom-book of the age, as Spengler's *Untergang des Abendlandes* would be seventy years later and Orwell's *Nineteen Eighty-four* another half a century after that.
10. G. de Maupassant, *Sur l'eau* (Paris, 1888). *Lettre d'un fou* (1885) is another superb autobiographical tale.
11. Alphonse Daudet, *L'Evangeliste* (Paris, 1883). A violently anti-religious as well as anti-morphine novel. But at the nursing home in Lamalou-les-Bains where the author spent his last years 'morphine flowed like water'. M. Taine, *La Comtesse Morphine* (Paris, 1885). Titillatingly damned both morphine and decadence.
12. M. Villars and Willy, *Les imprudences de Peggy* (Paris, *c.* 1885). Willy was Colette's husband and bloodsucker.
13. Quoted by P. Jullian, *Lorraine* (Paris, 1905), p. 65.
14. As described in B. Hodgson's brilliant depiction of the period, *In the Arms of Morpheus* (New York, 2001).
15. V. du Saussay, *La Morpheine* (Paris, *c.* 1895).
16. The famous words 'Is there a doctor in the house?' were spoken on the stage at Drury Lane Theatre on 17 February 1902 when Miss Kitty Smiles collapsed backstage. The traditional tale is that the only person to volunteer was a medical student from St Thomas's Hospital who not only correctly diagnosed morphine overdose but also instructed the manager not to pour hot coffee down the patient's throat until he ascertained that she could gag. If true (and why not?) he may have saved her life.
17. S. Bernhardt, *Memories of my Life* (New York, 1908), p. 165.
18. Dumas *père* turned the execution into a gruesome set piece in *The Count of Monte Cristo*.
19. The first more or less accurate measurement of morphine and heroin in blood and urine was based on the substances even in minute concentrations causing a painful erection of the tail of the mouse. The test, known as Straub's Mouse Test after the pharmacologist who described it, could even be made semi-quantitative. It has now been replaced by cheaper and kinder chromatographic measurements.
20. Mrs Ward was one of those formidable Victorian ladies whose manifold achievements would have been remarkable even in a physically robust person. In fact she suffered from a multiplicity of chronic ailments for which the only effective remedy was morphine.
21. Young Sigmund Freud took it up to earn enough money to marry his betrothed.
22. To P.G. Wodehouse's Bertie Wooster the eminent practitioner Sir Roderick Glossop was the 'loony doctor'. Unsurprisingly perhaps, the description did not catch on with the profession. 'Shrink' emerged in the army during the Second World War.
23. Eugene O'Neill put the words into the mouth of Edmund Tyrone in *A Long Day's Journey into Night*: 'It's pretty hard at times having a dope fiend for a mother'. The dramatist's own mother, Ella O'Neill, was hopelessly addicted.
24. Despite the recognition that morphine is the active principle in opium and that heroin is a more potent but direct derivative, the difference in public perception survives. In 1980 Yves St Laurent launched a successful new perfume called *Opium*. Only an entrepreneur with a death wish would launch a product named *Morpheine*, let alone *Héroïne*.
25. Quoted by M. Jay in *Emperors of Dreams* (Sawtry, Cambs, 2000), p. 77.
26. Quoted by M. Booth in *Opium. A History* (New York, 1996), p. 245.
27. T.C. Allbutt, 'On the abuse of hypodermic injected morphine', *Practitioner*, 1870 (5), 327. The author was the probable model for George Eliot's Dr Lydgate in *Middlemarch*

and a convert. While starting in practice in Yorkshire he was an enthusiastic user of morphine in his patients, especially in cardiac disease; but he changed his mind when he saw the abuse of the drug among fashionable physicians in London.

28. J.B. Mattison, 'The Treatment of Opium Addiction', *Journal of Nervous and Mental Diseases*, 1885 (12), 249.
29. O. Jennings, *On the Cure of the Morphia Habit* (London, 1880), p. 57.
30. L. Lewin, *Phantastica* (1924); trans. P.H.A. Wirth. Published in 1998 in Richmond, Vt. In the 1920s Lewin was regarded as the greatest authority on opium in Germany but even before Hitler his firm adherence to the Jewish faith prevented him from being appointed to a chair in any German university.
31. G.B. Shaw, *The Doctor's Dilemma*, Preface (London, 1906), p. 45.

20 AMERICAN VOICES

1. H.H. Kane, *Opium Smoking in America and China* (Philadephia, 1882), p. 68.
2. Quoted by D. Latimer and J. Goldberg, *Flowers in the Blood* (New York, 1981), p. 187.
3. W. Colton, 'Turkish sketches: effects of opium', *Knickerbocker*, 1837 (7), 421.
4. W. R. Cobbe, *Dr Judas: A Portrayal of the Opium Habit* (Chicago, 1895), p. 46.
5. F.H. Ludlow, *The Hasheesh Eater* (Philadelphia, 1857), p. 35.
6. Ibid., p. 46.
7. Ibid., p. 68.
8. Quoted by D.P. Dulchinos, *Pioneer of Inner Space, The Life of Fitz Hugh Ludlow, Hasheesh Eater* (New York, 1998), p. 124.
9. H. Day, *The Opium Habit with Suggestions as to the Remedy* (New York, 1868).
10. F.H. Ludlow, 'Outline of the opium care', *Harper's Monthly*, August 1864, p. 54.
11. Quoted in Dulchinos, *Pioneer of Inner Space*, p. 68.
12. Ibid., p. 89.
13. Ibid., p. 129.
14. F.H. Ludlow, 'What shall they do to be saved?' *Harper's New Monthly Magazine*, August 1867, p. 277. A cry of despair.
15. Quoted by H. Broun and M. Leech, *Anthony Comstock, Roundsman of the Lord* (London, 1928), p. 68. No mean feat to write with insight and sympathy but also convincingly about a monster.
16. Ibid., p. 346.
17. J. Parrish (Dr), *The Opium Appetite* (Philadelphia, 1870), p. 98.
18. A. Calkins, *Opium and the Opium-Appetite* (1871; Bibliobazaar, Charleston, SC, 2008), p. 68.
19. Ibid., p. 124.
20. Ibid., p. 312.
21. Fully reviewed by H. Wayne Morgan in *Drugs in America: A Social History, 1800–1980* (Syracuse, NY, 1981).
22. Cobbe, *Dr Judas*, p. 135.
23. *Knickerbocker*, February 1858, p. 197.
24. Quoted by Wayne Morgan in *Drugs in America*, p. 36.
25. H.H. Kane, *The Drugs that Enslave* (Philadelphia, 1881), p. 17.
26. Ibid., p. 138.
27. Ludlow, 'What shall they do to be saved?', p. 277.
28. Quoted by Wayne Morgan in *Drugs in America*, p. 268.

21 NERVOUS WASTE

1. G.M. Beard, 'American neurasthenia or nervous exhaustion', *Boston Medical and Surgical Journal*, 1869 (80), 217. The terms may be out of date but the concepts were pioneering.

2. That pompous term was not invented for another fifty years but Beard was clearly on to it. Beside 'neurasthenia' the similarly obscure term 'nervous exhaustion' was widely used in the opium literature.
3. M. Beard, *American Nervousness: Causes and Consequences* (New York, 1881), p. 133.
4. J.B. Mattison, 'The Treatment of Opium Addiction', *Journal of Nervous and Mental Diseases*, 1885 (12), 249.
5. N. Allen,, *The Opium Trade* (New York, 1850), p. 23.
6. W. Rosser Cobbe, *Dr Judas: A Portrayal of the Opium Habit* (Chicago, 1895), p. 245.
7. Ibid., p. 57.
8. H.H. Kane, *The Drugs that Enslave* (Philadelphia, 1881), p. 57.
9. Anon., 'Opium eating. An autobiographical sketch', *New York Times*, 1877 (30), 33.
10. Ibid., p. 34.
11. D. Starr Jordan, 'Drugs and character', *The Independent*, 1899 (52), 1059.
12. C.W. Earle, 'The responsibilities and duties of the medical profession regarding opium inebriety', *Transactions of the Illinois Medical Society*, 1886 (39), 56.
13. The American sanitarium never held quite the same connotation as the European, mainly tuberculosis, sanatorium. Perhaps the slight difference in derivation is significant. Sanitarium derives from *sanitas*, meaning health. Sanatorium derives from *sanare*, meaning healing
14. L.E. Keeley, *The Opium Habit: Its Proper Method of Treatment and Cure without Suffering and Inconvenience* (Dwight, Ill., 1882), p. 45.

    Gold had a long history as a therapeutic agent: it was an ingredient in most alchemists' mixes and almost invariably figured in the prescriptions recommended by Paracelsus. But it was not recommended as the only or main ingredient of a specific mixture until L.E. Keeley. What his 'Bichloride of Gold' was or how it was prepared was never revealed if indeed it existed.
15. His book, *A Popular Treatise on Drunkenness and the Opium Habit and their Successful Treatment with Double Chloride of Gold, the Only Successful Treatment* (Dwight, Ill., 1890) achieved 'countless editions' according to one dust jacket.

22 A HEROIC SUBSTANCE

1. In later life he published several popular books on chemical experimentation for the young.
2. The result would be aspirin.
3. Single-handed, according to Livy, the hero held up the enemy on the bridge leading to Rome. For a time Dreser considered naming the new drug *Coclein.*
4. Quoted in D. Latimer and J. Goldberg, *Flowers in the Blood* (New York, 1981), p. 235.
5. Terry Pellens, *The Opium Problem* (New York, 1936), p. 77.
6. Both effects are mediated through the central nervous system, not local actions on the gastrointestinal lining.
7. R.S. de Ropp, *Drugs and the Mind* (London, 1958), pp. 68–70.

23 THE BIRTH OF A CRUSADE

1. *Washington Post*, 2 June 1898, quoted in T.G. Paterson, J. Garry Clifford and K.J. Hagan, *American Foreign Policy: A History to 1914* (Lexington, Mass., 1983), p. 123.
2. 'To give them back to Spain would have been cowardly and dishonourable. We could not turn them over to France and Germany, our commercial rivals in the Orient: that would have been bad business and discreditable. Left to themselves they would soon have anarchy and misrule worse than under Spain.' Quoted in D. Reynolds, *America, Empire of Liberty* (London, 2010), p. 298.

3. B. McAllister Linn, *The Philippine War, 1899–1902* (Lawrence, Kan., 2000). The best and most balanced account of the conflict. See also Reynolds, *America*.
4. Ibid., p. 192.
5. Quoted by E. Marshall, *New York Times*, 12 March 1911, p. 32.
6. Ibid., p. 34
7. Ibid., p. 33.
8. Ibid., p. 35.
9. Ibid., p. 37.
10. *New York Times*, 23 April 1912, p. 65.
11. Ibid., p. 68.
12. Ibid., p. 69.
13. *Journal of the American Medical Association*, 1911 (34), 69.
14. E.H. Williams, *New York Times*, 8 February 1914, p. 5.
15. On the entry of the United States into the First World War he volunteered and was sent to France with the Medical Corps. He crashed his ambulance (carrying no patients) within a week and suffered severe injuries. Repatriated, he died in 1917 at the age of fifty.

24 WAR AND PEACE (OF SORTS)

1. Within a year of the signing of the peace treaties there was war in Poland, Turkey and Ireland. In 1931 Japan invaded China. Three years later Italy invaded Abyssinia. Two years later civil war erupted in Spain.
2. G. Csath, *The Diary of Geza Csath*, trans. Peter Reich (London, 2000), p. 56.
3. *Lancet*, editorial, 1916 (ii), p. 534.
4. *Lancet*, editorial, 1917 (i), p. 28.
5. Reggie was acquitted in the High Court: there was no evidence that he was responsible for Billie's death. A somewhat pathetic figure, after the case he disappeared from the social scene.
6. Quoted by M. Booth in *Opium. A History* (New York, 1996), p. 345.
7. Quoted by V. Berridge and G. Edwards in *Opium and the People* (London, 1981), p. 65.
8. F.P. Walters, *A History of the League of Nations*, 2 vols (Oxford, 1952), Vol. I, p. 32.
9. Summarised ibid., p. 128.
10. Deliberations at the League relating to opium are well described in J.M. Scott's *The White Poppy. A History of Opium* (London, 1971) and in a more gossipy vein by G. von Turkoczy, *Népszövetségi Krónika* [League of Nations Chronicle] (Budapest, 1942).
11. Quoted by J.M. Scott in *The White Poppy*, p. 243.
12. Von Turkoczy, *Népszövetségi Krónika* [League of Nations Chronicle], p. 176.
13. Quoted by Walters, *History of the League of Nations*, Vol. I, p. 324.
14. Count A. Apponyi, *Emlekezetek* [Remniscences] (Budapest, 1938), p. 432.

25 VICTIMS AND SURVIVORS

1. But there is now a good new translation by Susan Bennett into English available in paperback (London, 2009).
2. Hans was born Rudolf Ditzen in 1893.
3. Quoted in M. Seefelder, *Opium, eine Kulturgeschichte* (Hamburg, 1987), p. 168.
4. Professor Werner Goetze of Berlin University (died 2002 in Berlin); personal communication.
5. Sir T. Brunton, *Lectures on the Actions of Medicines* (London, 1897), p. 640.
6. Ibid., p. 642.
7. A. Birrel, *Sir Frank Lockwood* (London, 1898), p. 191.
8. A friend read the paper for him: Koller, a junior unpaid assistant in the University ophthalmology clinic was neither granted leave nor could he have afforded the train

fare. He later emigrated to the United States and became a respcted eye surgeon at Mount Sinai Hospital, New York. He died in 1944, aged eighty-seven.

9. Osler died in 1919. Among his books was a locked 'Secret History' which contained a detailed account of Halsted's addiction.
10. During the war cleaning and maintenance suffered from the depletion of staff and one horrible morning the screams of a nurse alerted the operating team, visitors and students: a mouse was scurrying across the floor. The ensuing hush was broken by the chief's enquiry: 'Has our rodent colleague been sterilised, Sister?'
11. The most important single feature of the operation was the dissection in one single 'block' of the breast with what were believed to be the commonest first sites of the spread of the cancer, the lymph nodes in the axilla. It was not till the 1970s that the mutilating procedure was replaced by a more limited removal of the cancer followed by radiotherapy.
12. Quoted in J.M. Scott, *The White Poppy. A History of Opium* (London, 1971), p. 145.
13. *League of Nations Publications* (Geneva, 1936), Vol. X, p. 564.
14. H.B. Spear, 'The early years of the "British System" in Practice' in J. Strang and M. Gossop (eds), *Heroin Addiction and Drug Policy. The British System* (Oxford, 1994), p. 142.

26 UNHOLY ALLIANCES

1. Goering too had a large stash of morphine when he was captured in 1945.
2. When Alfred W. McCoy published his carefully researched book the United States was once again at war, in Vietnam, and strenuous attempts were made to suppress it.
3. W. Burroughs, *The Naked Lunch* (1959; New York), p. 5.
4. See Chapter 32.
5. See Chapter 33.
6. As the North Vietnamese were closing in on Saigon, President Thieu, the last in the line, resigned, accusing Henry Kissinger and the United States of betraying Vietnam.

27 JUNKIES

1. P. Bailey, 'The heroin habit', *New Republic*, 12 April 1916.
2. D. Phear, 'The living dead', *Today*, 1928 (12), 24.
3. At one time Linda Lovelace claimed that she made *Deep Throat* while in an opium trance.
4. J. Kerouac, *On the Road* (1957; Penguin Classics, London, 1986). At first hard to read, it suddenly becomes compulsive.
5. Perhaps they were more comparable to the swarms of mendicant friars, mostly Franciscans and Dominicans, who, inspired by the preaching of St Francis, wandered over Christian Europe in the fourteenth century; or perhaps to the *Wandervögel* of German Romanticism in the nineteenth.
6. See 'Howl' in Allen Ginsberg, *Howl and Other Poems* (New York, 1956), p. 7.
7. But Burroughs could be as silly as Nancy. Opium, morphine and heroin have never been just 'great mind expanders'.
8. They included Joan Baez, Janis Joplin, The Who, Jefferson Airplane and Jimi Hendrix.
9. After his blistering first album, *Are You Experienced* (1967), Jimi Hendrix's career barely lasted four years (most of it spent in England) before he died of an overdose (?accidental ?suicidal ?murder) in 1970 aged twenty-seven.
10. The estimated figures in Millennium Year were 33.9 per cent brought in by lorry, 19.9 per cent by freight or cargo, 14.9 per cent in private luggage, 3.8 per cent on persons, half of that concealed internally, and the rest by other means, including the regular post, diplomatic bag services, sea drops and drops from light aircraft.
11. G. Greene, 'My own devil', *Vogue*, October 1973, p. 25.

28 GUARDIANS OF THE LAW

1. L. Katcher, *The Big Bankroll. The Life and Times of Arnold Rothstein* (New York, 1959), p. 324. Scott Fitzgerald used Rothstein as the model for Jay Gatsby's criminal associate Meyer Wolfsheim.
2. Quoted by D. Latimer and J. Goldberg, *Flowers in the Blood* (New York, 1981), p. 290.
3. H. Anslinger and C.R. Cooper, 'Marihuana, assassin of youth', *American Magazine*, 1937 (124) 150.4.
4. H. Anslinger, 'The psychiatric aspects of marihuana intoxication', *Journal of the American Medical Association*, 16 January 1943, p. 212.
5. R. Ellison, *Invisible Man* (Washington, 1952) p. 83.
6. H. Anslinger and W. Oursler, *The Murderer* (New York, 1961), p. 131.

29 ZERO TOLERANCE

1. Quoted by D. Latimer and J. Goldberg, *Flowers in the Blood* (New York, 1981), p. 298.
2. R.M. Nixon, *In the Arena* (New York, 1990), p. 199.
3. A. Summers, *The Arrogance of Power* (New York, 2000), p. 317.
4. G. Vidal, *Collected Essays, 1952–1972* (New York, 1974), p. 374.
5. Philippe Bourgeois, *In Search of Respect* (Cambridge, 1995), p. 345.
6. A. Efthimiou-Mordaunt, 'The user's voice', *Inside Out*, 2000 (14), 9.
7. But, unlike opium, pure heroin does not smoke well and generally continued to be injected, as it still is. For smoking it was usually mixed with another drug like caffeine or a barbiturate, the former mix commonly known as 'Chinese no. 3'. 'Brown opium' was Iranian heroin mixed with caffeine and was popular in MilleniumYear with smart young people. Other sophisticates added citric acid. 'Black tar heroin' contained an excess of acetic acid which made it comparatively easy to smoke (though irritating) and was fashionable in the United States and Britain in the 1980s.

   Smoking heroin is as addictive as the injection and can be as lethal. A few heroin smokers develop an ultimately fatal condition called leukoencephalopathy in which the white matter of the brain undergoes focal degeneration, resulting in progressively slurred speech, loss of coordination and other neurological signs.
8. Dale Beckett, 'Heroin, the gentle drug', *New Society*, 26 July 1979, p. 181.
9. Arthur Hawes, quoted by R. Davenport-Hines in *The Pursuit of Oblivion* (London, 2001), p. 367.
10. Quoted in Davenport-Hines, *Pursuit of Oblivion*. The London drug scene in the 1990s is particularly well described in this work.
11. R. Robertson, 'Epidemic of AIDS related virus infection among intravenous drug abusers', *British Medical Journal*, 1986 (i), 527.
12. A. Dally, *A Doctor's Story* (London, 1990). A key source for the period.
13. Royal College of Psychiatrists, *Drug Scenes* (London, 1987).
14. V. Berridge, *AIDS in the UK* (Oxford, 1996).

30 GOD'S OWN MEDICINE

1. Many charities shrivelled in the wake of the National Health Service, which was erroneously perceived as making private initiatives superfluous.
2. In retrospect her headmistress at Roedean did discern exceptional qualities of leadership in her.
3. In this as in many other respects there was a parallel between her and Florence Nightingale. Florence, coming from a county family, could write to 'Sidney' (Sidney Herbert, Secretary of State for War and cousin of the Earl of Pembroke) as a social equal and did so regularly from Scutari.
4. She was slightly disappointed that Pope John Paul II on a crowded four-day state visit to Britain planned in every detail over a period of months could not at a day's notice drop in at St Christopher's for tea. A cardinal was sent instead.

5. One standard textbook dismisses the disparity as 'for obvious reasons'. They have never been obvious to the present writer.
6. In principle the treatment of incurables preoccupied the great French clinicians of the nineteenth century. 'Guérír quelquefois, soulanger souvant, embrasser toujours' was an oft-quoted precept. But it was no Anglo-Saxon idea and even in France few knew how to put precept into practice.
7. Jean Martin Charcot was an exception to this as to many other rules. His monumental six-volume *Système de medecine*, one of the great medical texts of all times, has an excellent chapter devoted to the use of opiates.
8. J. Snow, 'Opium and cocaine treatment in cancerous disease', *British Medical Journal*, 1896 (21), 718. The author was unrelated to the John Snow of cholera and anaesthesia fame but an eminent physician.
9. Sir W. Osler in N. Sykes, P. Edmonds and J. Wiles (eds), *Management of Advanced Disease*, 3rd edition (London, 1905), p. 15.
10. A. Worcester, *The Care of the Aged, the Dying and the Dead* (Springfield, Ill., 1935; rep. Oxford, 1961).
11. R. Abrams, G. Jameson, M. Poehlman and S. Snyder, 'Terminal care in cancer. A study of 200 patients attending Boston clinics', *New England Journal of Medicine*, 1945 (232), 719.
12. C. Saunders, 'The treatment of intractable pain in terminal cancer', *Proceedings of the Royal Society of Medicine*, 1964 (4), 68.
13. St Christopher is no longer on the canonical register of saints and may have been a mythical person but as he was the traditional patron saint of those embarking on a long journey the choice was not inappropriate.
14. Cicely Saunders's personal achievements were also recognised. Honorary fellowships and doctorates rained on her; and she became a DBE as well as a member of the Order of Merit. The last honour she shared with Florence Nightingale who was among the first to be appointed when the Order was instituted by Edward VII. Dame Cicely died at St Christopher's Hospice on 5 July 2005.
15. J.N. Langley, *The Autonomous Nervous System* (Cambridge, 1921). A landmark book.
16. C. Pert and S.H. Snyder, 'Opium receptors: demonstration in nervous tissue', *Science*, 1973 (179), 1011.
17. A. Goldstein, *Addiction. From Biology to Drug Policy*, 2nd edn (Oxford, 2001). An excellent review by one of the protagonists.
18. A. Corbett, S. McKnight and G. Henderson, 'Opioid receptors': see http://opioids.com/receptors/index html, accessed July 2010.
19. S.A. Greenfield, *Brain Story* (London, 2000). The book of what must have been a fascinating television series.
20. J. Fichna, A. Janecka, M. Piestrzeniewicz, J. Costentin and J.C. do Rego, 'Antidepressant-like effects of Endorphin-1 and Endorphin-2 in mice', *Neuropsychopharmacology*, 2006 (6), 5.
21. Goldstein, *Addiction*, 2nd edn (Oxford, 2001), p. 79.
22. Ibid., p. 86.
23. W. Zhu, K. Mantione, R.M. Kream and G.B. Stefano, 'Alcohol-, nicotine-, and cocaine-induced release of morphine from human white blood cells', *Medical Science Monitor*, 2007 (12), 350.
24. A. Drewnonowski, D.D. Krahn, M.A. Demitrach, L. Nairn and P.A. Gosnell, 'Taste responses and preferences for sweet high-fat foods: evidence for opioid involvement', *Physiological Behaviour*, 1992 (51/2), 371.
25. The English novelist Enid Bagnold became a morphine addict during the First World War when she worked as a volunteer nurse; and, despite later and additional involvement with amphetamines, lived and worked till her death at the age of ninety-one.

31 TREATMENTS AND CURES

1. Quoted by D. Latimer and J. Goldberg, *Flowers in the Blood* (New York, 1981), p. 247.

2. Even serious textbooks suggest that the name was intended as a tribute to Adolf, the Führer, and was an abbreviation of Adolphine. The name in fact referred to *dolor*, pain, and was not publicised till after the war by the American branch of the pharmaceutical company Ely Lilly.
3. V. Dole, 'Addictive behaviour', *Scientific American*, December 1980.
4. J. Pullinger and A. Quickie, *Chasing the Dragon* (London, 1989), p. 65.
5. M. Booth, *Opium. A History* (New York, 1996), p. 195.

32 LIFE AND DEATH OF A DRUG LORD

1. The terms seem to have spread extraordinarily fast. The 'Golden Triangle' was introduced by United States Vice-Secretary of State Marshall Green during a press conference on 12 July 1971.
2. But according to Bertil Lintner it may have derived from the first traders of the border regions of the Triangle, especially those of the Thai–Burmese frontier towns, exchanging opium for 99 per cent pure gold ingots: B. Lintner, *Burma in Revolt* (Washington, DC, 1994).
3. Burma's new name was not as universally adopted as Sri Lanka, Bangladesh or Taiwan because it was proclaimed by an undemocratically installed military junta.
4. The mighty Shan Kingdom, dating back to the Mongol invasion in the thirteenth century, once extended far beyond northern Myanmar. It later declined but finally fell victim only to the cavalier map-drawing of colonial times and the hurried redrawing of maps after decolonisation.
5. At the top of the heap Marshall Sarit in the 1970s boasted a hundred mistresses, arbitrarily executed criminals as a public spectacle and died leaving an estate of $150 million. Thai police neither would nor perhaps could charge Bangkok's fifteen top narcotics brokers who insulated themselves from any direct contact with the traffic. Even with twenty United States agents and an annual budget of $6 million, the Drugs Enforcement Authority operations in Bangkok were still dependent on cooperation with the police; and 'the police are so corrupt it turns my stomach', said one Bangkok-based DEA agent in an interview in 1975. 'The border patrol police take pay-offs for letting the opium and heroin in and the provincial "antinarcotics" police transport the narcotics from Northern Thailand to Bangkok as instructed by the drug cartels' (A. W. McCoy's *The Politics of Heroin*, Chicago, 2003).
6. In 1974 a new Laotian ambassador was appointed to France and, as a mark of Laos's traditional esteem for French culture, the person appointed was Prince Sopsaisana. The prince was not only royalty but also vice-president of the Laotian national assembly, a member of numerous international bodies, president of the Laotian Academy and head of the Laotian judiciary. On alighting at Orly he was discovered through the agency of a sniffer dog to be carrying in his personal luggage 50 kilos of heroin on behalf of the Teochiu Triads.

   Despite sophisticated electronic equipment, sniffer dogs still win paws downs over machines. Active breeds like spaniels prance around and nothing can stop them. More phlegmatic breeds like setters or Labradors sit down next to the suspect and will not budge. Snag, a United States Customs sniffer, has notched up seizures worth nearly $1 billion.
7. A.F.B. (died in 2002) was a personal friend who visited the region professionally three times.
8. A.F.B., abridged personal account.
9. Jean-Louis Destruelle, personal communication.
10. John Newman, a politician who had spoken out against the rising menace, was gunned down outside his house in Cabramatta in Sydney. The assassins were themselves exterminated by a rival gang two weeks later. (See McCoy, *Politics of Heroin*, p. 354.)

11. A.T., personal communication. Professors of medicine do not wear ruby tiepins in the shape of dragons in England but happily the jewel could be converted into a lady's brooch.
12. A.W. McCoy pointed out the futility of vast sums of money being expended on trying to capture or kill top figures or names in the drug trade, a major aim of the United States War on Drugs. A replacement is always ready to step into his shoes (McCoy, *Politics of Heroin*, revised edn, Chicago, 2003).
13. Well described by B. Lintner and M. Black, *Merchants of Madness. The Methamphetamine Explosion in the Golden Triangle* (New York, 2009).

33 THE MAKING OF A MODERN NARCO-STATE

1. In what is today Pakistan and Iran both cultivation and consumption were more significant but neither country was among the world's leading suppliers.
2. 'The Great Game' is said to have been coined by Michael Conolly, a nineteenth-century intelligence officer but was introduced into common usage by Rudyard Kipling in *Kim* in 1901. Kipling's poem *White Man's Burden* dates from two years earlier.
3. An exhibition in the British Museum, London in 2010 devoted to Afghan art was an eye-opener to many.
4. On hearing of the ignominious retreat of the Duke of Brunswick after the cannonade of Valmy in 1792 the young Metternich noted in his diary: 'Never invade a revolution, never ever. However unlikely, the revolution will always prevail.' Countless historical examples bear this out.
5. The black pelt of newborn karakul lambs is said to be the softest but hardiest of all animal skins. It seems to be the material of President Karzai's customary headgear as it was Jinnah's.
6. The system involved payment at the time of sowing of half the amount the product would fetch after an average harvest.
7. Corruption in the current Western sense can exist only in societies in which financial probity is the norm. It was not always the norm even in the West. It was not in Britain's Indian Empire in the eighteenth century when a surge of buccaneering energy made established ethical notions obsolete. It was not the norm for Prince Talleyrand who would without compunction accept *douceurs* from his country's enemies in wartime. It is often not the norm when a foreign system of government is imposed on a traditional society unsympathetic to it. It is not the norm in Afghanistan today. Afghans often deplore corruption to foreign visitors but this is a polite ritual. It has effectively become a network of social cohesion, imperfect but the only one there is
8. Afghanistan is sometimes compared to Vietnam. One difference is that the heroin habit has not spread to the American and other Western troops. Or has it not? A news item on the BBC on 14 September 2010 for the first time mentioned allegations of 'large-scale trafficking' in heroin by British Army personnel using army mail and other facilities. The allegations are taken seriously and are being investigated.
9. See P.-A. Chouvy, *Opium. Uncovering the Politics of the Poppy* (Cambridge, Mass., 2010), p. 151.
10. D. Mansfield, '*Beyond the Metrics: understanding the nature of change in the rural livelihood of opium poppy growing households in the 2006/7 growing season*'. Report for the Afghan Drugs Interdepartmental Unit of the UK Government (London, 2007).
11. Quoted in Chouvy, *Opium*, p. 207.

34 NEW TRAILS: OLD TRIBULATIONS

1. Quoted by P.-A. Chouvy, *Opium. Uncovering the Politics of the Poppy* (Cambridge, Mass., 2010), p. 56.

2. This did not apply to the small artistic-literary circles of Moscow and St Petersburg. Ilya Repin's shattering portrait of Mussorgsky a few days before the composer's death is a representation of an addict as well as of an alcoholic.
3. The five victims were Tania Nicol (19), Anneli Alderton (24, mother of one), Annette Nicholls (29, mother of one), Paula Clennell (24, mother of three) and Gemma Adams (25; her partner too was a heroin addict).
4. It was second to Portugal until in Portugal the decriminalisation of drug possession significantly reduced it.
5. T. Rhodes, A. Sarang, P.Vickerman and M. Hickman, 'Why Russia must legalise methadone', *British Medical Journal*, 2010 (341), 129. An excellent and up-to-date review.
6. In London, Glasgow and Manchester it has been more stable over the past ten years than the price of bread.

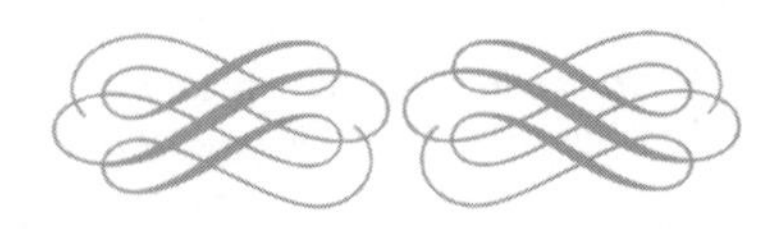

Selected Annotated Bibliography

WITH THE EXCEPTION of a few key papers, this Bibliography is of books and book-size documents. Most references to single articles, pamphlets, etc. are in the notes.

The literature on some famous opium users – Coleridge, Keats, Poe, Novalis and Baudelaire among others – is vast. An attempt has been made to select biographical works which shed light on the subject's opium habit.

A few particularly revealing works of fiction and recordings of music have been included.

* * *

Abraham, J., *Science, Politics and the Pharmaceutical Industry* (London, 1996).

Abrams, M.H., *The Milk of Paradise: The Effect of Opium Visions on the Works of De Quincey, Crabbe, Francis Thompson and Coleridge* (Boston, 1934). An interesting and pioneering work.

Accum, F., *A Treatise on Adulteration of Food and Culinary Products* (London, 1820). A ground-breaking, influential and still a somewhat blood-curdling book.

Adler, P.A., *Wheeling and Dealing: An Ethnography of an Upper-Level Drug Dealing and Smuggling Community* (New York, 1985). Not typical, but gripping.

Alagappa, M., ed., *Political Legitimacy in South-East Asia. The Quest for Moral Authority* (Stanford, 1995).

Alexander, R., *The Rise and Progress of British Opium Smuggling*, 3rd edn (London, 1886). Still an excellent read.

Allbutt, T.C., 'On the abuse of hypodermic injections of morphine', *Practitioner*, 1870 (5) 327. An important and in its time highly provocative paper.

Allbutt, T.C., *A System of Medicine* (London, 1897). The most famous textbook of its time with long and sensible sections on the use and abuse of morphine by injection.

Almeida, H. de, *Romantic Medicine and John Keats* (Oxford, 1991).

Alston, C., 'A Dissertation on Opium', in *Medical Essays and Observations*, Vol. V (Edinburgh, 1742), p. 110. The first cultivation of opium in Britain.

Anderson, F., *The Rebel Emperor* (London, 1958). One of the few sympathetic and imaginative accounts of the Taiping Rebellion.

Anderson, J., 'The Taliban Opium War', *New Yorker*, 9 July 2007.

Anderson, L., Captain, *A Cruise in an Opium Clipper* (London, 1935). A cruise off Taiwan in the 1850s; delightful.

Anderson, O., *Suicide in Victorian and Edwardian England* (London, 1987).

Anon., *Opium Eating. An Autobiographical Sketch by an Habituate* (Philadelphia, 1876). One of many such confessionals tracing the addiction to experiences in the Civil War.

Anstie, F.E., *Stimulants and Narcotics. Their Mutual Relations* (London, 1864).

Anstie, F.E., *Neuralgia and the Diseases that Resemble it* (London, 1871). Hopes that intraneural injections of morphine would cure neuralgia were later disappointed.

Ansubel, P.D., *Drug Addiction: Physiological, Psychological and Social Aspects* (New York, 1958).

Apponyi, A., Count, *Emlekezetek* [Reminiscences] (Budapest, 1938). Apponyi was the Hungarian chief delegate to the League of Nations and celebrated for his advocacy of a revision of the Paris Peace Treaties. He attended the third Opium Conference.

Arethaeus, *Opera* (mid-second century AD, trans. F. Adams; London, 1856).

Armstrong-Jones, R., 'Notes on some cases of morphinomania', *Journal of Mental Sciences*, 1902 (48) 478. The author was one of the leading 'alienists' in Britain.

Artaud, A., *Artaud Anthology* (San Francisco, 1965).

Ashley, R., *Heroin: The Myth and the Facts* (New York, 1972). A withering attack on legal drug control.

Assad, A.Z., and R. Harris, *The Politics and Economics of Drug Production on the Pakistan–Afghanistan Border* (Aldershot, 2003).

Auersperg, J., ed., *Erinnerungen des Prinzen Joachim von Auersperg, 1740–48* (Vienna, 1889). The author's tour of India was partly impelled by his creditors. The editor was his great-grandson whose fortunes were restored by judicious marriages. Not the most exciting of narratives as the prince's digestive ailments dominate the story.

Augustine, St, *Confessions*, trans. H. Chadwick, 3 vols (Oxford, 1992). The model for dozens of self-revelations, including De Quincey's.

Aung, Maung Htin *A History of Burma* (Cambridge, 1967). Balanced and informative on Shan aspirations.

Ball, J.C., and C.D. Chambers, eds, *The Epidemiology of Opium Addiction in the United States* (Springfield, Ill., 1970). A useful book even if the proper meaning of the term 'epidemiology' is stretched to its limits and beyond.

Ball, P., *The Devil's Doctor* (London, 2006). The best up-to-date introduction to the career of Paracelsus.

Balzac, H. de, *Opium*, in *Oeuvres Diverses*, Vol. II, ed. P. Castres, R. Chollet, R. Guise and C. Guise (first published Paris, 1830; Paris, 1996).

Barbeau, J.W., *Coleridge, the Bible and Religion* (London, 2007). Interesting.

Barker, J., *The Brontës* (London, 1994). The classic work on the family.

Barker, J., *The Brontës. A Life in Letters* (London, 1997). Complements the above.

Barry, M., *Ginsberg. A Biography* (London, 2001).

Barzun, J., *Berlioz and his Century* (New York, 1956). Good on the origin of the *Symphonie fantastique* and on the composer's background. The composer's father was an eminent physician who introduced acupuncture to Europe and later became an incurable addict.

Baselt, R., *Disposition of Toxic Drugs and Chemicals in Man*, 8th edn (Foster City, Calif., 2008).

Bateman, W., *Magnacopia. A Chemico-Pharmacological Library of Useful and Profitable Information for the Practitioner, Chemist and Druggist, Surgeon Dentist, etc.* (London, 1839). The medical Mrs Beeton of its time.

Bates, A., *Weeder in the Garden of the Lord: Anthony Comstock's Life and Career* (Lanham, Md., 1995). The testimony of an admirer.

Bateson, F.W., *Wordsworth: A Reinterpretation* (London, 1905).

Baudelaire, C., *Les Paradis artificiels* (Paris, 2002). An excellent edition of a classic.

Baudelaire, C., *Selected Poems* (London, 1986). A good selection of untranslatable poetry.

Baudelaire, C., *Intimate Journals*, trans. Christopher Isherwood (London, 1949).
Baum, D., *Smoke and Mirrors: The War on Drugs and the Politics of Failure* (New York, 1996). Lively and credible.
Baumler, A., *The Chinese and Opium under the Republic* (New York, 2007). Sound and well written.
Baumler, A., ed., *Modern China and Opium: A Reader* (New York, 2007).
Baxter, E., *De Quincey's Art of Autobiography* (Edinburgh, 1990).
Bayer, R., and Oppenheimer, G.M., eds, *Confronting Drug Policy: Illicit Drugs in a Free Society* (New York, 1993). A thoughtful discussion of decriminalisation by several experts.
Beard, G.M., *Stimulants and Narcotics: Medically, Philosophically and Morally Considered* (New York, 1871). Beard was the voice of the concerned medical profession in the 1870s and '80s, warning tirelessly against the 'creeping menace' that threatened the future of America.
Becher, J.R., *Über Hans Fallada* (Berlin, 1965). A sympathetic memoir by Fallada's Communist patron.
Becker, H.S., *The Outsiders: Studies in the Sociology of Deviance* (New York, 1963). Of great historical interest.
Beckett, D., 'Heroin the gentle drug', *New Society*, 1970 (4) 181.
Beeching, J., *The Chinese Opium Wars* (San Diego, 1975). Still, in the present writer's opinion, the best book on the opium wars – sound, evocative and an excellent read.
Behr, E., *Prohibition* (New York, 1997).
Bejerot, N., *Drugs and Society* (Springfield, Ill., 1980). Interesting.
Bello, D., *Opium and the Limits of the Empire: Drug Prohibition in the Chinese Interior, 1729–1850* (Cambridge, Mass., 2005).
Bellot, H.H., *The Pharmacy Acts, 1851–1908* (London, 1908). A good overview.
Bence-Jones, M., *Clive of India* (London, 1974). A good account of an empire-builder and addict.
Benedek, I., *Semmelweis Élete és Kora* (Budapest, 1960). The best book on Semmelweis.
Benedek, I., *Mandragora*, 2 vols (Budapest, 1962). A dull work by a good writer.
Benjamin, W., *Charles Baudelaire* (London, 1997).
Berlioz, H., *Symphonie fantastique* (Orchestre Nationale de France, Beecham); EMI 567972–2). The composer himself claimed that much of the music was composed in a haze of opium.
Berridge, V., 'Victorian opium eating: responses to opium eating in nineteenth-century England', *Victorian Studies*, 1978 (21) 25.
Berridge, V. 'Morality and medical science: concepts of narcotic addiction in Britain', *Annals of Science*, 1979 (36) 67.
Berridge, V., 'The origins of the British drug scene' *Medical History*, 1988 (32) 521.
Berridge, V., *AIDS in the UK: the Making of Policy, 1981–1994* (Oxford, 1996).
Berridge, V., and Edwards, G., *Opium and the People. Opiate Use and Drug Control Policy in Nineteenth and Early Twentieth Century England* (London, 1981). A ground-breaking book, interesting and readable.
Berridge, V., and P. Strong, eds, *AIDS and Contemporary History* (Cambridge, 1993). Contains an important article on British drug policy.
Bertram, E., Blackman, M., Sharpe, K., and Andreas, P., *Drug War Politics. The Price of Denial* (Berkeley, Calif., 1996).
Besant, W., *East London* (London, 1901).
Betts, W.R., 'The discovery of morphine', *Chemist and Druggist*, 1954 (162) 63.
Bewley-Taylor, D.R., *The United States and International Drug Control, 1909–1997* (London, 1999).
Black, D., *Triad Takeover* (London, 1991).
Blake, C., *Charles Elliot, RN* (London 1959). A sympathetic biography of a much-maligned character (London, 1959).
Blake, J.B., 'Mr Ferguson's hypodermic syringe', *Journal of the History of Medicine*, 1960 (17) 337.

Blanchard, C.M., *Afghanistan, Narcotics and US Policy* (New York, 2009). A short, up-to-date review.
Bliss, M., *William Osler. A Life in Medicine* (Oxford, 1999). The best and most recent life of a good doctor.
Blount, J., *The American Occupation of the Philippines, 1898–1912* (New York, 1912).
Blum, R.H., ed., *The Dream Sellers* (San Francisco, 1970).
Blum, R.H., *Society and Drugs* (San Francisco, 1980). Highly perceptive.
Blum, R.H. and H.H. Nowlis, *Drugs and the College Campus* (New York, 1969). An early study of opium in the universities.
Blunt, W., *Linnaeus. The Complete Naturalist* (London, 1971).
Bonavia, D., *China's War Lords* (Oxford, 1995).
Booth, M., *Opium. A History* (New York, 1996). An excellent work.
Booth, M., *The Dragon Syndicates. The Global Phenomenon of the Triads* (New York, 1999).
Boswell, J. *Boswell's Life of Johnson* (first published 1791; London, 1949).
Bourgeois, P., *In Search of Respect. Selling Crack in El Barrio* (Cambridge, 1995). A moving and gripping autobiographical novel about drug dealing in New York.
Bowra, C.M., *The Romantic Imagination* (Oxford, 1950).
Bowring, Sir J., *Autobiographical Recollections* (London, 1877). The surprisingly readable memoirs of a reputedly obnoxious character.
Bramall, C., *In Praise of Maoist Economic Planning: Living Standards and Economic Development in Sichuan since 1931* (Oxford, 1993). Useful.
Breasted, J.H., with medical notes by Dr A.B. Luckhardt, *The Smith Papyrus* (New York, 1930). A sumptuous facsimile edition with translation.
Brecher, E.M., *Licit and Illicit Drugs* (Boston, 1972). A landmark book marshalling the arguments for decriminalisation, much of it still relevant.
Brereton, W.H., *The Truth about Opium* (London, 1882). Not quite; but a well-informed apologia for British policy towards China.
Bressler, F., *The Chinese Mafia* (London, 1981). Remarkably insightful.
Brett, R.L., *George Crabbe* (London, 1956).
British Pharmacopoeia (London, 1858).
Brontë, A., *The Tenant of Wildfell Hall* (first published London, 1848; Panther edn, 1989). Contains an interesting discussion of 'hankering after forbidden things' and 'the evils of intemperance and abstinence' on p. 37. 'Is it prudent to allow a child a taste of forbidden substances?'
Brontë, C., *Villette* (London, 1853; Panther edition, 1987). Contains a haunting opium-like dream sequence and an incident of opium giving relief from pain and suffering.
Brook, T., and Wakabayashi, B., eds, *Opium Regimes: China, Britain and Japan, 1839–1952* (Los Angeles, 2000).
Brookner, A., *Romanticism and its Discontents* (London, 2000). Beautifully written; full of insights.
Broun, H., and M. Leech, *Anthony Comstock, Roundsman of the Lord* (London, 1928). An extraordinarily humane and balanced biography of a monster.
Brown, D., *Palmerston. A Biography* (London, 2010). The most recent and probably best biography of this flamboyant but still enigmatic character.
Browne, J.C., *Practical Instructions for the Treatment and Cure of Cholera and Diarrhoea by Chlorodyne* (London, n.d., approx. 1840). The beginning of one of the most popular opium preparations used by many generations. It saved lives in cholera epidemics.
Browne, T., Sir, *Works*, ed. H. Robbins, 2 vols (Oxford, 1986).
Bruun, K., P. Lynn, and R. Ingemar, *The Gentlemen's Club. International Control of Drugs and Alcohol* (Chicago, 1975).
Buddenberg, D., and W.A. Byrd, eds, *Afghanistan's Drug Industry: Structure, Functioning, Dynamics and Implications for Counternarcotics Policy* (New York, 2006). An authoritative and comprehensive review, containing some important papers.

Bulgakov, M., 'Morphine', in *A Country Doctor's Notebook*, trans. M. Glenny (first published 1925; London, 1975). The morphine-addicted young doctor Polyakov is one of the great tragic semi-fictional heroes of the morphine literature.
Bullivant, K., ed., *Culture and Society in the Weimar Republic* (Manchester, 1977). Contains some excellent articles and many references to drugs.
Bullock, H., *The Chinese Vindicated; or Another View of the Opium Question* (London, 1840). A splendidly robust attack on the opium trade.
Burke, T., *Limehouse Nights. Tales of Chinatown* (London, 1916). Superior tabloid journalism.
Burke, T., *The Ecstasies of Thomas de Quincey* (London, 1928).
Burne, P.G., *Addiction* (New York, 1974). A splendid collection of essays, including an outstanding chapter by D. F. Musto on the early spread of heroin.
Burnett, J., *Plenty and Want. A Social History of Diet in England* (London, 1968). Diet and opium use were often linked.
Burton, R., *Works*, including *Anatomy of Melancholy*, ed. T.E. Faulkner, N.K. Kiesling and R.L. Blair (Oxford, 1989–94).
Butel, P., *Opium: Histoire d'un fascination* (Paris, 1995). A good read.
Bynum, F.W., and Porter, R., eds, *Medical Fringe and Medical Orthodoxy, 1750–1850* (London, 1987). An invaluable collection.
Calkins, A., *Opium and the Opium-Appetite* (Philadelphia, 1871). An important publication which should have exploded the myth that morphine was a cure for opium addiction but sadly did not.
Camby, H.S., *Thoreau* (Boston, 1939).
Cameron, N., *An Illustrated History of Hong Kong* (Oxford, 1991).
Carey, J.T., *The College Drug Scene* (Englewood Cliffs, NJ, 1968). An early study of this important aspect of addiction.
Cartwright, F.F., *A Social History of Medicine* (London, 1977).
Celsus, *De Medicina*, translated by F. Marx (1915, Loeb edition, Havard, Vol. I, 1935, Vol. II, 1938).
Chang, Hsin-pao, *Commissioner Lin and the Opium War* (Cambridge, Mass., 1964). The only biography, based on Chinese as well as Western sources, of the outstanding Chinese personality in the opium wars. Fascinating.
Charters, A., *The Portable Kerouac* (first published 1995; London, 2007). An excellent Penguin Classic.
Charters, A., ed., *The Portable Beat Reader* (first published 1992; London, 2001). This Penguin paperback is probably the best introduction to the American Beat literature.
Checkland, S.G., *The Gladstones. A Family Biography, 1764–1851* (Cambridge, 1971).
Chesneaux, J., ed., *Popular Movements and Secret Societies in China, 1940–1950* (Stanford, Calif., 1972).
Chin, Ko-Lin, *The Golden Triangle. Inside South East Asia's Drug Trade* (Ithaca, NY, 2009). A first-hand account of the Wa Special Region of northern Burma by a brave traveller.
Chouvy, P.A., 'The ironies of Afghan opium production', *Asia Times*, 17 September 2003.
Chouvy, P.A., 'Afghanistan opium production in perspective', *China and Eurasia Forum Quarterly*, 2006 (February), 21.
Chouvy, P.-A., *Opium: Uncovering the Politics of the Poppy* (Cambridge, Mass., 2010). A survey of the modern opium trade with its historical background.
Christison, R., *Of the Action of Opium and the Symptoms it Excites in Man* (Edinburgh, 1827). A prophetic essay which did not get the response it deserved.
Christison, R., *On Poisons* (Edinburgh, 1829).
Christison, R., *A Treatise on Poisons* (Edinburgh, 1832). A pioneering textbook dealing with the dangers of laudanum.
Christison, R., *Cases and Observations in Medical Jurisprudence* (Edinburgh, 1834). An invaluable sourcebook.
Clarke, M., *The Secret Life of Wilkie Collins* (London, 1988). A good biography with a vivid background of the Victorian literary scene.

Clement-Jones, V., L. McLoughlin, S. Tomlin, G. Besser, L. Rees, and H. Wen, 'Increased beta-endorphin but not met-encephalin levels in human cerebrospinal fluid after acupuncture for recurrent pain', *Lancet*, 1980 (ii) 946.

Coates, A., *A Macao Narrative* (Hong Kong, 1978). Engaging.

Cobbe, W. R., *Dr Judas: A Portrayal of the Opium Habit* (Chicago, 1895). A sensationally successful book.

Cocteau, J., *Opium* (first published Paris, 1938; London, 1957). An iconic eulogy but also a warning by a fan and addict.

Cohen, N.M., *Health and the Rise of Civilisation* (New Haven, Conn., 1989).

Cohen, R., *Tough Jews: Fathers, Sons and Gangster Dreams* (London, 1999). Excellent on the Jewish gangster scene in New York following the Harrison Act.

Colantuoni, C., P. Rada, J. McCarthy, C. Patten, N.M. Avena, A. Chadeayne, and B.G. Hoebel, 'Evidence that intermittent, excessive sugar intake causes endogenous opioid dependence', *Obesity Research*, 2002 (0–6) 478.

Coleridge, E.H., ed., *The Complete Poetical Works of Samuel Taylor Coleridge*, 2 vols (London, 1912).

Coleridge, S., *Poppies* (London, 1834). Some affecting verse by the reputedly beautiful, accomplished and opium-addicted daughter of S.T.C. who first co-edited with her husband Henry Nelson Coleridge (a cousin) and her brother Derwent the poet's previously unknown work.

Coleridge, S.T., *Collected Letters*, ed. E.L. Griggs, 6 vols (Oxford, 1956–72).

Coleridge, S.T., *Notebooks of Samuel Taylor Coleridge*, ed. K. Coburn, 2 double vols (London, 1957–62). Kathleen Coburn as a young Oxford graduate came across some of Coleridge's notebooks almost by accident and devoted the rest of her life to collecting and editing the poet's work. Few poets have had such beautifully edited and commentated editions of virtually their entire written output.

Coleridge, S.T., and W. Wordsworth, *Lyrical Ballads* (1798), ed. D. Roper (London, 1968).

Collins, W., *No Name* (first serialised in *Harper's Magazine*, New York, 1862). A harrowing tale of suicide and murder with laudanum in what today would be described as a thoroughly dysfunctional family. Gripping.

Collins, W., *The Moonstone. A Romance* (first published London, 1868; Penguin, 1994). A captivating novel said to have been written in a daze of morphine.

Collis, M., *Foreign Mud* (London, 1952). An excellent account of the First Opium War and events leading up to it based on and summarising the Jardine Matheson Archive in the University Library, Cambridge.

Colton, W., 'Turkish sketches' (Philadelphia, 1836).

Comstock, A., *Traps for the Young* (New York, 1884). Probably Comstock's most readable outpouring.

Conrad, L.I., 'Arab-Islamic Medicine', in F.W. Bynum and R. Porter, eds, *Companion Encyclopedia of the History of Medicine* (London, 1993), 676. Magisterial.

Cook, M.C., *The Seven Sisters of Sleep* (London, 1867). Morphine in the factories: riveting.

Coomber, R., ed., *Drugs and Drug Use in Society, A Critical Reader* (Greenwich, UK, 1904).

Cooper, J.R., F.E. Bloom and R.H. Roth, *The Biochemical Basis of Neuropharmacology*, 7th edn (Oxford, 1996). The best of the standard texts dealing with neuroreceptors and psychotropic drugs.

Costa, A.M., *Making Drug Control Fit for Purpose. Building on the UNGLASS Decade* (New York, 2008). Important assessment of achievements and failures.

Cottle, J., *Reminiscences of Samuel Taylor Coleridge and Robert Southey* (London, 1847). Gossipy and unpleasant memoirs, a best-seller in its time.

Courtwright, D., *Dark Paradise. Opiate Addiction in America before 1940* (London, 1982).

Courtwright, D., *Forces of Habit: Drugs and the Making of the Modern World* (Cambridge, Mass., 2001). Insightful.

Courtwright, D., H. Joseph and D. Des Jarlais, eds, *Addicts who Survived. An Oral History of Narcotic Use in America, 1923–1965* (Knoxville, Tenn., 1989).

Crabbe, G., *The Works of the Rev. George Crabbe* (London, 1828). Contains the original 'Peter Grimes' novella.

Crabbe, G., *The Life of George Crabbe, by his Son*, introduced by E. Blunden (first published 1834; London, 1947). A charming and affectionate but revealing memoir by Crabbe's son, also a clergyman.

Crepon, T., *Leben und Tode des Hans Fallada* (Halle, 1978). A moving memoir.

Crothers, T.D., *Morphinism and Narco-Maniacs from Other Drugs* (Philadelphia, 1902). An example of the high-moral-ground censorious school. Interesting but missable.

Crowe, S.J., *Halsted of Johns Hopkins. The Man and his Men* (Springfield, Ill. 1957). Sympathetic and discreet.

Crumpe, S., *An Inquiry into the Nature and Properties of Opium* (London, 1793).

Csath, G., *The Diary of Geza Chath*, trans. P. Reich (Budapest, 2002). One of the memorable addict diaries by a doctor poet, now available in a good English translation.

Cull, J.G., and R.E. Hardy, eds, *Types of Drug Abusers and their Abuses* (Springfield, Ill, 1870). Useful pen portraits of the wide range of otherwise anonymous addicts in the United States at the time.

Curgenven, J.G., *The Waste of Infant Life* (London, 1867). A clear warning, largely ignored.

Curtis Smith, T., 'Hypodermic injection of morphine', *Kansas City Medical Journal*, 1872 (2) 73. One of a flood of papers, both hailing and warning against the perils of hypodermic injection. For good or ill but mainly for ill, 'the syringe transformed the drug menace'.

Cusket, W., and others, *Drug Trips Abroad: American Drug Refugees in Amsterdam and London* (Philadelphia, 1972). Interesting on the international and cosmopolitan ramifications of the drug world.

Cutting, W.C., 'Morphine addiction for 62 years', *Stanford Medical Bulletin* 1942 (1) 39. The story of the 'stable' addiction of a distinguished physician. This too happened.

Dally, A., *A Doctor's Story* (London, 1990). A fascinating book and a key source on the drug scene in the UK in the 1980s.

Daly, G., *Pre-Raphaelites in Love* (London, 1989). The tragic addiction of Lizzie Siddal and Dante Gabriel Rossetti.

Danchin, P., *Francis Thompson, la vie et l'oeuvre d'un poète* (Paris, 1959). An excellent book, as French biographies of English literary men usually are.

Daudet, L., *La Lutte* (first published 1863; Paris, 1935). The author was a doctor, journalist and novelist, intensely preoccupied with addiction; but, unlike his father, not an addict himself. In *La Lutte* a young doctor with a bright future who thinks he has tuberculosis becomes an addict.

Davenport-Hines, R., *The Pursuit of Oblivion: A Global History of Narcotics, 1500–2000* (London, 2001). Well researched and well written but perhaps covers too wide a ground. Most narcotics have something in common but their history is different. A valuable contribution nevertheless, especially about the late twentieth-century scenes in the United Kingdom and the United States.

Davies, A.M., *Clive of Plassey* (London, 1939). A sympathetic biography of an important historical character and an interesting human being.

Davis, J.F., *China during the War* (London, 1852). A vivid and relatively well-balanced contemporary account of the opium wars.

Day, H., *The Opium Habit with Suggestions as to the Remedy* (New York, 1868). One of the earliest DIY guides. Upbeat like most of the genre.

Debus, A.G., *The English Paracelsians* (New York, 1966). A meticulous study of a neglected field.

de Quincey, H.A., *Thomas de Quincey: A Biography* (Oxford, 1936).

de Quincey, T., *Confessions of an English Opium-Eater and Other Writings* (first published 1822; London, 2003). This Penguin edition of a still highly readable classic has a good introduction and postscript on the author's life, literary setting and opium by the editor, Barry Milligan.

de Quincey, T., 'The opium question with China', *Blackwood's Magazine*, May and June 1840.
de Quincey, T., *Reminiscences of the English Lake Poets*, ed. J.E. Jordan (London, 1961).
de Quincey, T., *The Works of Thomas de Quincey*, 21 vols. General ed. G. Lindop (London, 2000; BiblioBazaar, Vol. I, 2010). Fun to dip into.
Dickens, C., *Bleak House* (London, 1853). Contains a superb description of the discovery of a dead addict in Chapter 11.
Dickens, C., *Hard Times* (London, 1854). Contains the description of Coketown of opium addicts. Great.
Dickens, C., *The Mystery of Edwin Drood* (first published 1870; London, 1980). The last, unfinished novel.
Dikotter, F., L. Laaman and Zhy Xun, *Narcotic Culture. A History of Drugs in China* (Chicago, 2004). Excellent.
Dingle, A.E., *The Campaign for Prohibition in Victorian England* (London, 1980).
Disraeli, B., *Sybil, or The Two Nations* (London, 1845). Contains a splendidly barbed and only thinly disguised portrait of Jardine. It would send a publisher's legal advisers into hysterics today.
Dols, M.W., 'Diseases in the Islamic World,' in K.F. Kiple, ed., *The Cambridge World History of Human Disease* (Cambridge, 1993), 334.
Doré, G., and B. Jerrold, *A London Pilgrimage* (London, 1872). Vignettes by the appalled voyeur of London's underworld.
Dormandy, T., *The White Death. A History of Tuberculosis* (London, 2000).
Dormandy, T., *The Worst of Evils: The Fight against Pain* (New Haven and London, 2006).
Douglas, A., *Artists' Quarters. Reminiscences of Montmartre and Montparnasse* (London, 1941). Evocative.
Dover, T., *An Ancient Physician's Legacy to his Country* (London, 1732). Contains the historic formula of Dover's powder on p. 543.
Downes, D., *Contrasts in Tolerance: Post-war Penal Policy in the Netherlands and England and Wales* (Oxford, 1988). Highly revealing.
Downing, C. T., *The Fan-Qui in China* (London, 1838). An evocative account of early Canton by an intelligent surgeon.
Dryden, J., *Selected Poems*, ed. D. Hopkins (London, 1998).
Dubos, R., and I., *The White Plague. Tuberculosis, Man and Society* (London, 1953). A brilliant book.
Dubut de Laforest, J.-L., *Morphine* (Paris, 1891). One of the most powerful naturalistic novels of the *fin de siècle*.
Ducerf, L., *François de Menthon: Un Catholique au service de la République, 1900–1984*. De Menthon's activity in the Resistance in France was imperilled by drugs.
Duffin, J., *To See with a Better Eye. A Life of R.T.H. Laënnec* (Princeton, NJ, 1998).
Duffy, A., *The Sanitarians* (Ithaca, NY, 1953). An early history of American public health.
Dulchinos, D.P., *Pioneer of Inner Space. The Life of Fitz Hugh Ludlow, Hasheesh Eater* (New York, 1998). An excellent biography.
du Maurier, D., *The Infernal World of Bramwell Brontë* (London, 1960).
Dupouy, R., *Les Opiomanes* (Paris, 1912). Useful on the *fin-de-siècle* French scene.
Dupree, L., *Afghanistan* (Oxford, 1998). Probably the most authoritative book by an American authority.
Duster, T., *The Legislation of Morality: Law, Drugs and Moral Judgement* (New York, 1970). A pioneering and still relevant discussion of social and individual rights.
Dutton, M.R., *Policing and Punishment in China* (Cambridge, 1992).
Dyos, H.J., and M. Wolff, eds, *The Victorian City* (London, 1963). Contains an interesting chapter by G. Rosen on diseases, drugs and death.
Eames, J.B., *The English in China* (London, 1909).
Eaton, H.A., *Thomas de Quincey, a Biography* (Oxford, 1936). Still fresh.
Edwards, G., 'The British approach to the treatment of heroin addiction', *Lancet*, 1969 (i) 768.

Edwards, G., 'What drives British drug policies?' *British Journal of Addiction*, 1989 (84) 219. A historical overview.

Eldridge, W.B., *Narcotics and the Law: A Critique of the American Experiment in Narcotic Drug Control*, 2nd edn (Chicago, 1967). Now mainly of historical interest but still relevant.

Eliot, G., *Felix Holt, the Radical* (first published 1866; Penguin edn, London, 1887). A captive of opium, the hero consoles himself that 'if the sufferings ever became too frequent or intolerable, a simple increase in the dose would put an end to them altogether'.

Eliot, G., *Middlemarch* (first published 1871–72; Penguin edn, London, 1994). Dr Lydgate, the idealistic young surgeon, becomes an occasional opium user under pressure of 'foreseen difficulties'. Another character, Will Ladislaw, 'makes himself ill with opium'.

Eliot, G., *Daniel Deronda* (first published 1871–76; Penguin edn, London, 1976). One character, Hans Meyrick, is both an addict and a disapproving sufferer.

Elison, R.E., *Invisible Man* (New York, 1952) The youth of an African-American in the 1940s, partly autobiographical. One of the great novels of the century.

Elliot, C., Sir, *Memoirs of Admiral the Honourable Sir Charles Elliot* (London, 1863). Not the most exciting of memoirs but embellished with charming watercolours.

Elliott, G.J., *The Frontier. The Story of the North West Frontier of India* (London, 1968). Useful as background to the current Afghanistan conflict.

Ellis, A.S., *Ancient Anodynes* (London, 1936). An unjustly criticised compilation of numerous exotic salves.

Elvin, M., *The Pattern of China's Past* (London, 1973).

Endacott, G.O., *A History of Hong-Kong* (Oxford, 1958). A good but far from edifying tale.

Escayrac, de Lauture, E., comte de, *Mémoires sur la Chine* (Paris, 1865). Head of the scientific mission attached to the French Army during the Second Opium War and briefly a prisoner in Peking. Absorbing.

Fallada, H., *Kleiner Mann was Nun?* (Leipzig, 1932). A great novel set in Weimar Germany.

Fallada, H., *Every Man Dies Alone (Jeder stirbt für sich allein)* (London, 2009). First English translation of an important anti-Nazi German novel, the author's last.

Farin, K., *Hans Fallada: Welche sind, die haben kein Glück* (Munich, 1993).

Fay, P., *The Opium War, 1840–42* (Chapel Hill, NC, 1997).

Félice, P. de, *Poisons sacrés, ivresses divines* (Paris, 1936). Good on Baudelaire, his illness and his addiction.

Filan, K., *The Power of the Poppy. Harnessing Nature's Most Dangerous Ally* (Rochester, Vt., 2011). A brief up-to-date history with a good section on the beat generation and current preparations and techniques.

Fischer, L., 'The opium habit in children', *Medical Record*, 1894 (45) 194.

Fleming, D., *William H. Welch and the Rise of Modern Medicine* (Baltimore, Md., 1954).

Flow, K., *The Chinese Encounter with Opium* (Taipei, 2000). Expensive but sumptuously illustrated.

Forbes, R.B., *China and the China Trade* (Boston, 1844).

Ford, J., *Coleridge on Dreaming. Romanticism and the Medical Imagination* (Cambridge, 1998).

Foster, M., *Selected Poems and Letters of Elizabeth Barrett Browning*, 3rd edn (Baltimore, Md., 1988). Great.

Foxcroft, L., *The Making of Addiction. The Use and Abuse of Opium in Nineteenth-Century Britain* (Aldershot, 2007). A readable and scholarly work.

Freches, J., *Les merveilles du Palais d'Été* (Paris, 2006). An illustrated brief history of Peking's Summer Palace, its destruction and very partial reconstruction by the Dowager Empress Cixi. Heart-breaking.

Fry, E., 'China, England and Opium', *Contemporary Review*, 1875–76 (30) 1. A robust Quaker view.

Galagher, N., 'Islamic and Indian Medicine', in K. Kiple, ed, *The Cambridge World History of Human Diseases* (Cambridge, 1993), 27. Valuable.

Gaskell, E., *Mary Barton* (first published 1848; London, 1906). Haunting story of impoverished weavers narcotising away their cares.
Gavit, J.P., *Opium* (London, 1925).
Geller, M. L., *Ancient Babylonian Medicine* (London, 2011).
Gerin, W., *Bramwell Brontë* (London, 1961). The prodigal son and brother who never returned.
Ghodse, H., *Drugs and Addictive Behaviour. A Guide to Treatment* (Cambridge, 2002).
Ghosh, A., *Sea of Poppies* (London, 2008). A spacious slow-moving epic set against the background of the opium wars. Atmospheric; nominated for the Man-Booker Prize.
Giersch, C.P., *Asian Borderlands: The Transformation of Qing China's Yunan Frontier* (Cambridge, Mass., 2006). Excellent background to the Golden Triangle.
Gill, S., *William Wordsworth: A Life* (Oxford, 1989).
Gillman, J., *The Life of Samuel Taylor Coleridge* (London, 1834). A perceptive and illuminating memoir by the doctor who looked after Coleridge for the poet's last eighteen years.
Gilman, S., *Franz Kafka: The Jewish Patient* (London, 1995). Morphine alone relieved the terminal suffering of the writer dying from tuberculosis.
Glantz, M.D., and C.R. Hartel, eds, *Drug Abuse. Origins and Interventions* (Washington, DC, 2000). A cautious publication of the American Psychological Association.
Glendinning, V., *Anthony Trollope* (London, 1993).
Goldberg, J., *Anatomy of a Scientific Discovery* (New York, 1988). A well-written popular account of the discovery of the endogenous opioids.
Goldsmith, M., *The Trail of Opium. The Eleventh Plague* (London, 1939). Stylish and informative; but modern research (e.g. see R. Newman) has questioned some of her conclusions.
Goldstein, A., *Addiction. From Biology to Drug Policy*, 2nd edn (Oxford, 2003). Clear and authoritative by a leading neuroscientist.
Goode, E., *Drugs in American Society* (New York, 1972).
Goodman, J., P.E. Lovejoy and A. Sherratt, eds, *Consuming Habits. Drugs in History and Anthropology* (London, 1995).
Goodson, L., *Afghanistan's Endless War, State Failure, Regional Politics and the Rise of the Taliban* (New York, 2002).
Gray, M., *Drug Crazy* (New York, 1998). A vigorous anti-drugs-war treatise.
Greenberg, M., *British Trade and the Opening of China* (Cambridge, 1969). An excellent account of the early opium trade.
Greene, G., 'My own devil', *Vogue*, October 1973.
Griffiths, J.V., *Afghanistan, Land of Conflict and Beauty. A History of Conflict* (London, 2009). Excellent.
Grinspoon, L., and J. Bakalar, *Psychedelic Drugs Reconsidered* (New York, 1980). See V. Rubin.
Grinspoon, L., and P. Hedbloom, *The Speed Culture, Amphetamine Use and Abuse* (Cambridge, Mass., 1975). See V. Rubin.
Grohlman, W.H., *The Life of Ibn Sina* (London, 1930).
Gruner, O.C., *A Treatise on the Canon of Avicenna* (London, 1930).
'An Habituate', *Opium Eating, an Autobiographical Sketch* (Philadelphia, 1876). A revealing work with an American background.
Haddakin, L., *The Poetry of Crabbe* (London, 1955).
Halász, G., *Erdély Története* [History of Transylvania] (Budapest, 1932).
Hale-White, W., *Keats as Doctor and Patient* (Oxford, 1958).
Hall, M., 'The effects of the habit of giving opiates on the infantine constitution', *Edinburgh Medical and Surgical Journal*, 1816 (12) 423. The facts clearly explained: the warning totally ignored.
Han Suyin, *Eldest Son: Zhou Enlai and the Making of Modern China* (London, 1994).
Harding, G., *Opiate Addiction. Morality and Medicine* (New York, 1988).
Hardman, J.G., A. Goodman Gilman, L.E. Limbird et al., eds, *Goodman and Gilman's The Pharmacological Basis of Therapeutics*, 12th edn (New York, 2009). The most

authoritative current textbook on pharmacology, with sections on drug addiction and metabolism.

Harradan, B., *Ships that Pass in the Night* (London, 1893). A true sanatorium epic, beautifully told.

Harrison, B., *Drink and the Victorians. The Temperance Question in England, 1815–1872* (London, 1904). Attitudes to alcohol were inextricably bound up with attitudes to opium.

Hayter, A., *Mrs Browning* (London, 1952).

Hayter, A., *Opium and the Romantic Imagination* (London, 1968). A sparkling book but confined (apart from Baudelaire and Poe) to the British literary scene.

Heishman, S.J., K.J. Schuh, C.R. Schuster, J.E. Henningfield and S.R. Goldberg, 'Reinforcing and subjective effects of morphine in human opioid abusers: effect of dose and alternative reinforcer', *Psychopharmacology (Berlin)*, 2000 (148: 3) 272.

Hellbrand, M., *Narcotic Agent* (New York, 1941). A fascinating insider's account.

Helmer, J., *Drugs and Minority Oppression* (New York, 1975). Passionate and convincing.

Henig, R.B., ed., *The League of Nations. The Makers of the Modern World* (Edinburgh, 1973).

Hibbert, C., *The Dragon Wakens. China and the West, 1793–1911* (London, 1984).

Hill, J., *Family Herbal* (Bristol, 1812).

Hodgson, B., *Opium. The Portrait of a Heavenly Demon* (Oxford, 2000). A great read.

Hodgson, B., *In the Arms of Morpheus. The Tragic History of Laudanum, Morphine and Patent Medicines* (Buffalo, NY, 2001). A brilliant picture book with reproductions of old prints, film stills and photographs, more eloquent than words.

Holmes, R., *Coleridge: Early Visions* (New York, 1990). An authoritative biography.

Holmes, R., *Coleridge: Darker Reflections* (London, 1997). The sequel.

Holt, E., *The Opium Wars* (London, 1964). A fair popular history from the British point of view.

Homer, *The Odyssey* (Penguin edition, London 1948). What was the magic tranquillising ingredient in Aphrodite's secret potion?

Hosie, A., *On the Trail of the Opium Poppy*, Vol.I (London, 1914). A splendid travel journal; no second volume was ever published.

Hovanesia, R., *The Divine Answer of Commissioner Lin: The Story of History's Largest Narcotics Seizure and its Aftermath* (Baltimore, Md. 2007). An imaginative reconstruction of Commissioner Lin's activities during the First Opium War, written in the first person. A great read.

Howard Jones, N., 'Origins of hypodermic medication', *Scientific American*, 1971 (224) 96.

Hubbard, F.H., *The Opium Habit and Alcoholism* (New York, 1881).

Hughes, J.H., 'The autobiography of a drug fiend', *Medical Review of Reviews*, 1916 (22) 42.

Hunter, C., *On the Speedy Relief of Pain and Other Nervous Affections by Means of the Hypodermic Method* (London, 1865). The starry-eyed beginning.

Hurd, D., *The Arrow Incident* (London, 1967). An elegant diplomatic account of the absurd event which sparked the Second Opium War.

Huxtable, R.J., and S.K.G. Schwartz, 'The isolation of morphine. First principles in science and ethics', *Molecular Interventions*, 2001, 189.

Imber, G., *Genius on the Edge. The Bizarre Double Life of William Stewart Halsted* (New York, 2010). An excellent new biography.

Inglis, B., *The Forbidden Game. The Social History of Drugs* (London, 1975). Full of insights and useful generalisations.

Inglis, B., *The Opium War* (London, 1976). Not one of this writer's best.

International Opium Commission. *Report of the International Opium Commission, Shanghai, China, 1909*, 2 vols (Shanghai, 1909). The first attempt at international cooperation over opium.

Iverson, L.L., *The Science of Marihuana* (Oxford, 2000). Interesting and much of it relevant to opiates.

Jackson, H.J., 'Best order. How bigger pictures of Coleridge emerge from the detailed work of his editors', *Times Literary Supplement*, March 2011 (5632), 3. Coleridge continues to exercise literary minds and three books about him published in 2011 by A.D. Vardy, J. Worthen and J. Beer are reviewed in this article.

Jackson, J., *France, the Dark Years, 1940–44* (Oxford, 2001). Contains valuable information about drug trafficking in Occupied France.

Jacobi, J., ed., *Paracelsus. Selected Writings* (London, 1951). A good selection without too much mumbo-jumbo.

Jardine, L., *Ingenious Pursuits: Building the Scientific Revolution* (London, 1999). Excellent, with a good description of Sir Christopher Wren's historic experiment with intravenous opium.

Jay, M., *Artificial Paradises. A Drugs Reader* (London, 1999). Not a systematic history but full of interest.

Jay, M., *Emperors of Dreams: Drugs in the Nineteenth Century* (Sawtry, Cambs., 2000). A short but splendidly opinionated book.

Jay, M., *High Society. Mind-altering Drugs in History and Culture* (London, 2010). Excellent and lavishly illustrated catalogue to an exhibition of mind-altering drugs at the Wellcome Medical Historical Library, London. The title is clever but, appropriately perhaps, mind-bending.

Jeffreys, J., 'Observations on the improper use of opium in England', *Lancet*, 1840 (1) 382.

Jeffreys, J., *The Traffic in Opium in the East* (London, 1858).

Jelsma, M., T. Kramer and P. Vervest, eds, *Troubles in the Triangle: Opium and Conflict in Burma* (Silkworm Books, Chiang Mai, 2005). Contains some penetrating contributions.

Jennings, O., *On the Cure of the Morphia Habit* (London, 1880).

Jennings, O., *On the Cure of the Morphia Habit without Suffering* (London, 1901).

Jennings, O., *The Morphia Habit and its Renunciation* (London, 1909). Highly regarded in its time, not an easy read or very convincing today. Addicts who have successfully kicked the habit are rarely reliable general guides.

Johnstone, The Reverend, in *The Opium Trade in China* by an 'Eyewitness' (London, 1858). A comprehensive contemporary indictment of the opium trade, extensively quoted by more recent historians.

Jones, J., *Mysteries of Opium Revealed* (London, 1700). The pioneering first monograph in English about opium.

Judson, H.F., *Heroin Addiction in Britain* (New York, 1973).

Juillard, L.F., *Souvenirs d'un voyage en Chine* (Montbéliard, n.d.). Vivid and moving memoirs of a Protestant chaplain in the French Army during the Second Opium War, 1860.

Kabay, J.J., *Janos Kabay. The Life of an Inventor* (Australia, 1990). The affectionate biography of the Hungarian inventor of the method (still used in the pharmaceutical industry for producing medicinal heroin) for the extraction of opium from poppy straw, written by his son.

Kane, H.H., *The Drugs that Enslave* (Philadelphia, 1871). A powerful work by an authority at the time.

Kane, H.H., *The Hypodermic Injection of Morphine* (New York, 1880). A belated warning but widely quoted in Britain.

Kane, H.H., *Opium Smoking in America and China* (Philadelphia, 1882).

Keats, J., *Complete Poems*, ed. John Barnard (Penguin Classics, 2003). Includes many of the greatest opium-inspired poems in literature.

Keay, J., *The Honourable Company – History of the English East India Company* (London, 1991).

Keller, F., *The Lake Dwellings of Switzerland* (London, 1866).

Kelley, A.M., '*The School of Salernum, etc.*' (New York, 1970). A reprint of the translation by Sir John Harington.

Keug, W., ed., *Drug Abuse: Current Concepts and Research* (Springfield, Ill., 1972).
Kevran, R., *Laënnec, médecin breton* (Paris, 1955). A sympathetic biography of a great doctor by a fellow Breton.
King, R., *The Drug Hang-up: America's Fifty-Year Folly* (New York, 1972). A shrill but not negligible attack on opium law enforcement.
Kingsley, C., *Alton Locke* (London, 1850). Opium in the Fens.
Kirchner, L., *The Big Bankroll. The Life and Times of Arnold Rothstein* (New York, 1957).
Klein, A., *Drugs and the World* (New York, 2008). A balanced global view on the effects of the drugs war.
Knight, A.W., ed., *The Beat Vision* (New York, 1978).
Knopf, S.A., 'One million drug addicts in the United States', *Medical Journal and Record*, 1924 (119) 135.
Kobler, J., *Ardent Spirits: The Rise and Fall of Prohibition* (New York, 1972). Probably still the best book on the subject, highly relevant to the anti-drug war.
Kohn, M., *The Dope Girls. The Birth of the British Drug Underground* (London, 1992). A good portrait of the London drug scene during the Jazz Age.
Kuhnke, M., *Besuch bei Hans Fallada* (Neubrandenburg, 1996). Evocative.
Lancet, editorial, 'Poisoning with opium', 1828 (ii) 764.
Lancet, editorial, 'The Privy Council and the Pharmacy Act', 1869 (i) 469.
Lane Poole, S., *The Life of Sir Harry Parkes* (London, 1894). The troublesome hero of the Second Opium War.
Latimer, D., and J. Goldberg, *Flowers in the Blood. The Story of Opium* (New York, 1981). An outstanding, racily written and well researched history by two science journalists with numerous apt quotations.
Laurie, P., *Drugs* (London, 1967).
Lawn, B., *The Salernitan Question: An Introduction to the History of Medieval and Renaissance Problem Literature* (Oxford, 1963).
Lay, G.T., *The Chinese As They Are* (London, 1841). A surprisingly sympathetic book considering the off-putting title.
League of Nations, *First Opium Conference, Geneva, 1924–25* (Geneva, 1925).
Leask, N., *British Romantic Writers and the East: Anxieties of Empire* (London, 1993). Excellent.
Leavenworth, C.S., *The Arrow War with China* (London, 1901). Full of leisurely and somewhat irrelevant but occasionally illuminating reminiscences.
Leech, K., 'The junkies' doctors and the London drug scene in the 1960s', in D.K. Whynes and P.T. Bean, eds, *Policing and Prescribing. The British System of Drug Control* (London, 1991).
Lefebure, M., *Samuel Taylor Coleridge: A Bondage of Opium* (London, 1974). A beautifully written and sympathetic biography.
Lethbridge, H.J., *Hard Graft in Hong Kong* (Oxford, 1985). A great read.
Levine, S.M., *The Romantic Art of Confession* (London, 1998). Good on Rousseau and De Quincey.
Levinson, M.H., *The Drug Problem: A New View Using the General Semantics Approach* (New York, 2002).
Levinstein, E., *The Morbid Craving for Morphia* (trans. of *Die Morphinensucht*) (London, 1878).
Lewin, L., *Phantastica*, trans. P.H.A. Wirth (first published Berlin, 1924; trans. published Rochester, Vt., 1998). A pioneering book in its day surveying the toxicology of various inebriants including opium and its derivatives.
Lewin, L., *Die Gifte in der Weltgeschichte* [Poisons in World History] (Berlin, 1920). An impressively erudite book.
Li, *see* Xiaoxiong.
Light, A.B., and E.G. Torrence, *Opium Addiction* (Philadelphia, 1930).
Lindekerke, A. de, *La Belle Époque de l'opium* (Paris, 1984). A good read.

Lindley, A.F., *Taiping and Tien-kwoh: The History of the Taiping Revolution*, 2 vols (London, 1866). Not really a history but a good yarn about the Taiping Revolution.

Lindop, G., *The Opium-Eater: The Life of Thomas de Quincey* (London, 1981).

Lindsay, H., *Is the War with China a Just One?* (London, 1840). The answer, according to the author, is a resounding yes.

Lintner, B., *Burma in Revolt. Opium and Insurgency since 1948* (Washington, DC, 1994). Excellent.

Lintner, B., *Blood Brothers. The Criminal Underworld of Asia* (New York, 2003). A brilliant piece of in-depth reportage.

Lintner, B., and M. Black, *Merchants of Madness. The Methamphetamine Explosion in the Golden Triangle* (New York, 2009). Another riveting story, the aftermath of the opium-heroin boom.

Lodwick, K.L., *Crusaders against Opium. Protestant Missionaries in China* (Lexington, Mass., 1996). Not a success story but a moving one.

Lomax, E., 'The uses and abuses of opiates in nineteenth-century England', *Bulletin of the History of Medicine*, 1973 (47) 63. Valuable.

London, J., *The Little Lady in the Big House* (New York, 1916). Contains the haunting description of a morphine mercy-killing. The writer knew what he was talking about.

Long, M.T., A.M. Hailes, G.W. Kirby and N. C. Bruce, 'Transformations of morphine alkaloids by Pseudomonas putida M10', *Applied and Environmental Microbiology*, 1995 (61: 10) 3645. A seminal paper on the use of microorganisms to synthesise or consume morphine and chemical derivatives.

Lopez-Munoz, F., P. Garcia-Garcia and C. Alamo, 'The virtue of that precious balsam: approach to Don Quixote from the psychopharmacological perspective', *Actas Espagnolas de Psiquiatria*, 2007 (35) 149. A good introduction to the pharmacological speculations about Don Quixote.

Lowes, J.L., *The Road to Xanadu* (Boston, Mass., 1927). Slightly out of date but still a brilliant read.

Lowes, P.D., *The Genesis of International Narcotics Control* (Geneva, 1966). Contains many interesting details of the international concern about drugs and the frustrations arising from it.

Lubbock, B., *The Opium Clippers* (Glasgow, 1938). Comprehensive and technical.

Lucas, C., *The Fenman's World. Memories of a Fenland Physician* (Norwich, 1930). Great.

McAllister, W., *Drug Diplomacy in the Twentieth Century. An International History* (London, 2000). A readable survey of an extremely complex field.

Macauley, R., *Pleasure of Ruins* (London, 1953). Good on the romantic and opium-inspired art of John Martin and others.

MacCoun, R., and P. Reuter, *Drug War Heresies: Learning from Other Vices, Times and Places* (New York, 2001). Excellent and provocative.

McCoy, A.W., *The Politics of Heroin in South East Asia* (New York, 1972). One of the most explosive books published in the 1970s, exposing the involvement of the CIA with opium smuggling.

McCoy, A.W., *The Politics of Heroin: CIA Complicity in the Global Drug Trade* (Chicago, 2003). An update.

Macdonald, D., *Drugs in Afghanistan. Opium, Outlaws and Scorpion Tales* (London, 2007). Important and illuminating.

MacInnes, C., *Absolute Beginners* (London, 1959). A great novel about the drug-using African and Caribbean underworld of London in the 1950s.

Mackenzie, A., *A History of the Mathesons* (London, 1900). A narrative of family piety.

McMahon, K., *The Fall of the God of Money* (London, 2002). An excellent account of the nineteenth-century Chinese drug culture.

Maday, A., *Romai orvostudomány* [Roman Medicine] (Budapest, 1936).

Mahler, G., 'Das irdische Leben', from *Des Knaben Wunderhorn* (Anne-Sofie von Otter, Berliner Philharmoniker, Abbado). Referred to in Chapter 14.

Maimonides, *Correspondence*, ed. I. Twersy (New York, 1972). Inspiring.

Mann, J., *Murder, Magic and Medicine* (Oxford, 1994).

Mansfield, D., and A. Pain, *Alternative Livelihoods: Substance or Slogans?* Afghanistan Research and Evaluation Unit (AREU) briefing paper (Kabul, 2007).

Mansfield, D., and A. Pain, *Evidence from the Field: Understanding Changing Levels of Opium Poppy Cultivation*, briefing paper (Kabul, 2007). One of numerous important publications by the authors dealing with the opium problem in Afghanistan.

March, J., *The Pre-Raphaelite Sisterhood* (London, 1985).

Marez, C., *Drug Wars: The Political Economy of Narcotics* (Minneapolis, 2004).

Markus, J., *Dared and Done. The Marriage of Elizabeth Barrett and Robert Browning* (London, 1995). Morphine transformed into the language of love.

Marshall, E., 'Uncle Sam is the worst drug fiend in the world', *New York Times*, 22 March 1911.

Massing, M., *The Fix* (New York, 1998). A critical and readable review of drug policies over the past fifty years.

Matheson, D., *What is the Opium Trade?* (Edinburgh, 1857). A powerful anti-opium tract by the son of the co-founder of Jardine and Matheson.

Mattison, J.B., 'Opium addiction among medical men', *Medical Records*, 1883 (23) 621.

Mattison, J.B., 'The treatment of opium addiction', *Journal of Nervous and Mental Diseases*, 1885 (12) 249. A more detailed treatment of the earlier paper.

Mattison, J.B., 'Narcotic abuse and the public weal', *Medical News*, 1903 (82) 638.

Mauclair, C., A.H. Quinn, E. Wagenknecht, G. Rans, M. Phillips, *Edgar Allan Poe the Man*, 2 vols (Chicago, 1926). Poe's addiction to opium has been disputed. These authors doubt it; but most biographers, especially in France, follow Baudelaire and are in no doubt that he was at least a heavy user.

Maudsley, H., 'Opium in the treatment of insanity', *Practitioner*, 1869 (2) 1.

Maynard, W.P., *The Terror of Fu Manchu* (London, 2009).

Meynell, E., *The Life of Francis Thompson* (London, 1913). The tragic story of another opium-addicted tuberculous poet.

Mezey, A., *Muse in Torment* (Lewes, 2000). Many poets who were inspired by laudanum or morphine were considered mad by their contemporaries and some mad poets were undoubtedly hooked on the drug. In this brilliant book the writer, a psychiatrist, considers the relationship between mental illness and literary inspiration.

Mickel, E.J., *The Artificial Paradise in French Literature. The Influence of Opium and Hashish on the Literature of French Romanticism and* Les Fleurs du Mal (Chapel Hill, NC, 1969). Thorough and in parts enlightening.

Milligan, B., *Pleasures and Pains: Opium and the Orient in Nineteenth-Century British Culture* (Charlottesville, Va., 1995). Excellent.

Mills, J.H., and P. Barton, eds, *Drugs and Empires. Essays in Modern Imperialism and Intoxication* (New York, 2007). Contains some excellent essays.

Milne, L., *The Shans at Home* (London, 1910). Khun Sa's background.

Mirbeau, O., *Les Vingt et un jours d'un neurasthénique* (Paris, 1901). Brilliant fiction.

Mitcheson, M., and J. Hartnoll, 'Conflicts in deciding treatment within drug dependency clinics', in D.J. West, ed., *Problems of Drugs in Britain* (London, 1978). Illuminating.

Moges, marquis de (ed.), *Recollections of Baron Gros's Embassy to China and Japan in 1857–58* (London, 1860). A splendid memoir which includes Baron Gros's participation in the Second Opium War.

Moore, W.J., *The Other Side of the Opium Question* (London, 1882). A well argued defence of the official British policy by the Deputy Surgeon General of the Indian Army.

Moreau, F., *Thomas de Quincey, la vie, l'homme, l'oeuvre* (Paris, 1964). Perceptive.

Morel, B.A., *Traité sur la dégéneration* (Paris, 1857). An influential book in its day forecasting the dire consequences of the spreading drug habit.

Morgan, T., *Literary Outlaw: The Life and Times of William S. Burroughs* (New York, 1983).

Morison, J.L., *The 8th Earl of Elgin* (London, 1928). A slightly hagiographic biography of Elgin, the imperial proconsul. But he was a more complex and in many ways more interesting character.

Morison, R., *The English Opium Eater: A Biography of Thomas de Quincey* (London, 2009). A thorough and sympathetic modern biography.

Morris, M., ed., *The Book of Health* (London, 1883). Contains an insightful article by T.L. Brunton on narcotics.

Morse, H.B., *The International Relations of the Chinese Empire* (London, 1910). The best source book on the subject.

Mott, W., *Uber Opium in China. Erinnerungen eines alten Artztes* (Tübingen, 1880). A moving account by a missionary doctor.

Mulligan, B., 'Morphine-addicted doctors: The English opium eater and embattled medical authority', *Victorian Literature and Culture*, 1905 (33) 541. An important paper on a neglected aspect of opium addiction.

Musto, D., *The American Disease: Origins of Narcotics Control*, 3rd edn (Oxford, 1999). Splendid.

Nett, J.C., *Repression und Verhaltensanpassung in lokalen Heroin- und Kokainmärkten* [Repression and Behavioural Patterns in Local Heroin and Cocaine Markets] (Bern, 2006).

Newman, R., 'Opium smoking in late imperial China: a reconsideration', *Modern Asian Studies*, 1995 (29/4) 765. A controversial but ground-breaking revisionist study.

Nguyen te Duc, ed. G. Tredaniel, *Le Livre d'opium* (Paris, 2002).

Nicase, E., ed. and trans., *La Grande Chirurgie de Guy de Chauliac, composé en l'an 1363* (Paris, 1890).

Nisbet, J.F., *The Insanity of Genius and the General Inequality of Human Faculty Psychologically Considered* (London, 1881). An old chestnut, not so old at the time.

Norton, B.J., 'Experimental therapeutics in the Renaissance', *Journal of Pharmacology and Experimental Therapeutics*, 2003 (304) 489.

Nutton, V., ed., *Galen. Problems and Prospects* (London, 1981). As much as one wants to know about this great but unsympathetic character.

Oberman, J., *The Code of Maimonides* (New Haven, 1954).

Obersteiner, H., *Opium Dens in London* (London, 1906).

O'Brian, C.P., and J.H. Jaffe, eds, *Addictive States*, Research Publications, Vol. 70 (New York, 1992).

Oliphant, L., *Narrative of the Earl of Elgin's Mission to China*, 2 vols (London, 1859). The exuberant memoirs of Elgin's secretary.

O'Neill, E., *Long Day's Journey into Night* (New Haven, 1956).

Oppenheim, J., *'Shattered Nerves'. Doctors, Patients and Depression in Victorian England* (Oxford, 1991). Excellent.

Orczy, Baroness, *The Woman in the Big Hat* (London, 1910). Great yarn by the creator of the Scarlet Pimpernel centring on a fatal dose of morphine being administered in a cup of chocolate.

Osler, W., *The Principles and Practice of Medicine* (London, 1892). The most influential textbook of its time.

Ouchterloni, J., Lieutenant, *The Chinese War* (London, 1844). A good eyewitness account by a young Indian Army officer.

Owen, D.E., *British Opium Policy in China and India* (New Haven, 1934). Well researched but biased.

Pachter, H., *Paracelsus* (New York, 1951). Good but superseded by Pagel, *Paracelsus*.

Pacific Medical and Surgical Journal, editorial, 'Four hundred opium dens in San Francisco', 1881 (23) 523.

Page, H.A., *Life of Thomas de Quincey* (London, 1977). The first credible biography.

Pagel, W., *Paracelsus. An Introduction to Philosophical Medicine in the Era of the Renaissance* (London, 1982). A majestic survey by one of Paracelsus's most distinguished but slightly dotty champions.

Paludan, A., *A Chronicle of the Chinese Emperors* (New York, 1998).
Paoli, L., V.A. Greenfield and P. Reuter, *The World Heroin Market. Can Supply Be Cut?* (Oxford, 2009). An authoritative and up-to-date survey but not an easy read.
Parker, E.H., *Chinese Account of the Opium War* (first published Shanghai, 1888; Kessinger Publishing Legacy Reprints, 2006). A valuable reprint of the virtually unobtainable original, it covers the First Opium War from the Chinese angle.
Parkinson, J., *A Medical Admonition to Families Respecting the Preservation of Health and the Treatment of the Sick* (London, 1800). An excellent work that deservedly went through several editions. Opium is advocated 'in moderation'.
Parliamentay Papers, 1842 XXVII: On the Sanitary Conditions of the Labouring Population.
Parliamentary Papers 1894, XLII: Report of the Royal Commission on Opium.
Parliamentary Papers 1912–13 LXVIII: Instructions to the British Delegation to the International Opium Conference held at The Hague, December 1911.
Parliamentary Papers 1912–13, CXIX: Report of the British Delegates to the International Opium Conference.
Parry, N. and J., *The Rise of the Medical Profession. A Study of Collective Social Mobility* (London, 1976).
Parssinen, T.M., *Secret Passions, Secret Remedies: Narcotic Drugs in British Society, 1820–1930* (Philadelphia, 1983).
Parssinen, T., and K. Kerner, 'Development of the disease model of drug addiction in Britain, 1870–1926', *Medical History*, 1980 (24) 275. Important.
Pavel, E., *The Poet Dying. Heinrich Heine's Lost Years in Paris* (New York, 1995). Lost and given over to opium.
Pearsall, R., *Conan Doyle: A Biographical Solution* (London, 1977). The last word on the doctor's addiction.
Pereira, J., *Elements of Materia Medica and Therapeutics* (London, 1839). A respected text-book of the times.
Pert, C., *Molecules of Emotion: Why You Feel the Way You Feel* (New York, 1997). A racy, chatty, infuriating and exhilarating autobiographical canter around the beginnings of endorphins by one of their discoverers.
Peters, C., *The King of Invention. The Life of Wilkie Collins* (London, 1991).
Phayre, A.P., Lt Gen. Sir, *History of Burma*, 2nd edn (London, 1967).
Pichois, C., *Album Baudelaire* (Paris, 1974).
Pichois, C., and J. Ziegler, *Baudelaire* (Paris, 1996). New edition of a classic biography.
Pickering, G., *Creative Malady* (London, 1974). A great idea but disappointing.
Plant, S., *Writing on Drugs* (London, 1999) An imaginative, idiosyncratic anthology.
Plato, *The Last Days of Socrates* (Penguin edn, London, 1975). Was the hemlock laced with opium?
Poe, E.A., *The Portable Poe*, introduced by Philip van Doren Stern (first published 1959; Penguin edn, London, 1977). Was he or was he not an addict?
Poeaknapo, C., J. Schmidt, M. Brandsch, B. Drager, and M.H. Zenk, 'Endogenous formation of morphine in human cells', *Proceedings of the National Academy of Sciences of the USA*, 2004 (101–39) 14091.
Polachek, J., *The Inner Opium War* (Cambridge, Mass., 1993).
Pollock, J., *Wilberforce* (London, 1977).
Pomet, P., *l'Histoire générale des drogues, traitant des plantes, des animaux et des mineraux, etc.* (Paris, 1684), translated anonymously into English under the title *A Compleat History of Drugs* and published in London in 1712. A fascinating and very French mixture of the learned and the sensationally 'exotic', translated into German as well as into English and widely read and quoted in its day.
Porche, F., *Charles Baudelaire* (London, 1927). A neglected fine biography.
Porter, I.A., 'Thomas Trotter, MD, Naval Physician', *Medical History*, 1963 (7) 75.
Porter, R., *The Greatest Benefit to Mankind* (London, 1994). A one-volume, one-man history of medicine, a monumental and impossible task almost accomplished.

Porter, R., and M. Teich, *Drugs and Narcotics in History* (Cambridge, 1995). An important multi-authored work.
Posner, G., *Warlord of Crime* (New York, 1988).
Pratt, W.B., and P. Taylor, eds, *Principles of Drug Action: The Basis of Pharmacology*, 3rd edn (London, 1990).
Privy Council Papers, P.V. 8, 153, 1869 *Pharmacy Act, 1868. The Law Officers' Opinion.*
Proudfoot, A.T., *Acute Poisoning. Diagnosis and Management* (London, 1993).
Pullinger, J., *Crack in the Wall* (London, 1989).
Pullinger, J., and A. Quickie, *Chasing the Dragon* (London, 1989).
Qassem, A.S., *Afghanistan's Political Stability* (New York, 2009).
Raan, S., *Five Families. The Rise, Decline and Resurgence of America's Most Powerful Mafia Empires* (New York, 2005). Exciting.
Rahman, F., *Health and Medicine in the Islamic Tradition* (New York, 1987). Highly informative.
Ramon y Cajal, S., *Psicologia de Don Quijote y el quijotismo* (Madrid, 1905). The possible effect of drugs on Don Quixote has occupied the minds of Spain's foremost scientists for over a hundred years.
Rashid, A., *Taliban: Islam, Oil and the New Great Game in Central Asia* (London, 2000). Well written and illuminating.
Raskin, J., *American Scream: Allen Ginsberg's 'Howl' and the Making of the Beat Generation* (Berkeley, 2004) A good introduction.
Ray, O.S., *Drugs, Society and Human Behaviour* (St Louis, 1982). An insightful study.
Récamier, J.-P. de, *Voyage au Levant* (Paris, 1850). A useful but rather boring travelogue.
Reid, J.C., *Francis Thompson, Man and Poet* (London, 1959).
Renard, R., *The Burmese Connection. Illegal Drugs and the Making of the Golden Triangle* (London, 1996). An excellent history of the emergence of the Golden Triangle.
Rennie, D.F., *The British Arms in North China* (London, 1864). Includes a vivid account of the march on Peking.
Reuter, P., *Disorganised Crime: The Economics of the Visible Hand* (Cambridge, Mass., 1983).
Rimbaud, A., *Collected Poems*, ed. Oliver Bernard (first published 1962; Penguin edn, London, 1986).
Robb, G., *Rimbaud* (London, 2000). Excellent.
Roberts, C., and M. Cox, *Health and Disease in Britain* (Stroud, 2003).
C. Roberts, and K. Manchester, *The Archaeology of Disease*, 2nd edn (Ithaca, NY, 1995).
Robertson, F., *Triangle of Death* (London, 1977). Informative: ignore the awful title.
Robinson, K., *Wilkie Collins* (London, 1951). Not the latest but on Collins's opium habit the most informative biography.
Robinson, V., *Victory over Pain* (London, 1947). A prematurely triumphalist but well researched book.
Rohmer, S., *Dope, A Story of Chinatown and the Drug Traffic* (London, 1919). An excellent work by the disreputable creator of Fu Manchu.
Rolleston Report: *Report of the Departmental Committee on Morphine and Heroin Addiction* (London, 1926). The blueprint of the 'British Approach' to dealing with addicts.
Ropp, R.S. de, *Drugs and the Mind* (London, 1958). Excellent.
Rothstein, W., *American Physicians in the Nineteenth Century. From Sects to Science* (Baltimore, Md., 1973). A useful sourcebook.
Rowe, T.C., *Federal Narcotics Laws and the War on Drugs: Money down a Rat Hole* (New York, 2006). The subtitle is the message.
Rowell, E.A., *The Dope Adventures of David Dare* (Nashville, 1937). Victims of opium dealers are rescued by D.D.
Rowntree, T., *The Imperial Drug Trade* (London, 1905). A majestic indictment.
Royal Commission on Opium, *Report*, 7 vols (London, 1894–95).
Rubin, V., ed., *Cannabis and Culture* (The Hague, 1875). One of the most informative books on cannabis. Like amphetamines and hallucinogens marihuana has a large literature of its own, not included in the present Bibliography.

Rublowsly, J., *The Stoned Age: The History of Drugs in America* (New York, 1975). Journalistic and punchy, at times infuriating.

Rutkow, I. M., *Bleeding Blue and Gray. Civil War Surgery and the Evolution of American Medicine* (New York, 2005). An important aspect of the history of medicine, thoroughly researched.

Saikal, A., and W. Maley, *The Soviet Withdrawal from Afghanistan* (London, 1989). Illuminating background to the Afghan poppy 'explosion'.

Saimöng, S., *The Shan States and the British Annexation*, 2nd edn (Ithaca, NY, 1969). Useful background to Khun Sa and the Golden Triangle.

Sainsbury, H., *Drugs and the Drug Habit* (London, 1909).

Saussay, V. du, *La Morphine* (Paris, *c.* 1895). Among many memorable scenes in this *fin-de-siècle* novel Raoul, a terminal and sadistic addict, his beaten-down wife, his classy call-girl sister and his two prostitute sisters share a syringe in a Christmas orgy of morphine.

Scarborough, J., 'The opium poppy in Hellenistic and Roman medicine', in Porter and Teich, eds, *Drugs and Narcotics*, 4.

Schaffer, H., and Burglass, M.E., eds, *Classic Contributions to the Addictions* (London, 1981).

Schneider, E., *Coleridge, Opium and Kubla Khan* (Chicago, 1953).

Scholl, R., *Der Papyrus Ebers. Die grösste Buchrolle zur Heilkunde Altägyptens* (Leipzig, 2002). The best guide to the most important medical document of Ancient Egypt.

Scott, J.L., *Narrative of a Recent Imprisonment in China* (London, 1841). A fascinating account by a young naval officer.

Scott, J.M., *The White Poppy. A History of Opium* (London, 1971). A short but valuable history, at least a third of which is made up of contemporary quotations.

Seefelder, M., *Opium. Eine Kulturgeschichte* (Hamburg, 1987). More on the familiar Anglo-American than on the no less interesting German scene; but good on Fallada.

Senay, E.C., 'Drug abuse and public health – a global perspective', *Drug Safety* 6, Supplement 1, 1991 (1) 1.

Serturner, F., 'Darstellung der reinen Mohnsaure (Opiumsaure) nebst einer chemischen Untersuchung des Opiums mitvorzüglicher Hinsicht auf einen darin neu erdeckten Stoff und die dahin gehörigen Bemerkungen', *Journal der Pharmacie für Ärzte und Apotheker*, 1806 (14) 46. A landmark paper announcing the discovery of morphine, not yet so named.

Sévigné, Mme de, *Selected Letters*, trans., ed. and with an introduction by L. Tancock (London, Penguin Classics, 1982). Life and death at the court of Louis XIV would have been grim without opium.

Seznec, J., *John Martin en France* (London, 1964). The reciprocal influence of opium-inspired Romantic art and literature.

Sheridan, J.E., *China in Disintegration. The Republican Era in Chinese History, 1912–1949* (Ithaca, NY, 1975).

Simantov, R., and S. Snyder, 'Morphine-like peptides in mammalian brain: isolation, structure elucidation and interactions with the opiate receptor', *Proceedings of the National Academy of Sciences of the USA*, 1976 (73–7) 2515.

Sinclair, U., *The Jungle* (New York, 1906). A powerful novel about a poor immigrant to New York. Morphine flows in the brothel where she works.

Siraisi, N.G., *Medieval and Early Renaissance Medicine: An Introduction to Knowledge and Practice* (Chicago, 1990).

Sisman, A., *Wordsworth and Coleridge. The Friendship* (London, 2006). A unique poetic collaboration and friendship between two wholly dissimilar characters.

Smart, C., 'Social policy and drug addiction: a critical study of policy development', *British Journal of Addiction*, 1984 (79) 31.

Smith, T., 'How dangerous is heroin?' *British Medical Journal*, 1993 (1) 807.

Snyder, S.H., and Matthysse, S., *Opiate Receptor Mechanisms* (Cambridge, Mass., 1975). A landmark book.

Soheil, S.M., *Avicenna, his Life and Work* (London, 1958).

Sons of Wilberforce, *The Life of William Wilberforce*, 2 vols (London, 1838). The opium addiction is laid bare.

Spear, H.B., 'Management of drug addicts', *Lancet*, 1987 (i) 1322. Excellent.

Spence, J., 'Opium smoking in late imperial China', in F. Waterman and C. Grant, eds, *Conflict and Control in Late Imperial China* (Berkeley, 1975).

Starkie, E., *Baudelaire* (London, 1957). Still the best book on Baudelaire in English.

Stendhal, *La Chartreuse de Parme* (Paris, 1830). Contains one of the most vivid and harrowing descriptions of deliberate laudanum poisoning.

Sterne, M.B., *Louisa May Alcott* (New York, 1986). The tribulations of a lifelong addict. Sad.

Stevens, R., *American Medicine and the Public Interest* (New Haven, 1971). A valuable review of the social dimension of medicine, relevant to drug abuse.

Stimson, G. and E. Oppenheimer, *Heroin Addiction. Treatment and Control in Britain* (London, 1982). Valuable and objective.

Stirling, C., *Crime without Frontiers* (London, 1994).

Stoker, B., *Dracula* (1897; London, 2001). 'As if vampires weren't enough Stoker addles poor Lucy's brain with morphine as well'.

Stout, B., and M. Misra, eds, *Reports of the English East India Company* (Bombay, 1932).

Stowe, H.E.B., *Uncle Tom's Cabin* (1853). The despairing slave Cassie uses laudanum to kill her newborn child, 'I have made up my mind . . . I will never agin let a child live to grow up'.

Strang, J., and M. Gossop, eds, *Heroin Addiction and the 'British System': Understanding the Problem* (London, 2005). Important.

Strausbaugh, J., and D. Blaise, eds, *The Drug User* (New York, 1990). A good short anthology of literary pieces, some little known.

Sultzberger, H., ed., *All about Opium* (London, 1884). The eloquent reply of an opium merchant to the 'ever-increasing and unjust attacks of the Anti-Opium Society'.

Summers, A., *The Arrogance of Power* (New York, 2000).

Sydenham, T., *Opera Omnia* (1683), trans. as *Collected Works* by G. Greenhill in 2 vols (London, 1884).

Symons, J., 'Introduction' to Wilkie Collins, *The Woman in White* (London, 1860).

Szasz, T., *Ceremonial Chemistry: The Ritual Persecution of Drugs, Addicts and Pushers* (London, 1975). A challenging paper which concentrates on the role of doctors in creating an iatrogenic abnormality and epidemic.

Takano, H., *The Shore beyond Good and Evil: A Report from inside Burma's Opium Kingdom* (Tokyo, 2002). A first-hand account of life in the Golden Triangle.

Talmeyr, M., *La Comtesse Morphine* (Paris, 1885). One of the best of many violently anti-morphine novels of the *fin de siècle*.

Tandener, P.R., and A. Peleg, 'The medical Cervantes', *Journal of the Canadian Medical Association*, 2001 (165) 1123.

Tanner, S., *Afghanistan: A Military History from Alexander the Great to the War against the Taliban* (London, 2009). Excellent.

Taylor, A.H., *American Diplomacy and the Narcotics Traffic* (Durham, NC, 1969). A magisterial survey.

Taylor, A.S., *On Poisons in Relation to Medical Jurisprudence and Medicine* (London, 1848). A pioneering work by the first professor of medical jurisprudence in London. The opening sentence reads: 'The mid nineteenth century in England is an era soaked in poison.'

Temkin, O., *Galenism: Rise and Decline of a Medical Philosophy* (Ithaca, NY, 1973). The best introduction to this puzzling subject.

Teng, S.Y., *Chang Hsi and the Treaty of Nanking* (Chicago, 1944).

Teng, S.Y., *The Taiping Rebellion and the Western Powers* (Oxford, 1971). Excellent, partly based on Chinese sources.

Terenius, L., 'From opiate pharmacology to opioid peptide physiology', *Uppsala Journal of Medical Science*, 2000 (105–1) 1. A revealing personal account.

Terry, C.E., and M. Pellens, *The Opium Problem* (New York, 1928). A classic with valuable abstracts of the earlier literature.

Thackeray, W.M., *Vanity Fair* (1847). Contains many references to laudanum, including a particularly revealing scene of Emmy stopping her mother 'poisoning baby' with laudanum.

Thant, Myint-U, *The Making of Modern Burma* (Cambridge, 2001).

Thelwall, A.S., The Reverend, *The Iniquities of the Opium Trade with China* (London, 1839). A robust Christian view.

Thomson, C.J.S., *The Mystic Mandrake* (London, 1973). For centuries mandrake eclipsed poppy juice in Europe.

Thurman, J., *Secrets of the Flesh: A Life of Colette* (New York, 1999). Colette was not addicted but everybody around her seems to have been.

Toscher, N., *King of the Jews. The Arnold Rothstein Story* (London, 2005).

Toussaint, P., *Marie Dupléssis, la vrai Dame aux Camélias* (Paris, 1958).

Townshend, T.B., 'Neuralgia treated with enormous doses of sulphated morphia', *Ohio Medical and Surgical Journal*, 1863 (1) 87.

Transnational Institute (TNI), *Downward Spiral Banning Opium in Afghanistan and Burma* (Amsterdam, 2005). A useful discussion paper, as are later publications of the Institute.

Transnational Institute (TNI), *Losing Ground. Drug Control and War in Afghanistan* (Amsterdam, 2006).

Transnational Institute (TNI), *Missing Targets. Contemporary Drug Control Efforts in Afghanistan* (Amsterdam, 2007).

Transnational Institute (TNI), *Withdrawal Symptoms. Changes in the South East Asian Drugs Market* (Amsterdam, 2008).

Travis Hanes III, W., and F. Sanello, *The Opium Wars* (Naperville, Ill., 2002) Excellent: both readable and scholarly.

Trocki, C.A., *Opium and Empire: Chinese Society in Colonial Singapore, 1800–1910* (London, 1990).

Trocki, C.A., *The Opium Trade, 1750–1950* (London, 1999).

Trotter, T., *A View of the Nervous Temperament: Being a Practicle Inquiry into the Increasing Prevalence, Prevention, and Treatment of These Diseases Commonly Called Nervous, Bilious, Stomach and Liver Complaints* (Edinburgh, 1788). A critical contemporary view of the new 'Romantic Age'.

Trumbull, C.G., *Anthony Comstock, Fighter* (New York, 1913). Even in 1913 Comstock still had his champions.

Turner, F.S., *British Opium Policy and its Results in China* (London, 1876). An eloquent plea by an anti-opium campaigner.

UNAIDS, United Nations Programme on AIDS/HIV, *Report on the 2006 Global Epidemic* (Geneva, 2006).

UNAIDS, *Aids Epidemic Update* (Geneva, 2007).

UNODC, United Nations Office on Drugs and Crime, *Afghanistan, Annual Opium Poppy Survey, 2001* (Vienna, 2001).

UNODC, *Myanmar Opium Survey* (New York, 2005).

UNODC, *Opium Poppy Cultivation in the Golden Triangle, Laos PDR, Myanmar, Thailand* (New York, 2007).

UNODC, *Afghanistan Opium Survey, 2007. Executive Summary* (New York, 2007).

UNODC, *World Drug Report Seizures*, online at http://www.unodc.org. Accessed October 2008.

United States Department of State, *International Narcotics Control Strategy Report* (Washington, DC, 2005).

van Ash, C., *The Fires of Fu Manchu* (New York, 1987). An entertaining biography of Sax Rohmer by a close friend.

van Duyne, P.V., and M. Levy, *Drugs and Money: Managing the Drug Trade and Crime Money in Europe* (London, 2005).

Varin, P., *Expédition de Chine* (Paris, 1862). A prejudiced but interesting anti-British first-hand account of the Second Opium War.

Vicars, R.G., 'Laudanum drinking in Lincolnshire', *St. George's Hospital Gazette*, 1893 (1) 24.

Vickers, N., *Coleridge and the Doctors* (Oxford, 2004). Not a pleasant read for doctors.

Victoria, Queen, *Letters*, ed. E.C. Benson (London, 1907).

von Turkoczy, G., *Népszövetségi Krónika* [League of Nations Chronicle] (Budapest, 1942). A gossipy but informative account of the work of the League by a member of the Hungarian delegation.

Waldrond, T., ed., *Letters and Journals of James, 8th Earl of Elgin* (London, 1878).

Waley, A., *The Opium Wars through Chinese Eyes* (London, 1958). An indispensable account containing numerous translations of Chinese texts, including Commissioner Lin's letter to Queen Victoria.

Walshe, J., *Strange Harp, Strange Symphony. The Life of Francis Thompson* (New York, 1967).

Walters, F.P., *A History of the League of Nations*, 2 vols (Oxford, 1952). Essential for discussions of the opium conferences.

Ward, J.R., *Flashback. Drugs and Dealing in the Golden Age of the London Rave Scene* (Cullompton, Devon, 2010).

Watson, L. *Coleridge at Highgate* (London, 1925). An affectionate memoir by a descendant of Dr Gillman.

Watson, S., *The Birth of the Beat Generation. Visionaries, Rebels and Hipsters, 1944–1960* (New York, 1998). Good on the New York drug scene.

Wayne Morgan, H., *Drugs in America: A Social History, 1800–1980* (Syracuse, NY, 1981). Short but meticulously researched; an indispensable work.

Webster, C., Sir, *The Foreign Policy of Palmerston, 1830–41* (London, 1951).

Weir, H.C., 'The American Opium Peril', *Putnam's Magazine*, 1909 (7) 329. One of hundreds of articles in the popular press with apocalyptic warnings against morphine and heroin.

Weir-Mitchell, S. *Doctor and Patient* (Boston, 1887). A deeply insightful book.

Weir-Mitchell, S., *Autobiography of a Quack* (New York, 1899). A great read.

Weiss, R., and C. Butterworth, eds, *The Ethical Concepts of Maimonides* (New York, 1975).

Wells, W.A., *A Doctor's Life of John Keats* (New York, 1959). In the large Keats literature by literary scholars this deserves a hearing.

Welsh, F., *A History of Hong Kong* (London, 1993).

Westermeyer, J. *Poppies, Pipes and the People: Opium and its Use in Laos* (Berkeley, 1982).

Whitaker, B., *The Global Connection. The Crisis of Drug Addiction* (London, 1987). Excellent.

Whynes, D.K., and P.T. Bean, eds, *Policing and Prescribing. The British System of Drug Control* (London, 1991). Contains, among other good ones, an outstanding essay on junkies by K. Leach.

Wickens, G.M., ed., *Avicenna, Scientist and Philosopher. A Millenary Symposium* (London, 1952). A good sourcebook.

Wilde, O., *The Picture of Dorian Gray* (1891; Harmondsworth, 1966).

Williams, J., *More Lives than One. A Biography of Hans Fallada* (London, 1998). A meticulously researched biography of a great and tragic writer.

Willoughby, W.W., *Opium as an International Problem: The Geneva Conference, 1925* (Baltimore, 1926). A story of hope unfulfilled.

Wise, R.A., 'The neurobiology of craving: interactions for the understanding and treatment of addiction', *Journal of Abnormal Psychology*, 1988 (97) 118. An excellent paper on the possible mechanisms of craving for opiates and other drugs.

Wisotsky, S., *Breaking the Impasse in the War on Drugs* (London, 1986). Deeply pondered.

Wolmar, C., *Blood, Iron and Gold: How the Railways Transformed the World* (London, 2009). Contains a searing account of coolie labour and opium addiction on the railways in the United States and Panama.

Wood, A.C., *The History of the Levant Company* (Oxford, 1935).
Wordsworth, D., *The Grasmere Journals*, ed. P. Woof (Oxford, 1999).
World Bank, *World Development Report, 2007* (Washington, DC, 2008).
Xenophon, *The Banquet of Xenophon*, trans. J. Welwood (London, 1750).
Xiaoxiong Li, *Poppies and Politics in China, Sichuan Province, 1840–1940* (Newark, Del., 2010). A deeply researched work partly based on otherwise inaccessible Chinese sources.
Young, G., *A Treatise on Opium* (London, 1765). A pioneering book discussed in Chapter 10.
Yvorel, J.-J., *Les Poisons de l'esprit: drogues au XIX siècle* (Paris, 1992).
Zheng, Y., *The Social Life of Opium in China* (New York, 2005). Fascinating and somewhat revisionary history.
Ziegler, J., *Gautier, Baudelaire. Un carré de Dames* (Paris, 1978). Excellent on the French Romantic opium scene.
Zimmer, C., *Soul Made Flesh* (New York, 2004). Title apart, this is a riveting panorama of the early seventeenth-century scientific revolution, including early experiments with opium, centred on the great anatomist and doctor, Thomas Willis.

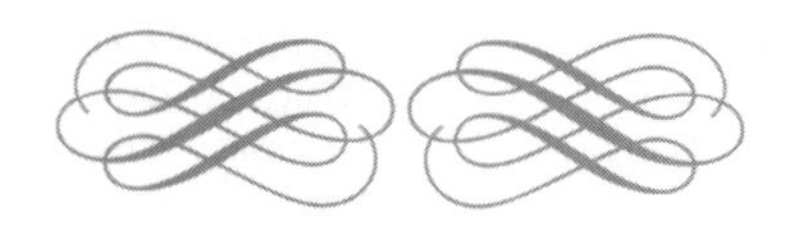

Index